Antennas for Base Stations in Wireless Communications

ABOUT THE EDITORS

DR. ZHI NING CHEN received his Ph.D. from the Institute of Communications Engineering and his DoE from the University of Tsukuba. Since 1988, he has worked at the Institute of Communications Engineering, Southeast University, City University of Hong Kong, University of Tsukuba, and the IBM T. J. Watson Research Center. He is currently Principal Scientist and Department Head for RF & Optical at the Institute for Infocomm Research in Singapore. He is concurrently holding guest/adjunct professor appointments at Shanghai Jiao Tong University, Southeast University, Nanjing University, Tongji University, and National University of Singapore. He is also technical advisor at Compex Systems.

As a key member of numerous international organizations, Dr. Chen has organized many events. He is the founder of the International Workshop on Antenna Technology (iWAT). He has published over 260 papers as well as authored and edited the following books: *Broadband Planar Antennas: Design and Applications* (Wiley, 2007), *Ultra Wideband Wireless Communication* (Wiley, 2006), and *Antennas for Portable Devices* (Wiley, 2007). He also contributed chapters to *Ultra Wideband Antennas and Propagation for Communications, Radar, and Imaging* (Wiley, 2006) as well as *Antenna Engineering Handbook, Fourth Edition* (McGraw-Hill, 2007). He holds 26 granted and filed patents with 15 licensed industry deals. He is the recipient of the CST University Publication Award 2008, IEEE AP-S Honorable Mention Student Paper Contest 2008, IES Prestigious Engineering Achievement Award 2006, I^2R Quarterly Best Paper Award 2004, and IEEE iWAT 2005 Best Poster Award.

Dr. Chen is a Fellow of the IEEE, Inc. and an IEEE AP-S Distinguished Lecturer (www1.i2r.a-star.edu.sg/~chenzn).

DR. KWAI-MAN LUK received his Ph.D. in electrical engineering from The University of Hong Kong. He is currently Head and Chair Professor of the Electronic Engineering Department at City University of Hong Kong. His recent research interests include design of patch, planar, and dielectric resonator antennas; microwave and antenna measurements; and computational electromagnetics. He is the author of two books, a contributing author to nine research books, and the author of over 250 journal papers and 200 conference papers. He has recently been awarded two U.S. patents and ten PRC patents on the design of a wideband patch antenna with an L-shaped probe. He received the Japan Microwave Prize at the 1994 Asia Pacific Microwave Conference held in Chiba in December 1994 and the Best Paper Award at the 2008 International Symposium on Antennas and Propagation held in Taipei in October 2008.

He was the Technical Program Chairperson of the 1997 Progress in Electromagnetics Research Symposium (PIERS 1997), the General Vice-Chairperson of the 1997 and 2008 Asia-Pacific Microwave Conference, and the General Chairman of 2006 IEEE Region Ten Conference. He is a deputy editor-in-chief of the *Journal of Electromagnetic Waves and Applications*.

Professor Luk is a Fellow of the IEEE, Inc.; a Fellow of the Chinese Institute of Electronics, PRC; a Fellow of the Institution of Engineering and Technology, UK; and a Fellow of the Electromagnetics Academy, U.S.

Antennas for Base Stations in Wireless Communications

Zhi Ning Chen

Kwai-Man Luk

New York   Chicago   San Francisco   Lisbon   London   Madrid
Mexico City   Milan   New Delhi   San Juan   Seoul
Singapore   Sydney   Toronto

Library of Congress Cataloging-in-Publication Data

Chen, Zhi Ning.
    Antennas for base stations in wireless communications / Zhi Ning Chen,
Kwai-Man Luk.
          p.   cm.
    Includes bibliographical references and index.
    ISBN 978-0-07-161288-3 (alk. paper)
    1. Antennas (Electronics) 2. Wireless communication systems.   I. Luk,
K. M. (Kwai Man), 1958-   II. Title.
    TK7871.6.C458   2009
    621.382'4—dc22   2009019119

McGraw-Hill books are available at special quantity discounts to use as premiums and sales promotions, or for use in corporate training programs. To contact a representative please e-mail us at bulksales@mcgraw-hill.com.

Antennas for Base Stations in Wireless Communications

1234567890   FGR / FGR   019

ISBN   978-0-07-161288-3
MHID   0-07-161288-2

Sponsoring Editor Wendy Rinaldi	**Proofreader** Susie Elkind	**Illustration** International Typesetting and Composition
Editorial Supervisor Janet Walden	**Indexer** Karin Arrigoni	**Art Director, Cover** Jeff Weeks
Project Editor LeeAnn Pickrell	**Production Supervisor** Jean Bodeaux	**Cover Designer** 12E Design
Acquisitions Coordinator Joya Anthony	**Composition** International Typesetting and Composition	
Copy Editor LeeAnn Pickrell		

Contents at a Glance

Contents

Contributors

Zhi Ning Chen   *Institute for Infocomm Research*
Brian Collins   *BSC Associates Ltd. and Queen Mary, University of London*
Anders Derneryd   *Ericsson AB Ericsson Research*
Martin Johansson   *Ericsson AB Ericsson Research*
Yasuko Kimura   *NTT DoCoMo*
Ahmed A. Kishk   *University of Mississippi*
Ka-Leung Lau   *City University of Hong Kong*
Kwai-Man Luk   *City University of Hong Kong*
Xianming Qing   *Institute for Infocomm Research*
Shie Ping See   *Institute for Infocomm Research*
Wee Kian Toh   *Institute for Infocomm Research*
Hang Wong   *City University of Hong Kong*

Preface

Wireless communication has undergone rapid development during the past four decades. Starting with the first generation wireless communications in the 1970s, the industry progressed to the second generation in the 1990s, the third generation in the 2000s, and at present, is entering the fourth generation. Applications range from cellular mobile phones to healthcare and radio frequency identifications, to name just a few.

Base station antennas are important components of any wireless communication system. The wide range of applications poses stringent specifications in power efficiency, sectorial directivity, polarization diversity, and adaptability. In addition, factors such as visual attractiveness, electromagnetic impact on environment, and, above all, cost are important considerations. The art of designing base station antennas is thus at the forefront of antenna technology and presents challenging problems requiring innovative solutions.

This book, *Antennas for Base Stations in Wireless Communications*, edited by Zhi Ning Chen and Kwai-Man Luk, is a timely and unique contribution that addresses the background, fundamental principles, and practical solutions to many challenges in base station antenna design. Both Dr. Chen and Dr. Luk are well known and have done distinguished work in the field. They have assembled an international group of experts to contribute to the book, which covers all of the relevant issues. The authors, who are from industry, research institutions, and universities, have extensive experience in the research, development, and application of antennas in wireless communications. They provide perspectives from diverse backgrounds and, in many instances, describe their own solutions to the challenges of designing base station antennas.

This book will be a welcome addition to the library of anyone interested in the forefront of antenna technology.

—Kai Fong Lee
University of Mississippi
Oxford, Mississippi
May 2009

Acknowledgments

We are very excited about the publication of *Antennas for Base Stations in Wireless Communications*. This project has presented many challenges, such as gathering the ideas needed to write a unique book, organizing the content effectively, finding the appropriate contributors, working on agreements, consolidating all the chapters and moderating the book's contents, and writing our own chapters. Our grateful thanks are due to all the authors for contributing chapters to this book. They have expended much effort and time to complete this hard task within a very short period. Also, we would like to take this opportunity to thank all the people involved in this project, especially Wendy Rinaldi, Editorial Director at McGraw Hill, for her professional support and patience to make this project a huge success. Finally, we would like to thank our families, namely, Chen's wife Lin Liu and his twin sons, Shi Feng and Shi Ya, as well as Luk's wife Jennifer and his sons, Alec and Lewis, for their understanding and support when we had to work on this project past midnight and during many weekends and holidays.

Introduction

Over the past two decades, developments in very large-scale integration (VLSI) or ultra large-scale integration (ULSI) technologies for electronic circuits and lithium batteries have been revolutionary. In parallel, huge progress has been made in the fields of computer science and information theory. These innovations are the ingredients, in terms of both hardware and software, for the rapid growth of modern mobile communication systems, networks, and services. As antenna engineers, we have been challenged by the extremely fast changes in technical requirements and strong demands in the application market. This book presents the latest advances in antenna technologies for a variety of base stations in mobile wireless communication systems.

Mobile Wireless Communications and Antenna Technologies

The development of the cellular mobile phone is an excellent example of a modern mobile wireless communication technology. Modern personal mobile wireless communications started with the first commercial mobile phone network—the *Autoradiopuhelin (ARP)*, designated as the *zero generation (0G)* cellular network—launched in Finland in 1971.

After that, several commercial trials of cellular networks were carried out in the United States before the Japanese launched the first successful commercial cellular network in Tokyo in 1979. Throughout the 1980s, mobile cellular phones were progressively introduced in commercial operations. At that time, the cellular network consisted of many base stations located in a relatively small number of cells that covered the service areas. Within the network and by using efficient protocols, automated handover between two adjacent cells could be achieved seamlessly when a mobile phone was moved from one cell to another. All the cellular systems were based on analog transmissions. Due to low-degree integration and high-power consuming circuits as well as bulky batteries, mobile phones at that time were too large to carry until Motorola, Inc., introduced the first portable mobile phone.[1] Later, analog systems, known as the *first generation (1G)* mobile phone systems, were accepted as true personal mobile wireless communication systems.

In the 1990s, digital technology was employed for the development of mobile phone systems, which were rapidly advancing and quickly replacing the analog systems to become the *second generation (2G)* personal mobile phone systems. Benefiting from the huge progress seen in integrated circuits (IC) and batteries as well as the deployment of more base stations, the bricks (mobile phones) shrunk to become actual hand-held devices. Meanwhile, cellular networks began to provide users with additional new services such as text messaging (*short message service or SMS*) and media content such as downloadable ring tones.

After the success of the 2G cellular network, extended 2G systems, such as the CDMA2000 1xRTT and GPRS featuring multiple access technology, were developed to enhance network performance and provide significant economic advantages. These systems are known as the *2.5 generation (2.5G)*. In theory, the 2.5G CDMA2000 1xRTT network can achieve the maximum data rate of up to 307 kbps for delivering voice, data, and signaling data.

The 2G and/or 2.5G networks provided users with quality of service (QoS). With so many different standards, however, different technologies had to be developed. To achieve a worldwide standard, the *third generation (3G)* systems have been standardized in the International Telecommunication Union (ITU) family of standards, or the IMT-2000, with a set of technical specifications such as a 2-Mbps maximum data rate for indoors and 384-kbps maximum data rate for outdoors, although this process did not standardize the technology itself. The first commercial 3G CDMA-based network was launched by NTT DoCoMo in Japan on October 1, 2001. Following this, many 3G networks have been set up worldwide. By the end of 2008, global subscriptions to 3G networks exceeded 300 million. For example, in China alone, the number of subscribers had reached 118 million by the end of 2008.

After the 3G systems, the market is looking toward the next-generation systems, which will be the *fourth generation (4G)* or *beyond 3G (B3G)*. The development of the 4G systems targets QoS and increasingly high data rates to meet the requirements of future applications such as wireless broadband access, multimedia messaging service (MMS), video chat, mobile TV, high-definition TV (HDTV) content, digital video broadcasting (DVB-T/H), and so on.

Before the 4G systems are realized, however, many B3G systems are being developed. For example, the 3GPP Long Term Evolution (LTE), an ambitious project called the "Third Generation Partnership Project," is designed to improve the Universal Mobile Telecommunications System (UMTS) in terms of spectral efficiency, costs, service quality, spectrum usage, and integration with other open standards.

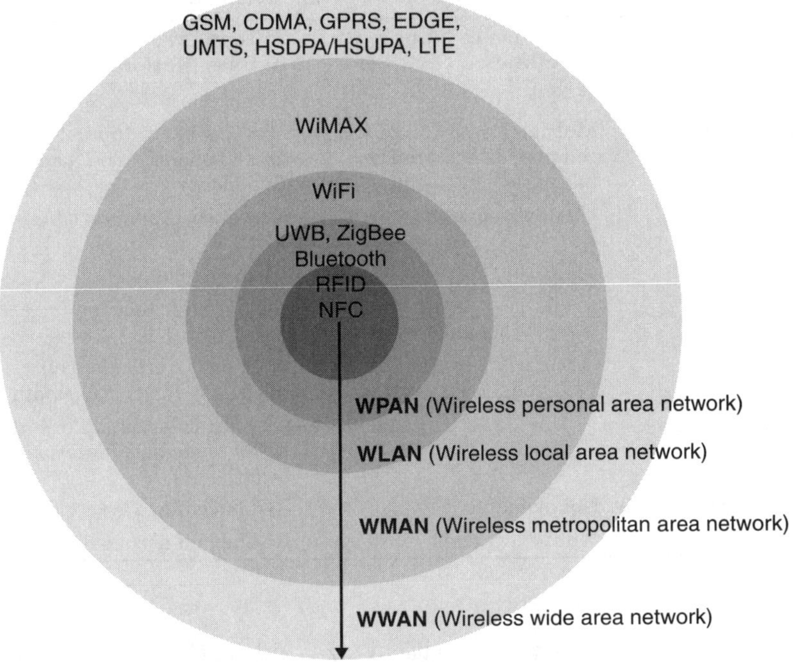

Figure 1   Coverage of modern personal wireless networks

Our lifestyle has changed significantly compared to 20 years ago. Besides the proliferation of mobile phones, wireless communication technologies have penetrated many aspects of life from business to social networks to healthcare and medical applications. A variety of wireless communication systems are now available to connect almost all people and premises, as shown in Figure 1.

In these systems, antennas play a vital role as one of the key components or subsystems. The rapid proposal of new applications has also spurred strong demands for new high-performance antennas. Although the fundamental physical principles of antennas have not changed, antenna engineers have been experiencing the rapid advancement of antenna technologies. For example, antenna technologies for the terminals and base stations in mobile communications have changed remarkably since the first mobile phones, which had long whip antennas, appeared in the market, as detailed in Table 1.

Conventionally, an antenna or array is designed as a radiating RF component. By optimizing the radiators' shapes or configurations, or by combining different radiating elements, the antenna or array will achieve the performance needed to meet system requirements. Such a methodology has been used for a long time in the design of commercial

TABLE 1   Features of Antennas in Personal Mobile Terminals and Base Stations

	Antennas in Mobile Terminals	Antennas/Arrays in Base Stations
Size	Very small	Compact
Operating bands	Multiple bands (up to six) for cell phones and laptops	Multiple bands (up to four) for cell phones, WiFi, and Bluetooth
	Ultra-wideband	Universal UHF bands for RFID
		Ultra-wideband
Diversity	Available in laptop and UWB wireless USB dongles	Available if needed
Polarization	Circularly polarized in RFID handheld readers	Dual polarization for cell phones, WiFi, and bluetooth
		Circular polarization in RFID readers and location beacons
Adaptive beamforming	Not available	Available if needed
MIMO	Will be available	Available if needed

mobile terminal antennas where critical size and cost constraints are imposed. For base stations, the requirements for high-performance antennas or arrays have pushed antenna engineers to implement more complicated RF, electrical, and/or mechanical designs to control antenna and array performance, in particular for the radiation patterns of base station antennas or arrays. Therefore, the antenna or array is designed as a subsystem rather than only an RF component.

Furthermore, the antenna or array will be intelligent (*smart* or *adaptive*) if powerful signal processing functions are applied to controlling antenna performance or digitally processing signals from antennas to form a closed feedback loop. The concept of antenna design is, therefore, expanded to cover not only RF radiators, controlling circuits, or subsystems but also signal processing algorithms, as shown in Figure 2. Such a concept has been used substantially in the development of antennas for base stations in wireless communications, the details of which will be found in this book.

In This Book

Antennas have become one of the most important technologies in modern mobile wireless communications. Many excellent books discuss the fundamental concepts or typical designs for general applications. Most of them are textbooks for students,[2–5] whereas the practical issues in antenna engineering are usually included in engineering handbooks

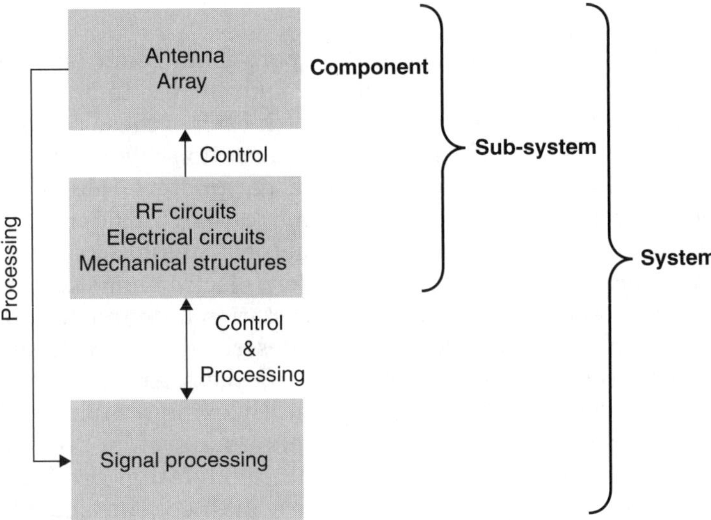

Figure 2   Antenna technology: RF radiator, controlling subsystem, and
signal processing

and monographs.[1,6–11] This book will focus on antenna technologies
for base stations in modern mobile wireless communication systems,
including the most popular applications in WPAN (UWB and RFID),
WLAN (Bluetooth and WiFi), WMAN (WiMAX), WWAN (cellular phones
such as GSM, CDMA, and WCDMA), and so on. The discussion will
cover all aspects of antenna technologies, from antenna element design
to antenna design in systems. Key design considerations and practical
engineering issues will be highlighted in the following chapters, all
written by highly experienced researchers, issues not readily found in
other existing technical literature. The contributors are from industry,
research institutes, and universities; all of them, in addition to ample
engineering experience, have worked for years in the research, development, and application of antennas in wireless communications.

Chapter 1

Chapter 1 defines antenna parameters and their fundamentals. Unlike
other books, we concentrate on the practical aspects of antennas and
the measurement techniques for different parameters such as input
impedance and radiation patterns. Measurements in far-field and near-
field ranges are addressed. Techniques for characterization of linearly
and circularly polarized antennas are also described. One important
parameter, which is usually not discussed in most antenna books, is
the intermodulation distortion of the antenna, which is considered here.
This chapter will be well received by antenna engineers.

Chapter 2

Chapter 2 begins with a description of the operational requirements for base station antennas and relates those requirements to antenna specification parameters and the means by which each of these parameters are controlled. The use of space- and polarization-diversity systems is described, and the required antenna parameters related to dual-polarized antennas are reviewed. As a major design consideration, the description of a variety of construction methods emphasizes both antenna performance and cost. The role of modern electromagnetic simulation tools is discussed in the context of base station antenna design. A variety of practical radiating elements are described to provide the designer with potential starting points for new designs. Array designs are also examined. The general practice of multiplexing a number of transmitters and receivers into a common antenna system creates severe requirements for avoiding passive intermodulation products (PIMs). Methods for minimizing PIMs are thus examined. This chapter includes descriptions of a variety of mechanical design approaches and provides a guide to good design practice.

It has become general practice to control the elevation beam direction (*beamtilt*) of base station antennas, either locally or remotely, with the expectation being that this control will be integrated with the management of traffic loading and the optimization of system capacity. Remote control is also being applied to the azimuth radiation pattern for changing the beamwidth and pointing the beams to desired directions. These developments are described, together with some of the design considerations involved in their realization. The measurement of base station antenna arrays is also discussed, along with techniques for optimizing and diagnosing common problems encountered during design. The installation of base stations is often the occasion for objections related to potential visual and electromagnetic impact on the local environment. These topics and alternative methods for reducing their impact are discussed and examples are provided. In the last part of the chapter, some suggestions are made about possible future developments of base station antennas.

Chapter 3

Chapter 3 describes the commercial requirements for the performance of both indoor and outdoor base station antennas for various mobile phone systems. Conventional techniques for developing base station antennas are also reviewed, including directed dipoles and aperture-coupled patch antennas. The L-probe fed patch antenna is a wideband design that has attracted much interest in the antenna community in recent years due to its simple structure and low cost. By examining five different antenna designs with simulations and measurements, this chapter demonstrates that this new antenna structure is highly suitable for developing base station antennas for both 2G and 3G mobile wireless communication systems.

Chapter 4

Chapter 4 introduces advanced antenna technologies for the fixed GSM-base stations in mobile wireless communication systems. It starts by describing the benefits obtained from advanced antennas in GSM base stations. Then we briefly review advanced antenna technologies, including antenna beamtilt, higher order sectorization, fixed multibeam array antennas, steered beam array antennas, receiver-diversity, coverage concepts, three-sector omnidirectional patterns, modular high-gain antennas, amplifier integrated sector antennas, and amplifier integrated multibeam array antennas. After that, the case studies related to the antenna technologies just mentioned are included. This chapter highlights many practical engineering issues from a systems point of view.

Chapter 5

Chapter 5 first presents the special aspects of mobile communication systems operating in Japan, which occupy so many frequency bands that antenna arrays with multiple operating frequency bands or multiple antenna arrays operating in different frequency bands are required. Several designs are available to meet such practical requirements. The TDMA-based PDC and W-CDMA systems are briefly introduced for understanding the requirements of base stations. Then the practical design considerations for antenna design are highlighted from a system point of view. In 3G networks, one of the most important considerations in antenna configuration is increasing the number of subscribers or system capacity. Ways to select the antenna structures and design the actual antennas for increasing system capacity are suggested. The technique to control vertical radiation patterns is elaborated from an engineering point of view.

This chapter looks at five typical antenna designs, which have been applied in practical wireless access systems. First, slim antennas with small widths, which may be single-band or multiple-band antennas, are preferred due to limited space for, and the high cost of, antenna installations in Japan. A detailed case study is presented. Second, this chapter demonstrates that the horizontal radiation pattern of an antenna can be controlled by adding metallic conductors in the vicinity of the existing base station antenna. Third, an omnidirectional vertical array antenna of width 0.09λ for spot cells is described. Fourth, booster antennas that have the high FB ratio and low sidelobe levels needed for the realization of a reradiation system in a shadow area are presented. Fifth, we will show that the vertical radiation pattern due to a number of vertical array elements can be controlled by changing the phases of each array element.

Chapter 6

Chapter 6 introduces several wideband unidirectional antenna designs based on microstrip antenna technology. All designs employ electrically thick substrates with a low dielectric constant for achieving wide impedance bandwidth performance. Moreover, these antennas using the twin L-probe feed, meandering probe feed, or differential-plate feed not only achieve wide impedance bandwidths, but also possess excellent electrical characteristics such as low cross polarization, high gain, and symmetrical E-plane radiation. After that, the chapter proceeds to illustrate a new type of wideband unidirectional antenna element—a complementary antenna. This novel wideband unidirectional antenna is composed of a planar electric dipole and a shorted patch antenna that is equivalent to a magnetic dipole. A new Γ-shaped feeding strip, comprising an air microstrip line and an L-shaped coupled strip, is selected for exciting the dipole and the shorted patch. This configuration of antenna structure accomplishes excellent electrical characteristics, such as wide impedance bandwidth, low cross polarization, low backlobe radiation, nearly identical E- and H-plane patterns, a stable radiation pattern, and steady antenna gain across the entire operating frequency band. In addition, two alternative feeding structures, T-strip and square-plate coupled lines, demonstrate the flexibility of antenna feed design. All antennas presented in this chapter find practical applications in many recent wireless communication systems like 2G, 3G, WiFi, ZigBee, and so on.

Chapter 7

Chapter 7 provides a general description of the standards and deployment scenarios of WLAN (WiFi). Designs are considered from a system perspective, including materials, fabrication process, time-to-market, as well as deployment and installation. The application of MIMO technology in WLAN systems in order to provide reliability and high-speed wireless links is also discussed. In MIMO systems, antenna performance will greatly impact capacity through the cross-correlation of the signals in transmission and reception. The mutual coupling between the antennas will, therefore, play a critical role in antenna design, which includes element selection and array configuration. The optimized antenna designs with low mutual coupling will enhance the diversity performance of the MIMO systems. Furthermore, the MIMO systems are also advantageous for various types of diversity techniques, for instance, space, pattern, and polarization diversity when applied simultaneously.

Several state-of-the art antenna design solutions are presented as well. This includes outdoor point-to-point antennas, outdoor point-to-multiple point antennas, and indoor point-to-point antennas. The various design challenges and tradeoffs of these antennas are also discussed. Client-device antennas have very critical constraints in terms of cost and size, which severely limit antenna performance. As for the base station antennas in point-to-point (P2P) and/or point-to-multipoint (P2MP) systems, the challenges faced include the tradeoffs among the performance, cost, size, integration of multiple functions into a single antenna design, as well as integration of antennas into radios.

Based on the specifications and the antenna design considerations for WLAN systems, several antenna designs and their practical challenges and tradeoffs are highlighted as case studies. The performance, simplicity, cost effectiveness, and manufacturability of the antenna designs are emphasized. Several applied design innovations are highlighted, for example, embeddable antenna on multilayered substrates, dual broadband antennas, integrated dual-band arrays, and arrays with broadband feeding network technologies.

A wireless personal area network (WPAN) is a short range network for wirelessly interconnecting devices centered around an individual person's workspace. Typically, a WPAN uses the technology that allows communication within about 10 meters. There are many technologies such as Infrared, Bluetooth, HomeRF, ZigBee, ultra-wideband (UWB), radio frequency identification (RFID), and near-field communication (NFC), that have been used for WPAN applications. Some of them—Infrared, Bluetooth, and RFID, for example—are mature commercial products that have been developed for years. The others such as UWB and NFC are still being developed.

Chapter 8

Chapter 8 addresses the antenna designs for two WPAN technologies: RFID for assets tracking and UWB for target positioning.

RFID technology has been developing rapidly in recent years, and its applications can be found in many service industries, distribution logistics, manufacturing companies, and goods flow systems. The reader antenna is one of the key components in an RFID system. The detection range and accuracy of RFID systems depend directly on reader antenna performance. The optimal antenna design always offers the RFID system long range, high accuracy, reduced fabrication cost, as well as simplified system configuration and implementation. The frequencies for RFID applications, which span from a low frequency of 125 KHz to the microwave frequency of 24 GHz, the selection of the antenna, and the design considerations for specific RFID applications feature distinct differences. Loop antennas have been the popular choice of reader antennas for LF/HF RFID systems. For the RFID applications at UHF and MWF bands, a number of antennas can be used as reader antennas, whereas the circularly polarized patch antenna has been the most widely used antenna.

The technology of UWB positioning is still being developed. UWB radios employ very short pulses that spread energy over a broad range of the frequency spectrum. Due to the inherently fine temporal resolution of UWB, arrived multipath components can be sharply timed at a receiver to provide accurate time of arrival (TOA) estimates. This characteristic makes the UWB ideal for high-precision radio positioning applications. Mono-station UWB positioning technology features a highly accurate simple system configuration and has great potential for indoor positioning applications. The six-element sectored antenna array shown in this chapter demonstrates the requirements and design considerations for such a system. The antenna configuration, the antenna element design, and the method for controlling the antenna element's beamwidth are expected to be beneficial when designing sectored antenna arrays for indoor UWB positioning applications.

References

1. Z. N. Chen and M. Y. W. Chia, *Broadband Planar Antennas: Design and Applications*, London: John Wiley & Sons, February 2006.
2. W. L. Stutzman and G. A. Thiele, *Antenna Theory and Design*, 2nd edition, New York: Artech House, 1997.
3. J. D. Kraus and R. J. Marhefka, *Antennas*, 3rd edition, New York: McGraw-Hill Education Singapore, 2001.
4. R. S. Elliott, *Antenna Theory & Design, in IEEE Press Series on Electromagnetic Wave Theory*, Rev sub-edition, New York: Wiley-IEEE Press, 2003.
5. C. A. Balanis, *Antenna Theory: Analysis and Design*, 3rd edition, New York: Wiley-Interscience, 2005.
6. J. J. Carr (ed.), *Practical Antenna Handbook*, 4th edition, New York: McGraw-Hill, 2001.
7. K. M. Luk and K. W. Leung (eds.), *Dielectric Resonator Antennas, in Antennas Series*, Cambridge, UK: Research Studies Press, 2002.
8. K. L. Wong, *Compact and Broadband Microstrip Antennas*, New York: Wiley-Interscience, 2002.
9. R. C. Hansen, *Electrically Small, Superdirective, and Superconducting Antennas*, New York: Wiley-Interscience, 2006.
10. J. Volakis (ed.), *Antenna Engineering Handbook*, 4th edition, New York: McGraw-Hill Professional, 2007.
11. Z. N. Chen (ed.), *Antennas for Portable Devices*, London: Wiley, 2007.

Fundamentals of Antennas

Ahmed A. Kishk

Center of Electromagnetic System Research (CEDAR)
Department of Electrical Engineering
University of Mississippi

An antenna is a device that is used to transfer guided electromagnetic waves (signals) to radiating waves in an unbounded medium, usually free space, and vice versa (i.e., in either the transmitting or receiving mode of operation). Antennas are frequency-dependent devices. Each antenna is designed for a certain frequency band. Beyond the operating band, the antenna rejects the signal. Therefore, we might look at the antenna as a bandpass filter and a transducer. Antennas are essential parts in communication systems. Therefore, understanding their principles is important. In this chapter, we introduce the reader to antenna fundamentals.

There are many different antenna types. The isotropic point source radiator, one of the basic theoretical radiators, is useful because it can be considered a reference to other antennas. The isotropic point source radiator radiates equally in all directions in free space. Physically, such an isotropic point source cannot exist. Most antennas' gains are measured with reference to an isotropic radiator and are rated in decibels with respect to an isotropic radiator (dBi).

1.1 Basis Parameters and Definitions of Antennas

Some basic parameters affect an antenna's performance. The designer must consider these design parameters and should be able to adjust, as needed, during the design process the frequency band of operation, polarization, input impedance, radiation patterns, gain, and efficiency. An antenna in the transmitting mode has a maximum power acceptance. An antenna in the receiving mode differs in its noise rejection properties. The designer should evaluate and measure all of these parameters using various means.

1.1.1 Input Impedance and Equivalent Circuits

As electromagnetic waves travel through the different parts of the antenna system, from the source (*device*) to the feed line to the antenna and finally to free space, they may encounter differences in impedance at each interface. Depending on the impedance match, some fraction of the wave's energy will reflect back to the source, forming a standing wave in the feed line. The ratio of maximum power to minimum power in the wave can be measured and is called the *standing wave ratio (SWR)*. An SWR of 1:1 is ideal. An SWR of 1.5:1 is considered to be marginally acceptable in low-power applications where power loss is more critical, although an SWR as high as 6:1 may still be usable with the right equipment. Minimizing impedance differences at each interface will reduce SWR and maximize power transfer through each part of the system.

The frequency response of an antenna at its port is defined as *input impedance* (Z_{in}). The input impedance is the ratio between the voltage and currents at the antenna port. Input impedance is a complex quantity that varies with frequency as $Z_{\text{in}}(f) = R_{\text{in}}(f) + jX_{\text{in}}(f)$, where f is the frequency. The antenna's input impedance can be represented as a circuit element in the system's microwave circuit. The antenna can be represented by an equivalent circuit of several lumped elements, as shown in Figure 1.1. In Figure 1.1, the equivalent circuit of the antenna is connected to a source, V_s, with internal impedance, $Z_s = R_s + jX_s$. The antenna has an input impedance of $Z_{\text{in}} = R_a + jX_a$. The real part consists

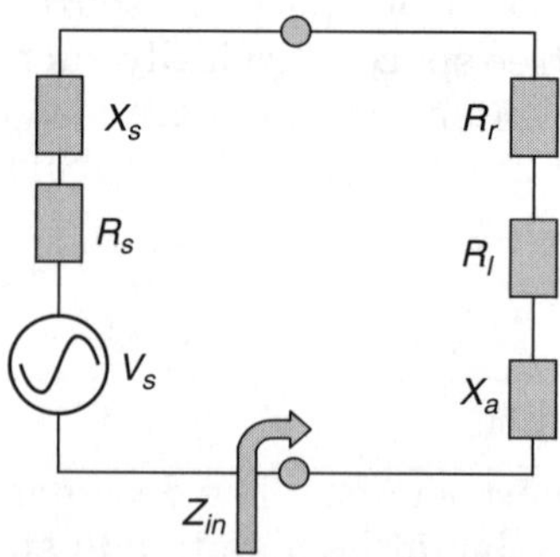

Figure 1.1 Equivalent circuit of an antenna

of the radiation resistance (R_r) and the antenna losses (R_l). The input impedance can then be used to determine the reflection coefficient (Γ) and related parameters, such as voltage standing wave ratio (VSWR) and return loss (RL), as a function of frequency as given in[1-4]

$$\Gamma = \frac{Z_{in} - Z_o}{Z_{in} + Z_o} \tag{1.1}$$

where Z_o is the normalizing impedance of the port. If Z_o is complex, the reflection coefficient can be modified to be

$$\Gamma = \frac{Z_{in} - Z_o^*}{Z_{in} + Z_o} \tag{1.2}$$

where Z_o^* is the conjugate of the nominal impedance. The VSWR is given as

$$\text{VSWR} = \frac{1 + |\Gamma|}{1 - |\Gamma|} \tag{1.3}$$

And the return loss is defined as

$$\text{RL} = -20 \log |\Gamma| \tag{1.4}$$

Input impedance is usually plotted using a *Smith chart*. The Smith chart is a tool that shows the reflection coefficient and the antenna's frequency behavior (inductive or capacitive). One would also determine any of the antenna's resonance frequencies. These frequencies are those at which the input impedance is purely real; conveniently, this corresponds to locations on the Smith chart where the antenna's impedance locus crosses the real axis.

Impedance of an antenna is complex and a function of frequency. The impedance of the antenna can be adjusted through the design process to be matched with the feed line and have less reflection to the source. If that is not possible for some antennas, the impedance of the antenna can be matched to the feed line and radio by adjusting the feed line's impedance, thus using the feed line as an impedance transformer.

1.1.2 Matching and Bandwidth

In some cases, the impedance is adjusted at the load by inserting a matching transformer, matching networks composed of lumped elements such as inductors and capacitors for low-frequency applications, or implementing such a matching circuit using transmission-line technology as a matching section for high-frequency applications where lumped elements cannot be used.

The *bandwidth* is the antenna operating frequency band within which the antenna performs as desired. The bandwidth could be related to the antenna matching band if its radiation patterns do not change within this band. In fact, this is the case for small antennas where a fundamental limit relates bandwidth, size, and efficiency. The bandwidth of other antennas might be affected by the radiation pattern's characteristics, and the radiation characteristics might change although the matching of the antenna is acceptable. We can define antenna bandwidth in several ways. Ratio bandwidth (BW$_r$) is

$$\mathrm{BW}_r = \frac{f_U}{f_L} \tag{1.5}$$

where f_U and f_L are the upper and lower frequency of the band, respectively. The other definition is the percentage bandwidth (WB$_p$) and is related to the ratio bandwidth as

$$\mathrm{BW}_p = 200\,\frac{f_U - f_L}{f_U + f_L}\,\% = 200\,\frac{\mathrm{WB}_r - 1}{\mathrm{WB}_r + 1}\,\% \tag{1.6}$$

1.1.3   Radiation Patterns

Radiation patterns are graphical representations of the electromagnetic power distribution in free space. Also, these patterns can be considered to be representative of the relative field strengths of the field radiated by the antenna.[1-4] The fields are measured in the spherical coordinate system, as shown in Figure 1.2, in the θ and ϕ directions. For the ideal isotropic antenna, this would be a sphere. For a typical dipole, this would be a toroid. The radiation pattern of an antenna is typically represented by a three-dimensional (3D) graph, as shown in Figure 1.3, or polar plots of the horizontal and vertical cross sections. The graph should show sidelobes and backlobes. The polar plot can be considered as a planer cut from the 3D radiation pattern, as shown in Figure 1.4.

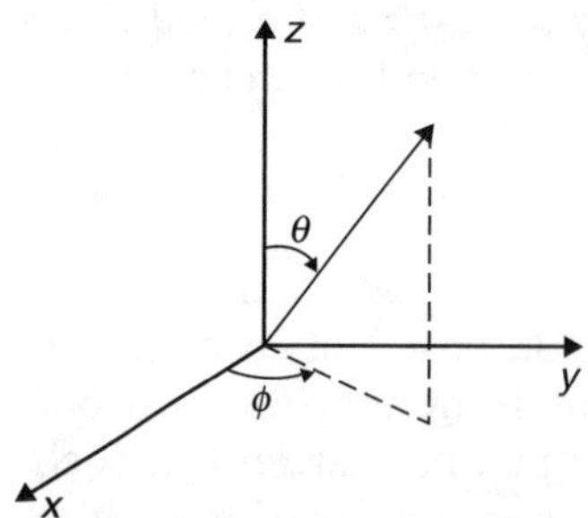

Figure 1.2   Spherical coordinates

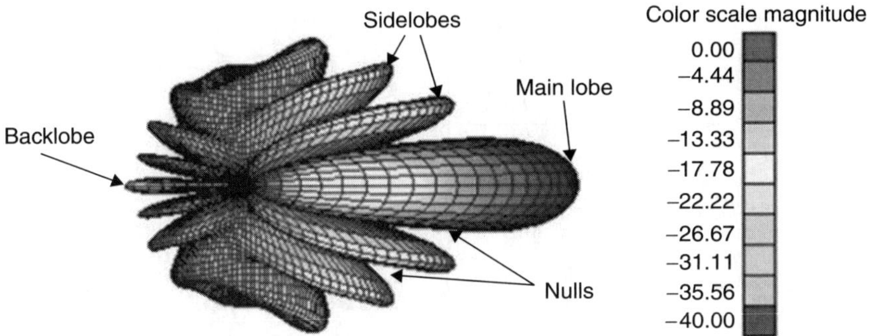

Figure 1.3 3D radiation pattern

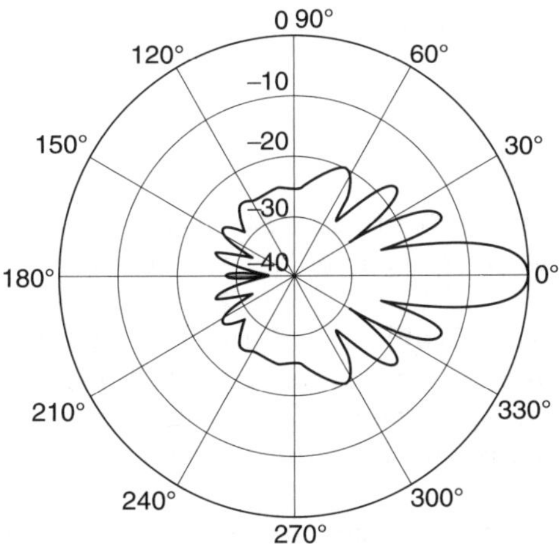

Figure 1.4 Polar plot radiation

The same pattern can be presented in the rectangular coordinate system, as shown in Figure 1.5. We should point out that these patterns are normalized to the pattern's peak, which is pointed to $\theta = 0$ in this case and given in decibels.

1.1.3.1 Beamwidth   The beamwidth of the antenna is usually considered to be the angular width of the half power radiated within a certain cut through the main beam of the antenna where most of the power is radiating. From the peak radiation intensity of the radiation pattern, which is the peak of the main beam, the half power level is −3 dB below such a peak where the two points on the main beam are located; these points are on two sides of the peak, which separate the angular width of the half power. The angular distance between the half power points is defined as the *beamwidth*. Half the power expressed in decibels is −3 dB, so the

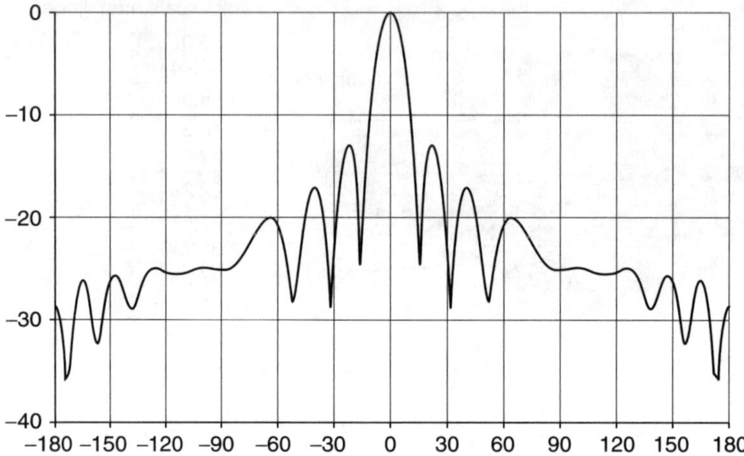

Figure 1.5   Rectangular plot of radiation pattern

half-power beamwidth is sometimes referred to as the 3-dB beamwidth. Both horizontal and vertical beamwidths are usually considered.

1.1.3.2 Sidelobes and Nulls   No antenna is able to radiate all the energy in one preferred direction. Some energy is inevitably radiated in other directions with lower levels than the main beam. These smaller peaks are referred to as *sidelobes*, commonly specified in dB down from the main lobe.

In an antenna radiation pattern, a null is a zone in which the effective radiated power is at a minimum. A null often has a narrow directivity angle compared to that of the main beam. Thus, the null is useful for several purposes, such as suppressing interfering signals in a given direction.

Comparing the *front-to-back ratio* of directional antennas is often useful. This is the ratio of the maximum directivity of an antenna to its directivity in the opposite direction. For example, when the radiation pattern is plotted on a relative dB scale, the front-to-back ratio is the difference in dB between the level of the maximum radiation in the forward direction and the level of radiation at 180°. This number is meaningless for an omnidirectional antenna, but it gives one an idea of the amount of power directed forward on a very directional antenna.

1.1.4   Polarization of the Antenna

The *polarization* of an antenna is the orientation of the electric field (*E-plane*) of the radio wave with respect to the Earth's surface and is determined by the physical structure and orientation of the antenna. It has nothing in common with the antenna directionality terms: horizontal, vertical, and circular. Thus, a simple straight wire antenna will have one polarization when mounted vertically and a different polarization when mounted horizontally. Electromagnetic wave polarization filters are structures that can be employed to act directly on the electromagnetic wave to filter out wave energy of an undesired polarization and to pass wave energy of a desired polarization.

Reflections generally affect polarization. For radio waves, the most important reflector is the ionosphere—the polarization of signals reflected from it will change unpredictably. For signals reflected by the ionosphere, polarization cannot be relied upon. For line-of-sight communications, for which polarization can be relied upon, having the transmitter and receiver use the same polarization can make a huge difference in signal quality; many tens of dB difference is commonly seen, and this is more than enough to make up the difference between reasonable communication and a broken link.

Polarization is largely predictable from antenna construction, but especially in directional antennas, the polarization of sidelobes can be quite different from that of the main propagation lobe. For radio antennas, polarization corresponds to the orientation of the radiating element in an antenna. A vertical omnidirectional WiFi antenna will have vertical polarization (the most common type). One exception is a class of elongated waveguide antennas in which a vertically placed antenna is horizontally polarized. Many commercial antennas are marked as to the polarization of their emitted signals.

Polarization is the sum of the E-plane orientations over time projected onto an imaginary plane perpendicular to the direction of motion of the radio wave. In the most general case, polarization is elliptical (the projection is oblong), meaning that the polarization of the radio waves emitting from the antenna is varying over time. Two special cases are *linear polarization* (the ellipse collapses into a line) and *circular polarization* (in which the ellipse varies maximally). In linear polarization, the antenna compels the electric field of the emitted radio wave to a particular orientation. Depending on the orientation of the antenna mounting, the usual linear cases are horizontal and vertical polarization. In circular polarization, the antenna continuously varies the electric field of the radio wave through all possible values of its orientation with regard to the Earth's surface. Circular polarizations (CP), like elliptical ones, are classified as right-hand polarized or left-hand polarized using a "thumb in the direction of the propagation" rule. Optical researchers use the same rule of thumb, but point it in the direction of the emitter, not in the direction of propagation, and so their use is opposite to that of radio engineers. Some antennas, such the helical antenna, produce circular polarizations. However, circular polarization can be generated from a linearly polarized antenna by feeding the antenna by two ports with equal magnitude and with a 90° phase difference between them. From the linear field components in the far zone, the circular polarization can be presented as

$$E_c(\theta, \phi) = \frac{E_\phi(\theta, \phi) + jE_\theta(\theta, \phi)}{\sqrt{2}} \tag{1.7}$$

$$E_x(\theta, \phi) = \frac{E_\phi(\theta, \phi) - jE_\theta(\theta, \phi)}{\sqrt{2}} \tag{1.8}$$

where E_c is the copolar of the circular polarization, in this case, the left-hand CP, and E_x is the cross-polarization, or the right-hand CP.

In practice, regardless of confusing terminology, matching linearly polarized antennas is important, or the received signal strength is greatly reduced. So, horizontal polarization should be used with horizontal antennas and vertical with vertical. Intermediate matching will cause the loss of some signal strength, but not as much as a complete mismatch. Transmitters mounted on vehicles with large motional freedom commonly use circularly polarized antennas so there will never be a complete mismatch with signals from other sources. In the case of radar, these sources are often reflections from rain drops.

In order to transfer maximum power between a transmit and receive antenna, both antennas must have the same spatial orientation, the same polarization sense, and the same axial ratio. When the antennas are not aligned or do not have the same polarization, power transfer between the two antennas will be reduced. This reduction in power transfer will reduce the overall system efficiency and performance as well. When transmit and receive antennas are both linearly polarized, physical antenna misalignment will result in a polarization mismatch loss, which can be determined using the following formula:

$$\text{Loss (dB)} = 20 \log (\cos\phi) \tag{1.9}$$

where ϕ is the difference in alignment angle between the two antennas. For 15°, the loss is approximately 0.3 dB; for 30°, the loss is 1.25 dB; for 45°, the loss is 3 dB; and for 90°, the loss is infinite.

In short, the greater the mismatch in polarization between a transmitting and receiving antenna, the greater the apparent loss will be. You can use the polarization effect to your advantage on a point-to-point link. Use a monitoring tool to observe interference from adjacent networks, and rotate one antenna until you see the lowest received signal. Then bring your link online and orient the other end to match polarization.

This technique can sometimes be used to build stable links, even in noisy radio environments.

1.1.5 Antenna Efficiency

Antenna efficiency is the measure of the antenna's ability to transmit the input power into radiation.[1-4] Antenna efficiency is the ratio between the radiated powers to the input power:

$$e = \frac{P_r}{P_{in}} \tag{1.10}$$

Different types of efficiencies contribute to the *total* antenna efficiency. The *total* antenna efficiency is the multiplication of all these efficiencies. Efficiency is affected by the losses within the antenna itself and the reflection due to the mismatch at the antenna terminal. Based on the equivalent circuit on Figure 1.1, we can compute the radiation efficiency of the antenna as the ratio between the radiated powers to the input power, which is only related to the conduction losses and the dielectric losses of the antenna structure as

$$e_r = \frac{P_r}{P_{in}} = \frac{R_r}{R_{in}} = \frac{R_r}{R_r + R_l} \tag{1.11}$$

Due to the mismatch at the antenna terminal, the reflection efficiency can be defined as

$$e_{ref} = (1 - |\Gamma|^2) \tag{1.12}$$

Then the total efficiency is defined as

$$e = e_r e_{ref} \tag{1.13}$$

In this formula, antenna radiation efficiency only includes conduction efficiency and dielectric efficiency and does not include reflection efficiency as part of the total efficiency factor. Moreover, the IEEE standards state that "gain does not include losses arising from impedance mismatches and polarization mismatches."[5]

Efficiency is the ratio of power actually radiated to the power input into the antenna terminals. A dummy load may have an SWR of 1:1 but an efficiency of 0, as it absorbs all power and radiates heat but not RF energy, showing that SWR alone is not an effective measure of an antenna's efficiency. Radiation in an antenna is caused by radiation resistance, which can only be measured as part of total resistance, including loss resistance. Loss resistance usually results in heat generation rather than radiation and reduces efficiency. Mathematically, efficiency is calculated as radiation resistance divided by total resistance.

1.1.6 Directivity and Gain

The *directivity* of an antenna has been defined as "the ratio of the radiation intensity in a given direction from the antenna to the radiation intensity averaged over all directions." In other words, the directivity of a nonisotropic source is equal to the ratio of its radiation intensity in a given direction, over that of an isotropic source[1-4]:

$$D = \frac{U}{U_i} = \frac{4\pi U}{P_r} \tag{1.14}$$

where D is the directivity of the antenna; U is the radiation intensity of the antenna; U_i is the radiation intensity of an isotropic source; and P_r is the total power radiated.

Sometimes, the direction of the directivity is not specified. In this case, the direction of the maximum radiation intensity is implied and the maximum directivity is given as

$$D_{max} = \frac{U_{max}}{U_i} = \frac{4\pi U_{max}}{P_r} \tag{1.15}$$

where D_{max} is the maximum directivity and U_{max} is the maximum radiation intensity.

A more general expression of directivity includes sources with radiation patterns as functions of spherical coordinate angles θ and ϕ:

$$D = \frac{4\pi}{\Omega_A} \tag{1.16}$$

where Ω_A is the beam solid angle and is defined as the solid angle in which, if the antenna radiation intensity is constant (and maximum value), all power would flow through it. Directivity is a dimensionless quantity because it is the ratio of two radiation intensities. Therefore, it is generally expressed in dBi. The directivity of an antenna can be easily estimated from the radiation pattern of the antenna. An antenna that has a narrow main lobe would have better directivity than the one that has a broad main lobe; hence, this antenna is more directive. In the case of antennas with one narrow major lobe and very negligible minor lobes, the beam solid angle can be approximated as the product of the half-power beamwidths in two perpendicular planes:

$$\Omega_A = \Theta_{1r}\Theta_{2r} \tag{1.17}$$

where, Ω_{1r} is the half-power beamwidth in one plane (radians) and Ω_{2r} is the half-power beamwidth in a plane at a right angle to the other (radians). The same approximation can be used for angles given in degrees as follows:

$$D \approx 4\pi \frac{\left(\dfrac{180}{\pi}\right)^2}{\Theta_{1d}\Theta_{2d}} = \frac{41253}{\Theta_{1d}\Theta_{2d}} \tag{1.18}$$

where Ω_{1d} is the half-power beamwidth in one plane (degrees) and Ω_{2d} is the half-power beamwidth in a plane at a right angle to the other (degrees). In planar arrays, a better approximation is[6]

$$D \approx \frac{32400}{\Theta_{1d}\Theta_{2d}} \tag{1.19}$$

Gain as a parameter measures the directionality of a given antenna. An antenna with low gain emits radiation with about the same power in all directions, whereas a high-gain antenna will preferentially radiate in particular directions. Specifically, the *gain, directive gain,* or *power gain* of an antenna is defined as the ratio of the intensity (power per unit surface) radiated by the antenna in a given direction at an arbitrary distance divided by the intensity radiated at the same distance by a hypothetical isotropic lossless antenna. Since the radiation intensity from a lossless isotropic antenna equals the power into the antenna divided by a solid angle of 4π steradians, we can write the following equation:

$$G = \frac{4\pi U}{P_{\text{in}}} \tag{1.20}$$

Although the gain of an antenna is directly related to its directivity, antenna gain is a measure that takes into account the efficiency of the antenna as well as its directional capabilities. In contrast, directivity is defined as a measure that takes into account only the directional properties of the antenna, and therefore, it is only influenced by the antenna pattern. If, however, we assume an ideal antenna without losses, then antenna gain will equal directivity as the antenna efficiency factor equals 1 (100% efficiency). In practice, the gain of an antenna is always less than its directivity.

$$G = \frac{4\pi U}{P_{\text{in}}} = e_{cd}\frac{4\pi U}{P_r} = e_{cd}D \tag{1.21}$$

Equations 1.20 and 1.21 show the relationship between antenna gain and directivity, where e_{cd} is the antenna radiation efficiency factor, D the directivity of the antenna, and G the antenna gain. We usually deal with *relative* gain, which is defined as the power gain ratio in a specific direction of the antenna to the power gain ratio of a reference antenna in the same direction. The input power must be the same for both antennas while performing this type of measurement. The reference antenna is usually a dipole, horn, or any other type of antenna whose power gain is already calculated or known.

$$G = G_{ref} \frac{P_{max}}{P_{max}\,|_{ref}} \qquad (1.22)$$

In the case that the direction of radiation is not stated, the power gain is always calculated in the direction of maximum radiation. The maximum directivity of an actual antenna can vary from 1.76 dB for a short dipole to as much as 50 dB for a large dish antenna. The maximum gain of a real antenna has no lower bound and is often −10 dB or less for electrically small antennas.

Antenna absolute gain is another definition for antenna gain. However, absolute gain does include the reflection or mismatch losses:

$$G_{abs} = e_{eff}G = e_{refl}e_{cd}D \qquad (1.23)$$

As defined before, e_{refl} is the reflection efficiency, and e_{cd} includes the dielectric and conduction efficiency. The term e_{eff} is the total antenna efficiency factor.

Taking into account polarization effects in the antenna, we can also define the partial gain of an antenna for a given polarization as that part of the radiation intensity corresponding to a given polarization divided by the total radiation intensity of an isotropic antenna. As a result of this definition for the partial gain in a given direction, we can present the total gain of an antenna as the sum of partial gains for any two orthogonal polarizations:

$$G_{total} = G_\theta + U_\phi \qquad (1.24)$$

$$G_\theta = \frac{4\pi U_\theta}{P_{in}} \quad \& \quad G_\phi = \frac{4\pi U_\phi}{P_{in}} \qquad (1.25)$$

The terms U_θ and U_ϕ represent the radiation intensity in a given direction contained in their respective E-field component.

The gain of an antenna is a passive phenomenon; power is not added by the antenna but simply redistributed to provide more radiated power in a certain direction than would be transmitted by an isotropic antenna. An antenna designer must take into account the antenna's application when determining the gain. High-gain antennas have the advantage of longer range and better signal quality but must be aimed carefully in a particular direction. Low-gain antennas have shorter range, but the orientation of the antenna is inconsequential. For example, a dish antenna on a spacecraft is a high-gain device (must be pointed at the planet to be effective) whereas a typical wireless fidelity (WiFi) antenna in a laptop computer is low-gain (as long as the base station is within range, the antenna can be in any orientation in space). Improving horizontal range at the expense of reception above or below the antenna makes sense.

1.1.7 Intermodulation

Generally, an antenna is considered a passive linear device. However, when such a device is excited by high enough power, it acts slightly as a nonlinear device. The nonlinearity is normally caused by metal-to-metal joints and nonlinear materials in the antenna structure. Therefore, when signals with multiple frequencies are fed into nonlinear devices, intermodulation product terms whose frequencies are different to those of the input signal are generated. A typical passive intermodulation signal level is from −180 to −120 dBc (dBc relative to carrier power).[7–8]

An antenna's intermodulation degrades a wireless system's performance if the system has the following features:

- High transmitted power is adopted.
- The system is equipped with high receiver sensitivity.
- One antenna is used for both transmitting and receiving.
- Signals at more than one frequency are transmitted.

Base stations normally have this entire feature set. Base stations, therefore, suffer from passive intermodulation (PIM). High-power signals excite the antenna of the base station, and intermodulation components cause back reflection to the receiver due to the antenna's PIM. Since the receiver is highly sensitive and is able to sense very weak signals, the intermodulation signals cause interference. The problem becomes worse if the intermodulation term falls inside the receiving band because the interference cannot be removed by filtering. For example, for the P-GSM-900 system whose downlink band is from 935 MHz to 960 MHz and whose uplink band is from 890 MHz to 915 MHz, the 3rd order intermodulation term at the base station side may be $2 \times 935 - 960 = 910$ MHz, which falls inside the uplink band. On the other hand, the PIM problem is not that serious at a client terminal, such as a cell phone, a personal digital assistant (PDA), or a laptop with wireless capability, and is normally ignored. On the client side, the transmitted power is not that high due to limited battery capacity and for electromagnetic safety reasons, and thus the PIM reflected to the receiver is weaker than that at the base station. The relatively low-power transmission does not reduce the quality of uplink as the base station is equipped with a highly sensitive receiver. In addition, the receiver sensitivity is not high at a client terminal, and the reflected PIM level is thus lower than the noise level. Similarly, the relatively lower receiver sensitivity does not degrade the downlink performance as a high-power signal is transmitted from the base station.

An antenna's PIM can be measured by a dedicated analyzer. For example, Summitek Instruments provides such an analyzer. Figure 1.6 shows the block diagram of a PIM analyzer that measures the PIM of a two-port device. It has two measurement modes called reverse measurement and forward measurement. As shown in Figure 1.6, a two-tone high-power signal is fed into Port 1 of the device under test (DUT). The RF switch is in the "Rev" position for the reverse measurement mode or in the "Fwd" position for the forward measurement. For the measurement of an antenna's PIM, the reverse measurement is used not only because an antenna is a one-port device but also because the reverse measurement corresponds to the operation condition of a base station antenna.

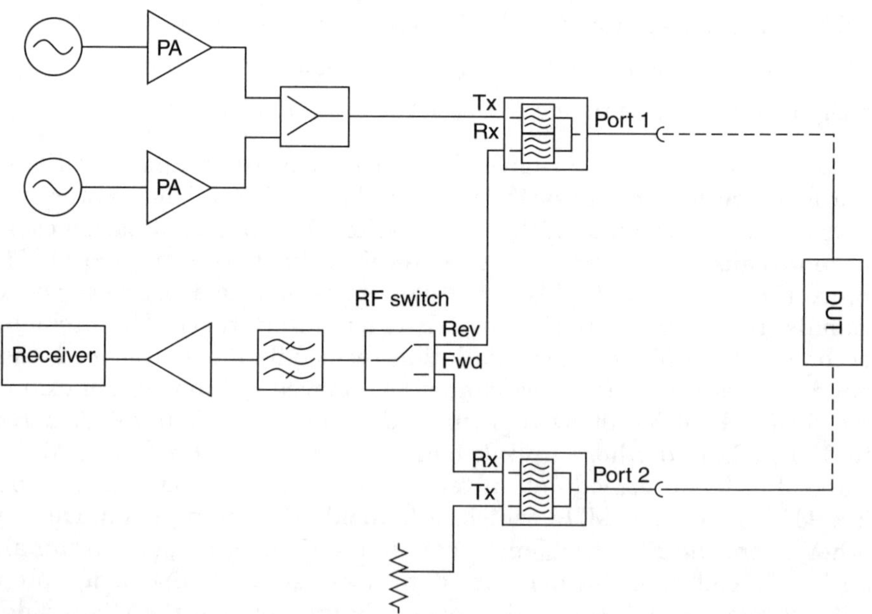

Figure 1.6 Block diagram of a PIM analyzer

1.2   Important Antennas in This Book

Here we introduce several antennas that are recently developed and could be considered as relatively new. These antennas are conventional microstrip antennas as narrow band planar printed antennas, suspended planar antennas as wideband antennas, and planar monopole as an ultra-wideband antenna (UWB).

1.2.1   Patch Antennas

The microstrip patch antenna is a popular printed resonant antenna for narrow-band microwave wireless links that require semihemispherical coverage. Due to its planar configuration and ease of integration with microstrip technology, the microstrip patch antenna has been studied heavily and is often used as an element for an array.

Common microstrip antenna shapes are square, rectangular, circular, ring, equilateral triangular, and elliptical, but any continuous shape is possible.[9] Figure 1.7 shows the parameters of circular and rectangular patches. Some patch antennas eschew a dielectric substrate and suspend a metal patch in air above a ground plane using dielectric spacers; the resulting structure is less robust but provides better bandwidth.

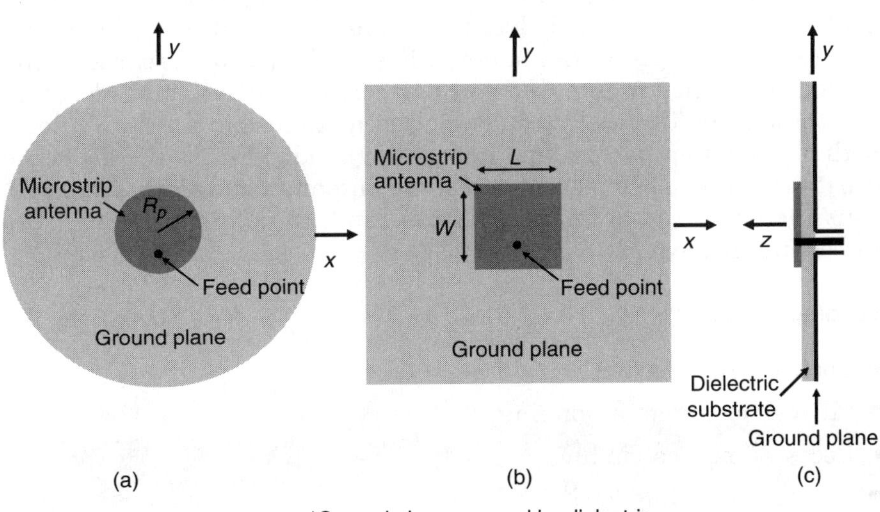

Figure 1.7   (*a*) Circular patch, (*b*) rectangular patch, and (*c*) side view

Advantages:

- Planer (and can be made conformal to shaped surface)
- Low profile
- Ease of integration with microstrip technology
- Can be integrated with circuit elements
- Ability to have polarization diversity (can easily be designed to have vertical, horizontal, right-hand circular (RHCP), or left-hand circular (LHCP) polarizations)
- Lightweight and inexpensive

Disadvantages:

- Narrow bandwidth (typically less than 5%), requiring bandwidth-widening techniques
- Can handle low RF power
- Large ohmic loss

The most common microstrip antenna is a rectangular patch. The rectangular patch antenna is approximately a one-half wavelength long section of rectangular microstrip transmission line. When air is the antenna substrate, the length of the rectangular microstrip antenna is approximately one-half of a free-space wavelength. If the antenna is loaded with a dielectric as its substrate, the length of the antenna decreases as the relative dielectric constant of the substrate increases. The resonant length of the antenna is slightly shorter because of the extended electric *fringing fields*, which increase the antenna's electrical length slightly. The dielectric loading of a microstrip antenna affects both its radiation pattern and impedance bandwidth. As the dielectric constant of the substrate increases, the antenna bandwidth decreases. This increases the antenna's Q factor and, therefore, decreases the impedance bandwidth.

Feeding Methods:

- Coaxial probe feeding
- Microstrip transmission line
- Recessed microstrip line
- Aperture coupling feed[10–11]
- Proximity-coupled microstrip line feed (no direct contact between the feed and the patch[12])

Bandwidth can be increased using the following techniques:

- Using thick and low permittivity substrates
- Introducing closely spaced parasitic patches on the same layer of the fed patch (15% BW)
- Using a stacked parasitic patch (multilayer, BW reaches 20%)
- Introducing a U-shaped slot in the patch (to achieve 30% BW)[13]
- Aperture coupling (10% BW, high backlobe radiation)[10–11]
- Aperture-coupled stacked patches (40–50% BW achievable)[14]
- L-probe coupling[15]

The size of the patch antenna can be reduced by using the following techniques:

- Using materials with high dielectric constants
- Using shorting walls
- Using shorting pins[16]

To obtain a small size wide-bandwidth antenna, these techniques can be combined.

1.2.2 Suspended Plate Antennas

A suspended plate antenna (SPAs) is defined as a thin metallic conductor bonded to a thin grounded dielectric substrate, as shown in Figure 1.8. Suspended plate antennas have thicknesses ranging from $0.03\ \lambda_1$ to $0.12\ \lambda_1$ (λ_1 is the wavelength corresponding to the minimum frequency of the well-matched impedance bandwidth) and a low relative dielectric constant of about 1. SPAs have a broad impedance bandwidth and unique radiation performance.[17]

The use of thick dielectric substrates is a simple and effective method to enhance the impedance bandwidth of a microstrip patch antenna by reducing its unloaded Q-factor. As the impedance bandwidth increases, however, surface wave losses also increase, which reduces radiation efficiency. To suppress the surface waves a low permittivity of the substrate is required.

Advantages:

- Easy to fabricate
- Not expensive
- Large bandwidth
- No surface waves

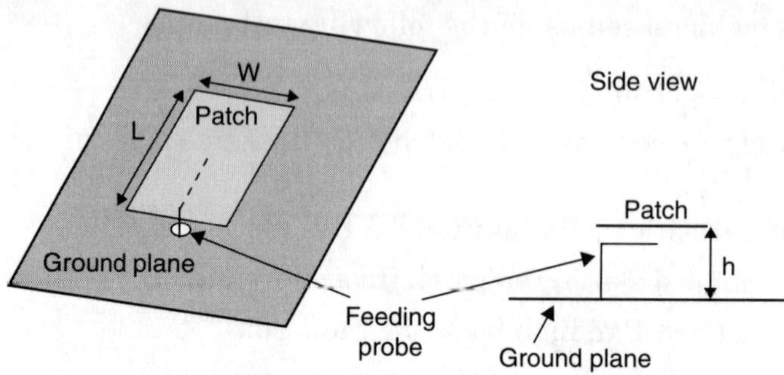

Figure 1.8 Suspended plate antenna patch fed by an L-shaped probe

Disadvantages:

- High cross polarization
- Thick Antenna compared to conventional microstrip antenna

Feeding Methods:

- Coaxial probe (poor matching BW is only 8%)

  Bandwidth can be improved using these techniques:[17]

- A dual probe-feeding arrangement consisting of a feed probe and a capacitive load[18]
- Long U-shaped slot, cut symmetrically from the plate (BW is 10%–40%)
- L-shaped probe (BW reaches 36%)[19]
- T-shaped probe (BW reaches 36%)
- A half-wavelength feeding strip
- A center-fed SPA with a symmetrical shorting pin
- Stacked suspended plate antenna[20]

1.2.3 Planer Inverted-L/F Antennas

A planar inverted-L/F antenna is an improved version of the monopole antenna. The straight wire monopole is the antenna with the most basic form. Its dominant resonance appears at around one-quarter of the operating wavelength. The height of quarter-wavelength has restricted their application to instances where a low-profile design is necessary.[17]

Figure 1.9 shows the geometry of a narrow-strip monopole with a horizontal bent portion, and the planer inverted-F antenna (PIFA) is shown in Figure 1.10.

A PIFA can be considered a kind of linear inverted-F antenna (IFA), with the wire radiator element replaced by a plate to expand bandwidth.

Advantages:

- Reduced height
- Reduced backward radiation
- Moderate to high gain in both vertical and horizontal polarizations

Disadvantages:

- Narrow bandwidth

The shorting post near the feeding probe of the PIFA antennas is a good method for reducing the antenna size, but this results in the narrow impedance bandwidth.

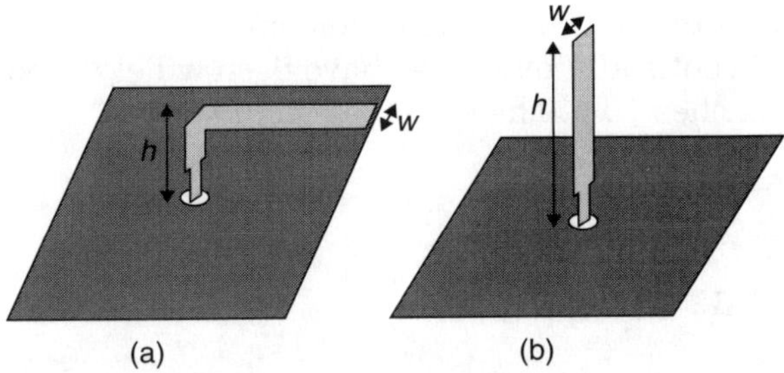

(a)   (b)

Figure 1.9 Geometry of a (*a*) narrow-strip monopole with a horizontal bent portion (*b*) monopole

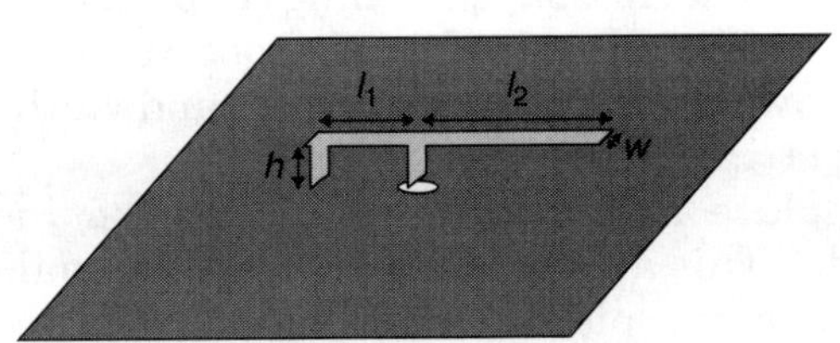

Figure 1.10 Geometry of the PIFA

Bandwidth can be improved using these techniques:

- Using thick air substrate
- Using parasitic resonators with resonant lengths close to the resonant frequency[21]
- Using stacked elements[22]
- Varying the size of the ground plane[23]

The size of the PIFA can be reduced using these techniques:

- Using an additional shorting pin[24]
- Loading a dielectric material with high permittivity[25]
- Capacitive loading of the antenna structure[26]
- Using slots on the patch to increase the antenna's electrical length[27]

1.2.4 Planer Dipoles/Monopoles

Dipoles and monopoles are the most widely used antennas. The monopole is a straight wire vertically installed above a ground plane; it is vertically polarized and has an omnidirectional radian in the horizontal plane. To increase the impedance bandwidth of the monopole antenna, planer elements can be used to replace the wire elements.[28]

Planer designs with different radiator shapes have been widely used in which the bandwidth reaches 70%. These shapes include[17]

- Circular (BW from 2.25–17.25 GHz)
- Triangular
- Elliptical (BW from 1.17–12 GHz)
- Rectangular (BW of 53%)
- Ring
- Trapezoidal (80% BW)
- Roll monopoles (more than 70% BW)

The square planar monopole with trident-shaped feeding strip (shown in Figure 1.11) was introduced[29] with a bandwidth of about 10 GHz (about 1.4–11.4 GHz). This bandwidth is three times the bandwidth obtained using a simple feeding strip.

A compact wideband cross-plate monopole antenna (shown in Figure 1.12) has been proposed.[30] This antenna has a cross-sectional area only 25% that of a corresponding planar cross-plate monopole antenna and can generate omnidirectional or near omnidirectional

Figure 1.11 Geometry of the planar monopole antenna with a trident-shaped (three-branch) feeding strip

Figure 1.12 A compact wideband cross-plate monopole antenna

radiation patterns for frequencies across a wide operating bandwidth of about 10 GHz (1.85–11.93 GHz).

1.3 Basic Measurement Techniques

Performing measurements of the antenna parameters to verify the simulated design is very important. Measurements are also needed to verify that the antenna achieves its requirements. The parameters of the antenna that need to be measured are the input impedance, radiation pattern, directivity, gain, and efficiency. Here, we will briefly state the techniques used for these parameters. We are going to avoid some of the laborious old techniques that were used for lack of modern equipment and instead concentrate on the techniques that use modern equipment.

1.3.1 Measurement Systems for Impedance Matching

Antenna impedance can be measured using a *vector network analyzer*. A vector network analyzer is able to separate the forward wave, V^+, and the reflected wave, V^-, from an antenna at a reference plane where calibration is done, and thus the reflection coefficient is provided, which is the ratio between the reflected to incident waves as $G = V^-/V^+$. Then, the impedance at the reference plane can be computed as $Z = Z_c\,(1+\Gamma)/(1-\Gamma)$, where Z_c is the characteristic impedance of the cable connecting to the antenna. Consequently, before starting to measure the DUT, the network analyzer has to be calibrated using the slandered techniques of standard short, open, and matching loads. One of the most important parameters that calibration establishes is the reference plane, which is critical for phase information. This is a single port measurement.

In some cases, two port measurements are needed to measure the reflection coefficients for the two ports and the mutual coupling between them. These are referred to as the *S-parameter measurements*. This measurement determines the mutual coupling between two antennas or, for some antennas, determines the isolation between two ports of a dually polarized antenna.

1.3.2 Measurement Setups for Far-Zone Fields

For far-field measurements, the distance between the transmitter and receiver has to be large enough to be sure that the transmitter is in the far zone of the antenna under test. To perform these measurements indoors, you have to provide an environment that ensures the antenna does not interact with the surroundings and operates within the environment as if in free space. To achieve this, an anechoic chamber is used with its walls covered with proper absorbing materials that reduce or eliminate the reflections from the walls. The absorbing materials have a certain bandwidth or, in other words, a certain lower frequency bound. The lower frequency is reduced as the absorbing material size is increased.

1.3.2.1 Far-Field System To measure the far-zone field, the transmitting and receiving antennas are put into an anechoic chamber, using a spacing that satisfies each other's far-zone requirement. Figure 1.13 shows an anechoic chamber instrumentation block diagram. This distance guarantees the wave impinging on the receiving antenna can be approximated as a plane wave. Generally, the far-field distance, d, is considered to be

$$d = \frac{2D^2}{\lambda} \tag{1.26}$$

where D is the antenna diameter and λ is the wavelength of the radio wave. Separating the antenna under test (AUT) and the instrumentation antenna by this distance reduces the phase variation across the AUT enough to obtain a reasonably good antenna pattern.

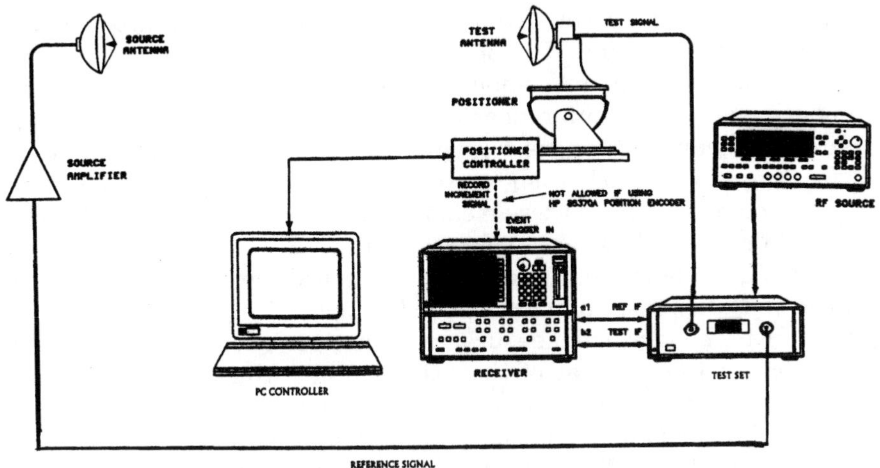

Figure 1.13   Anechoic chamber instrumentation block diagram

The antenna under test will rotate while the other antenna, referred to as the *testing probe*, is fixed. If the measurement is carried out in a transmitting mode, the antenna under test transmits signals while rotating two dimensionally or three dimensionally depending on the rotating mechanism. At the same time, the testing probe receives signals and the measurement system records the data. On the other hand, if the measurement is carried out in a receiving mode, the testing probe transmits signals and the antenna under test receives signals while rotating. This measurement gives the radiation pattern of the antenna under test, if the polarization is the same as that of the testing probe, which could be vertically polarized, horizontally polarized, or circularly polarized. In case the far-zone requirement is difficult to satisfy, a parabolic reflector antenna is used to generate a plane wave. Such a system is referred to as a *compact range*.

1.3.2.2 Near-Field System   To measure the near-zone field, the antenna under test is normally fixed and operates in either a transmitting or receiving mode, whereas the testing probe scans on a surface. The scanning surface could be a flat plane, a cylindrical surface, or a spherical surface depending on the mechanical scheme of the measurement system. Near the antenna under test, the system records the measured data.[5,31]

Planar near-field measurements are conducted by scanning a small probe antenna over a planar surface, as in Figure 1.14*a*. These measurements are then transformed to the far-field by use of a *Fourier Transform*, or more specifically, by applying a method known as *stationary phase* *to the Laplace Transform*. Three basic types of planar scans exist in near-field measurements. The probe moves in the Cartesian coordinate system, and its linear movement creates a regular rectangular sampling grid with a maximum near-field sample spacing of $\Delta x = \Delta y = \lambda/2$. Cylindrical near-field ranges measure the electric field on a cylindrical surface close to the AUT, as shown in Figure 1.14b. Cylindrical harmonics are used to transform these measurements to the far-field. Spherical near-field ranges measure the electric field on a spherical surface close to the AUT, as shown in Figure 1.14c. Spherical harmonics are used to transform these measurements to the far-field.

Similar to the far-zone measurement, the obtained field data has the same polarization as that of the testing probe. Normally, the testing probe is linearly polarized, and thus the measurement is often carried out twice for the two orthogonal polarization components on the sampling surface. This offers a convenient way to measure the radiation patterns of an electrically large antenna, which you may not be able to measure in an anechoic chamber.

Agilent provides high-quality microwave instrumentation, and combining Agilent's microwave instrumentation with NSI's software and systems expertise provides a solution that is unrivaled in the antenna measurement industry. The basic near-field range system block diagram shown in Figure 1.15 is similar to the University of Mississippi's available planar near-field system.

1.3.2.3 Circularly Polarized Systems   By choosing different testing probes in the far-zone measurement, there are three ways to measure the far-zone circularly polarized field. If the measurement system could get

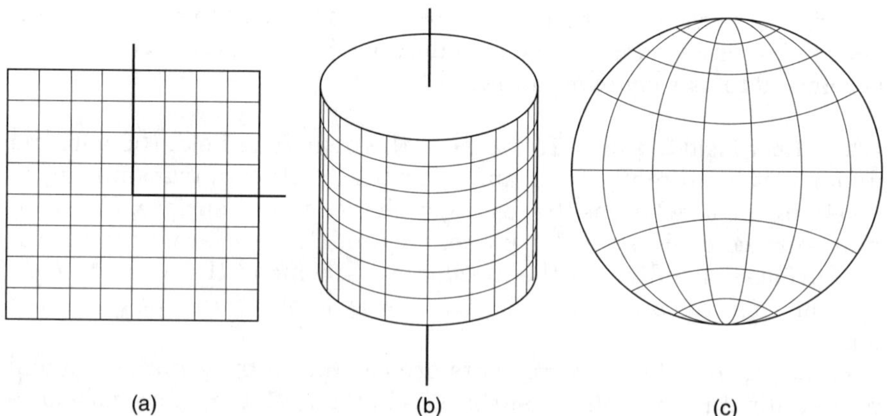

(a)   (b)   (c)

Figure 1.14   Near-field scanning: (a) planar scanning, (b) cylindrical scanning, and (c) spherical scanning

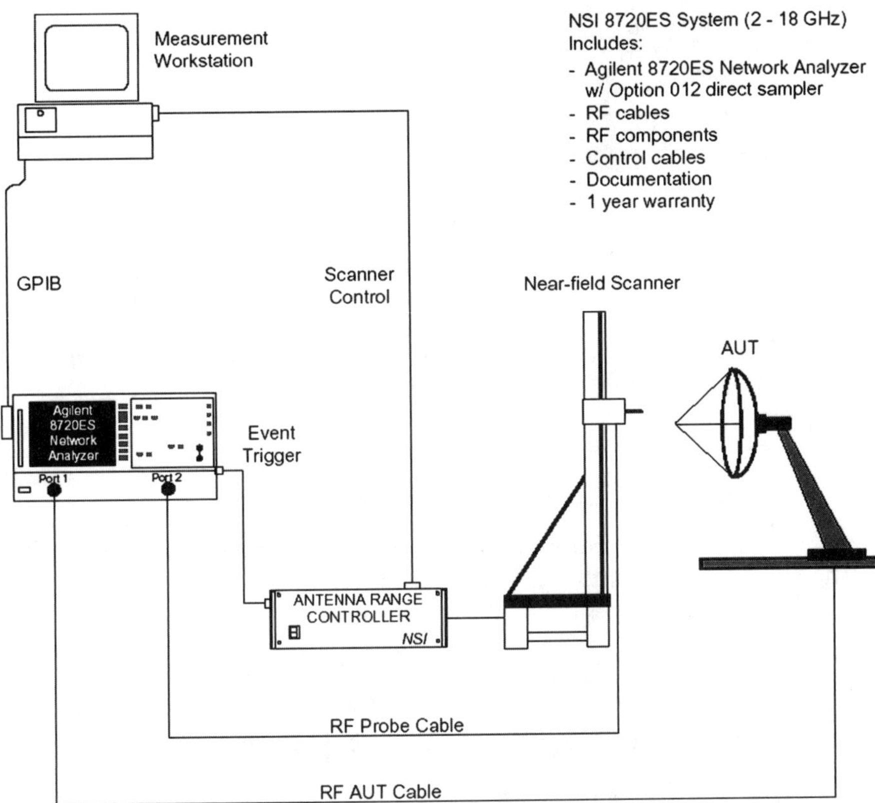

Figure 1.15 NSI VNA antenna measurement system based on the NSI Model 200 V- 5'×5' Near-Field Vertical Scanner and the Agilent 8720ES Vector Network Analyzer

both the magnitude and phase of the received data, the testing probe can be linearly polarized. After the measurement for the two orthogonal polarization components, the left-hand or right-hand circular polarization component can be computed using the complex data collected as given in Eq. 1.7 and Eq. 1.8. If the measurement system can only get the magnitude but not the phase of the data, the measurement may be carried out twice, once using the left-hand polarized testing probe and then once using the right-hand polarized probe, to get the two circularly polarized components (copular and cross-polar, respectively). These two methods require two measurements. Another convenient way of avoiding having to perform two measurements is to use a linearly polarized testing probe rotating at a rate much faster than that of the antenna under test. The resulting pattern is an oscillating pattern called an *axial ratio pattern*, as shown in Figure 1.16. The difference of the two envelops gives the axial ratio. This method can be applied to a system

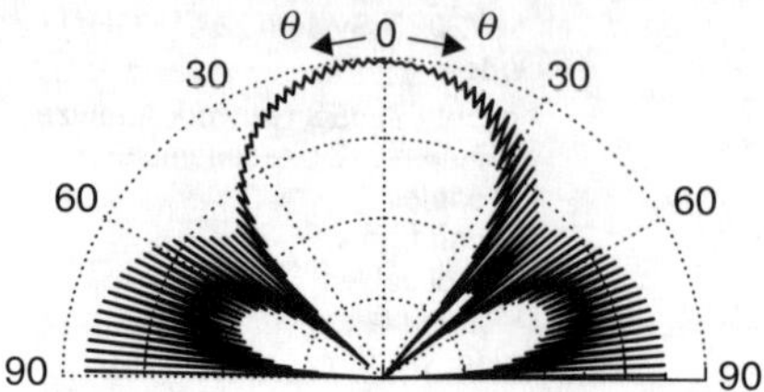

Figure 1.16 Example of an axial ratio pattern

that is able to get only the magnitude of the data. The two envelops are presented by E_1 and E_2 as

$$E_1(\theta,\phi) = E_c(\theta,\phi) + E_x(\theta,\phi)$$
$$E_2(\theta,\phi) = E_c(\theta,\phi) - E_x(\theta,\phi)$$

$$(1.27)$$

From E_1 and E_2 the axial ratio that is used as a measure of the quality of the circular polarization is given as

$$|\mathrm{AR}(\theta,\phi)| = 20\log\left|\frac{E_1(\theta,\phi)}{E_2(\theta,\phi)}\right|$$

$$(1.28)$$

Usually the AR is measured for the main beam and is given as

$$|\mathrm{AR}| = |\mathrm{AR}(0,0)|$$

$$(1.29)$$

1.3.3 Measurement Systems for Intermodulation

The intermodulation phenomenon exists in nonlinear devices working in a high-power environment. Intermodulation is widely discussed for the design of power amplifiers. Similarly, for a base station transmitting antenna, which emits high power, the intermodulation exists due to the metal/metal joint and material nonlinearity. The intermodulation is a kind of interference that should be suppressed. Suppose two closely located fundamental frequencies, f_1 and f_2, are transmitted by the antenna; due to nonlinearity, the radiated field has components at the frequencies of $f_1, f_2, 2f_1, 2f_2, 3f_1, 3f_2, f_1 + f_2, f_1 - f_2, 2f_1 - f_2, 2f_2 - f_1$, and so on. Since f_1 and f_2 are close to each other, the 3rd order intermodulation components at $2f_1 - f_2$ and $2f_2 - f_1$ are very close to the fundamental frequencies and are difficult to remove using filters, thus introducing interferences.

To measure the 3rd order intermodulation, the antenna under test operates at a transmitting mode and the testing probe operates at a receiving mode. The testing probe should have a very good linearity within a large dynamic range of power levels in order not to introduce measurement errors. When measuring, the antenna under test is excited by a given power level at two close fundamental frequencies, and the fundamental and 3rd order intermodulation components are measured and recorded on the receiving side. The operation is repeated for different input power levels, and finally it gives two curves—one is the received power level of the fundamental component versus input power, and the other one is the received power level of the 3rd order intermodulation component versus input power. From the two curves, we can obtain the 3rd order intercept point, which indicates the power dynamic range of the antenna under test for linear operation. Figure 1.17 shows an example of measurement systems for intermodulation in an anechoic chamber, which is installed at the City University of Hong Kong, SAR, China.

Figure 1.17 Example of measurement systems for passive intermodulation (Facility is available at the State Key Laboratory of Millimeter Waves at City University of Hong Kong.)

1.4   System Calibration

Gain and Polarization Calibrations of Standard Antennas   This calibration service is offered primarily for determining the absolute on-axis gain and polarization of standard gain horns, which, in turn, are used as reference standards in determining the gain and polarization of other antennas by the gain comparison technique. The antennas need not be identical. This method is the most accurate technique known for absolute gain and polarization measurements. For gain measurements, the uncertainties are typically 0.10–0.15 dB. Uncertainties of 0.05 dB/dB for polarization axial ratio measurements are typical.

Near-Field Scanning Techniques   With this technique, gain, pattern, and polarization parameters are calculated from near-field amplitude and phase measurements taken over a surface close to the test antenna. The absolute gain can be determined to within about 0.2 dB, the polarization axial ratio to within about 0.10 dB/dB, and sidelobe levels down to −50 dB or −60 dB. The exact uncertainties in these parameters will depend on such factors as the frequency, type, and size of antenna. Calibrated probes are normally required for these measurements. In order to achieve accurate results with the planar, cylindrical, or spherical near-field method, the transmitting or receiving properties of the probe must be known. With this information, the measured data can be corrected for the nonideal pattern and polarization properties of the probe. Probes are characterized by a three-step process: (1) The on-axis gain and polarization properties are measured using the technique described; (2) the far-field amplitude and phase patterns are measured for two nominally orthogonal polarizations of the incident field; and (3) the on-axis and pattern data are combined to obtain the probe correction coefficients at the desired lattice points for the measurement surface specified.[5, 31]

1.5   Remarks

In this chapter, we briefly presented the fundamental parameters of the antenna so beginning engineers can quickly grasp the meaning of antenna parameters. We also gave examples of different printed and simple antennas and their advantages and disadvantages. In addition, we touched on the measurement techniques for antenna parameters. We indicated the significance of the calibration of these systems to provide reliable measurements. The reader who needs more detail should refer to the references provided.

References

1. C. A. Balanis, *Antenna Theory Analysis and Design*, New York: John Wiley & Sons, Inc., 2005; C. A. Balanis, *Modern Antenna Handbook*, New York: John Wiley & Sons, Inc., 2008.
2. S. Drabowitch, A. Papiernik, H. D. Griffiths, and J. Encinas, *Modern Antennas,* 2nd edition, New York: Springer, 2005.
3. J. de Kraus and R. J. Marhefka, *Antennas for All Applications,* 3rd edition, New York: McGraw-Hill Science/Engineering/Math, 2001.
4. W. L. Stutzman and G. A. Thiele, *Antenna Theory and Design,* New York: John Wiley & Sons, Inc., 1981.
5. *ANSI / IEEE Standard Test Procedure for Antennas,* ANSI/IEEE Std 149, New York: John Wiley Distributors, 1979.
6. Richard C. Johnson and Henry Jasik, *Antenna Engineering Handbook*, 2nd edition, New York: McGraw-Hill Book Company, 1984.
7. Summitek Instruments, www.summitekinstruments.com.
8. S. Hienonen, "Studies on microwave antennas: passive intermodulation distortion in antenna structures and design of microstrip antenna elements," Dissertation, Helsinki University of Technology Radio Laboratory, Espoo Finland, March 2005.
9. Ramesh Garg, R. B. Garg, P. Bhartia, Prakash Bhartia, Apisak Ittipiboon, and I. J. Bahl, *Microstrip Antenna Design Handbook*, New York: Artech House Inc., 2000.
10. P. L. Sullivan and D. H. Schaubert, "Analysis of an aperture coupled microstrip antenna," *IEEE Transaction on Antennas and Propagation*, vol. 34, no. 8 (August 1986).
11. D. M. Pozar, "A reciprocity method of analysis for printed slot and slot-coupled microstrip antennas," *IEEE Transaction on Antennas and Propagation*, vol. 34, no. 12 (August 1986).
12. K. Wu, J. Litva, and R. Fralich, "Full wave analysis of arbitrarily shaped line-fed microstrip antennas of arbitrary shape," *Proc. IEE, Pt. H*, vol. 138 (1991): 421–428.
13. A. K. Schackelford, K. F. Lee and K. M. Luk, "Design of small-size wide-bandwidth microstrip patch antenna," *IEEE Antennas and Propagation Magazine*, vol. 45, no. 1 (February 2003).
14. P. Besso, D. Finotto, D. Forigo, and P. Gianola, "Analysis of aperture-coupled multi-layer microstrip antennas," *IEEE Antennas and Propagation Society International Symposium Digest*, vol. 3 (28 June to 2 July 1993): 122 –1227.
15. K. M. Luk, C. L. Mak, Y. L. Chow, and K. F. Lee, "Broadband microstrip patch antenna," *Electronics Letters*, vol. 34 (1998): 1442–1443.
16. K. F. Lee and R. Chair, "On the use of shorting pins in the design of microstrip patch antennas," *Transactions—Hong Kong Institution of Engineers,* vol. 11, no. 4 (2004).
17. Z. N. Chen and M. Y. W. Chia, *Broadband Planar Antennas,* New York: John Wiley & Sons, Inc., 2006.
18. G. Mayhew-Ridgers, J. W. Odendaal, and J. Joubert, "Single-layer capacitive feed for wideband probe-fed microstrip antenna elements," *IEEE Transactions on Antennas and Propagation,* vol. 51, no. 6 (June 2003): 1405–1407.
19. Z. N. Chen, "Impedance characteristics of probe-fed L-shaped plate antenna," *Radio Science*, vol. 36, no. 6 (2001): (1377–1383).
20. Z. N. Chen, M. Y. W. Chia, and C. L. Lim, "A stacked suspended plate antenna," *Microwave and Optical Technology Letters,* vol. 37, no. 5 (June 5, 2003): 337–339.
21. P. K. Panayi, M. Al-Nuaimi, and L. P. Ivrissimtzis, "Tuning techniques for the planar inverted-F antenna," *IEE National Conference on Antennas and Propagation,* no. 461 (30 March–1 April 1999).
22. M. Tong, M. Yang, Y. Chen, and R. Mittra, "Design and analysis of a stacked dual-frequency microstrip planar inverted-F antenna for mobile telephone handsets using the FDTD," *IEEE Antennas and Propagation Society Internatíonal Symposium,* vol. 1 (2000): 270–273.

23. M. C. Huynh and W. Stutzman, "Ground plane effects on the planar inverted—F antenna (PIFA) performance," *IEE Proc. Microwave Antennas and Propagation,* vol. 150, no. 4 (August 2003): 209–213.
24. L. Zaid and R. Staraj, "Miniature circular GSM wire patch on small ground plane," *Electronic Letters,* vol. 38 (February 2002): 153–154.
25. T. K Lo and Y. Hwang, "Bandwidth enhancement of PIFA loaded with very high permittivity material using FDTD," *IEEE Antennas and Propagation Society International Symposium,* vol. 2 (June 1998): 798–801.
26. T. K Lo, J. Hoon, and H. Choi, "Small wideband PIFA for mobile phones at 1800 MHz," *IEEE Vehicular Technology Conference,* vol. 1 (May 2004): 27–29.
27. C. R. Rowell and R. D. Murch, "A capacitively loaded PIFA for compact mobile telephone handsets," *IEEE Trans. Antennas and Propagation,* vol. 45, no. 5 (May 1997): 837–842.
28. N. P. Agrawall, G. Kumar, and K. P. Ray, "Wide-band planer monopole antenna," *IEEE Transactions on Antennas and Propagation,* vol. 46, no. 2 (1998): 294–295.
29. K. Wong, C. Wu, and S. Su, "Ultra-wideband square planar metal-plate monopole antenna with a trident-shaped feeding strip," *IEEE Transactions on Antennas and Propagation,* vol. 53, no. 4 (April 2005): 1262–1268.
30. K. Wong, C. Wu, and F. Chang, "A compact wideband omnidirectional cross-plate monopole antenna," *Microwave and Optical Technology Letters,* vol. 44, no. 6 (March 20, 2005).
31. G. E. Evans, *Antenna Measurement Techniques,* New York: Artech House Inc., 1990.

Base Station Antennas for Mobile Radio Systems

Brian Collins

BSC Associates Ltd. and Queen Mary, University of London

As well as their obvious function of providing a link between a base station and a mobile station, base station antennas provide an essential and increasingly important tool for the control of frequency re-use and the optimization of channel capacity in a mobile radio network. In this chapter, we examine the way in which these considerations contribute to the specification of the performance parameters required and the engineering means by which these parameters can be provided.

The descriptions are of general application to many frequency bands, so to avoid tedious repetition, the following bands will be defined and usually referred to by the short names shown. In the context of multiband arrays, 850-MHz and 900-MHz bands will be referred to as the *low band* and 1710–2170-MHz as the *high band*. The nomenclature for base stations and mobile stations differs between air interface specifications, but for convenience the terms *base station* (BS) and *mobile station* (MS) will be used here without implying reference to any particular air interface. To avoid repetition, the following comments apply throughout this chapter:

- *Reciprocity* The principle of reciprocity applies to most of the performance parameters of an antenna. Gain, radiation pattern, efficiency, and many other characteristics have the same value whether an antenna is transmitting or receiving a signal, so in the discussion that follows the direction of transmission is chosen to suit the simplest understanding of the phenomena described.

TABLE 2.1   Frequency Bands, Nomenclatures, and Uses

Frequency band	Short reference	Service
450–470 MHz	450 MHz	Phone + data
824–890 MHz	850 MHz	Phone + data
870 (880)–960 MHz	900 MHz	Phone + data
824–960 MHz (850 and 900 MHz)	Low bands	Phone + data
1710–1880 MHz	1800 MHz	Phone + data
1850–1990 MHz	1900 MHz	Phone + data
1900–2170 MHz	2100 MHz	Phone + data
1710–2170 MHz (1800, 1900, and 2100 MHz bands)	High bands	Phone + data

- ***Intellectual property***   Many of the topics discussed in this chapter relate to areas in which a wide variety of patents have been granted; it should not be assumed that because a technique or physical structure is described here it is free of patents or other commercial IP.

- ***Frequency bands***   Table 2.1 includes most major worldwide assignments but other frequency bands are allocated to mobile radio services in some countries. Future bands for UMTS or additional 3G services are not included. Following the transfer of broadcast TV services to digital format, a significant amount of the present analog TV spectrum will be reassigned to mobile services, although the extent to which these may be common on an international basis is not clear at the time of writing.

2.1   Operational Requirements

The dominant concerns of mobile radio operators are to optimize network capacity within the allocated number of channels and to obtain the maximum revenue generation from the installed equipment base. These economic objectives imply securing the most intensive possible frequency re-use and controlling the geographical distribution of channel capacity to match the distribution of user demand in space and time. The physical location of base stations and the engineering of their antenna systems are invaluable tools in this optimization. The development of the network and the antenna systems employed will be determined both by current usage patterns and also by those foreseen in the future. An important feature of modern antenna design is the increasing availability of techniques that permit the adaptation of antenna characteristics to current usage patterns in real time.

In the discussion that follows, we assume the reader has a general understanding of how mobile radio air interfaces operate and is familiar with the general principles of network and cell planning. Special attention is given to the ways in which antenna specifications are chosen and achieved to optimize network coverage and capacity.

2.2 Antenna Performance Parameters

In order to provide the required azimuth beamwidth and gain, a BS antenna typically comprises a vertical array of radiating elements. The design of the individual elements provides the required azimuth radiation pattern characteristics, whereas the vertical extent of the array and the number of radiating elements in the array is chosen to provide the necessary gain.

The *azimuth beamwidth* of a base station antenna is chosen to suit the frequency re-use plan chosen for the network. Base stations typically support three cells spaced 120° apart in azimuth, although this plan is by no means universal especially where coverage may be limited by buildings or hills or where usage patterns are nonuniform, as in the case of a BS alongside a major highway. Code division multiple access (CDMA) systems require careful control of the overlap between cells to avoid excessive loss of capacity in soft/softer handoff modes, so a 3-dB beamwidth of 65° is the norm. Global System for Mobile Communication (GSM) systems typically use antennas with 65° beamwidth in urban areas but often use wider beamwidths in rural areas. The apparent gap between 65° sectors is filled from an adjacent BS. In dense urban areas the planning of most networks is determined by the need to provide sufficient capacity rather than to cover the largest possible area, so different considerations may apply to the design of BS antennas.

For azimuth angles beyond the edges of the served cell, allowing some overlap for handoff control and optimization, the signal radiated by a BS antenna has no functional use and serves only to impair the C/I ratio in other cells. For this reason, it is desirable that beyond ±60° from boresight the azimuth pattern falls monotonically at as high a rate as practicable and that rearward radiation is suppressed.

Although we are accustomed to seeing rooftop BS antennas in many cities, a substantial number of cells use antennas mounted at lower levels; these make use of the built environment to limit the area they cover—allowing more intensive frequency re-use—and they can more easily be disguised to reduce problems with planning consent.

The use of omnidirectional antennas is usually limited to small in-fill base stations because their ability to support only a single cell limits their capacity and frequency re-use capabilities.

The main BS network is usually provided with antennas having the highest gain economically possible; this reduces the number of stations needed, improves in-building penetration, and helps to resolve some shadowing problems. Severe shadowing problems are better resolved by changing the location or height of the antenna; where a higher antenna causes problems with frequency re-use, adding a microcell to cover the unserved coverage hole may be a better option.

Reducing the *elevation beamwidth* and selecting the elevation angle of the beam maximum provide further methods for controlling the field strength over the intended service area of a cell. If a base station antenna has the maximum of its elevation radiation pattern aligned precisely in the horizontal plane, in many situations the elevation of the antenna over the surrounding area—and at longer distances the curvature of the earth—will ensure the maximum signal will pass over the heads of most users. By directing the beam slightly downward, the field strength for most users within the intended coverage area will be increased and at the same time the power radiated into neighboring cells will be reduced, significantly improving the C/I ratio in neighboring cells that share the same frequency. The effect may be compared with dipping the headlights of an automobile to avoid dazzling approaching drivers.

The elevation pattern of a column of uniformly excited radiating elements is characterized by a succession of minor lobes and nulls both above and below the main beam. Nulls below the main beam—especially the null closest to it—can cause areas of poor coverage close to the BS; side lobes immediately above the main beam can cause interference with neighboring cells if the main beam is downtilted or if the terrain rises between one BS and another. *Elevation pattern shaping* is commonly used to fill at least the first null below the main beam and to suppress the level of sidelobes for some chosen range of elevation angles above the main beam.

In order to serve street-level users from antennas mounted on the roofs of nearby high buildings, antennas with large elevation beamwidths (and therefore with low gain) may be desirable. In this situation many users are typically located well below the horizontal as seen from the BS antenna, and large beamtilts may be used both to correctly illuminate the intended users as well as to reduce interference levels in surrounding cells.

Figure 2.1 shows one way in which we can envisage the frequency re-use situation. Cell A uses frequency f_1 and provides a usable signal as far as distance d_1 in some azimuth direction. Beyond this distance, the signal provided from Cell A is too low to provide a reliable link, but is too high to allow the frequency to be re-used, whereas at some distance d_2 the level of interference from Cell A has fallen to a level that allows f_1 to be re-used by Cell B. In order to provide the maximum network capacity, we need to ensure that signal intensity falls as rapidly as possible with distance and we assist this both by site placement and by downtilting the antenna at Cell A.[1]

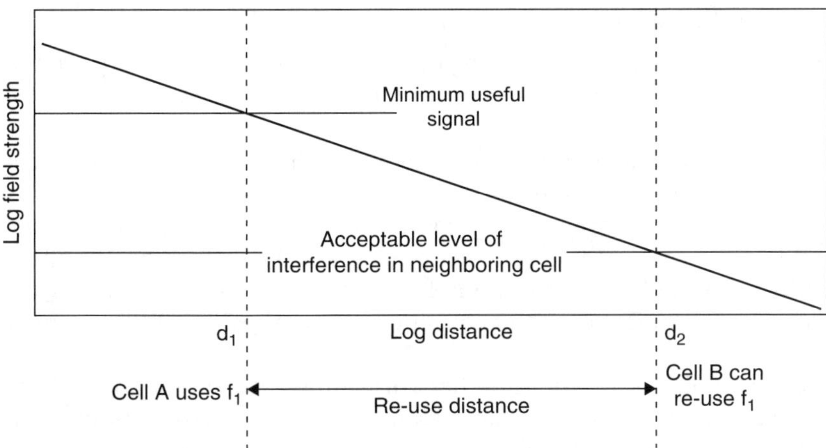

Figure 2.1 Re-use distance. The propagation loss is typically proportional to about $d^{3.8}$. The limiting radius of the cell represents the distance at which there is a reasonable chance the SINR ratio will be sufficient to provide an adequate BER. Beyond this the use of the same frequency will not be possible because of mutual interference. At a sufficient distance the signal level has fallen enough for another cell to re-use the same frequency. The elevation beamwidth of a base station antenna is typically only 5°–7°. By tilting the beam downward significant reduction of the re-use distance is possible.

The angle by which the elevation pattern maximum is placed below the horizontal is known as the *beamtilt* of the antenna. Beamtilt can be provided by two mechanisms. *Mechanical tilt* is provided by angling the base station antenna physically downward, whereas *electrical tilt* is provided by controlling the phases of the radiating currents in each element of the array so the main beam is moved downward. An antenna may have both electrical and mechanical beamtilt, the net beamtilt being the sum of both.

Comparing Mechanical and Electrical Tilt

The effective azimuth pattern of an antenna with electrical tilt diminishes in range as the tilt increases, but it has an essentially constant shape irrespective of the applied tilt. If an antenna is mechanically tilted, there is a tendency for the pattern to become both shorter and wider as the tilt increases, particularly for antennas having azimuth beamwidths in excess of 60°, because the tilt has little effect at azimuth angles far from boresight. In a number of situations, for example an antenna mounted on a wall and firing obliquely from it, an antenna with mechanical tilt is visually much more obtrusive than an antenna mounted vertically and provided with electrical tilt.

2.2.1 Control of Antenna Parameters

We now examine how each of the performance parameters required by the user can be provided by the antenna design engineer.

2.2.1.1 Azimuth Radiation Pattern   The azimuth beamwidth of an array is determined by the design of the radiating element(s) used for each vertical unit (*tier*), together with its relationship to the reflecting surface behind it, used to create a unidirectional beam. The width and shape of the reflector will also control the front-to-back ratio and to some extent the rate of roll-off off the pattern beyond the -3 dB points of the azimuth pattern.

In order to obtain predictable network hand-off behavior, the shape of the azimuth radiation pattern should change as little as possible over the operating frequency band. The MS measures the level of the Broadcast Control Channel (BCCH) GSM or Pilot Channel (CDMA) received from different cells, both of which are functions of the beamwidth of the BS antenna at the BS transmit (TX) frequency; the system assumes that a handoff made by comparative measurements on these channels will be accompanied by a matching change in the BS receive (RX) channel gain, and a significant difference between antenna gain on the TX and RX frequencies may result in problems with uplink quality, unstable handoff performance, or dropped calls. In a CDMA system, the overlap between cells has a critical effect on the proportion of traffic that uses soft/softer handoff modes, a parameter that directly affects network capacity. For these reasons, it is usual to specify tight limits on the azimuth beamwidth as a function of frequency, typically $65°\pm3°$ or $90°\pm5°$ over the whole band. The beamwidth at -10 dB is sometimes also specified, but this is relatively difficult to control by antenna design.

The typical front-to-back (F/b) ratio of a $65°$ antenna is 30 dB, and this is specified to control the frequency re-use characteristics of the base station. Whatever antenna is chosen for use, the network engineer should use measured pattern data in system coverage modeling and should check on the coverage implications of any change in radiation patterns and gain over the band.

If the required F/b ratio is greater than is offered in an otherwise suitable antenna, the effective ratio can be increased by selecting an antenna with a larger electrical beamtilt than is required and mounting it in an uptilted attitude. This places the main beam with the desired beamtilt in the forward direction while tilting the unwanted rearlobe well below the horizontal in the rearward direction.

The usual convention is that the azimuth pattern is specified and measured at an elevation angle that contains the elevation beam maximum, so for an antenna with an electrical downtilt of $5°$, the azimuth pattern is plotted on the surface of a cone with a half angle $85°$ from the vertical. A beamtilt of 5° is understood to mean a *downtilt* of 5°. When plotting elevation patterns in Cartesian form, the x-axis is arranged with angles below the horizon to the right of the y-axis and angles above to the left; although universally labeled *angle of elevation*, the scale is really the *angle of depression*.

2.2.1.2 Gain Gain is a function of both the azimuth beamwidth of the antenna—which will be selected to suit the manner in which cells are fitted together to meet frequency re-use requirements—and its electrical length in the vertical plane. The maximum length of a base station antenna is determined either by the physical size that can be accepted without becoming too obtrusive (as in the low bands) or by the minimum acceptable vertical beamwidth, which diminishes as the length and gain increase. As the antenna length increases, the length and attenuation of internal transmission lines increase; for this reason doubling the antenna length halves the elevation beamwidth and doubles the directivity, but does not double the available gain.

A broadside array delivers the maximum possible directivity if all its elements are excited with equal co-phased currents. In practice the application of beam-shaping requires nonuniform element currents and the directivity obtained falls short of that of a uniform array; this shortfall is known as *pattern shaping loss* and it increases as the extent of applied null fill and sidelobe suppression is increased. (The nonuniformity of the element currents results in a wider elevation beamwidth than would be obtained with a uniform distribution and the reduced directivity is a direct consequence of the increased beamwidth.)

2.2.1.3 Elevation Beamwidth The elevation beamwidth of a BS array is a function of its electrical length in the vertical plane. There is a practical limit to the extent to which the elevation beamwidth can be reduced in order to obtain increased gain because an antenna with a very small vertical beamwidth is physically large and, because of its size and narrow beamwidth, it requires a very rigid support to avoid unacceptable deflection by high winds. In hilly terrain it may not provide good coverage of the intended service area because users may occupy an elevation arc of larger angular extent than the antenna can illuminate. Standard practice is to use antennas up to 8 wavelengths long in the low bands, and up to 12, or occasionally 16, wavelengths long in the high bands. These array lengths provide elevation beamwidths of around 7°, 5°, and 3.5°, respectively.

2.2.1.4 Beamtilt For an array with a fixed beamtilt, electrical tilt is usually effected by adding a linear phase shift across the whole array. This ensures that only a small change occurs in both gain and elevation pattern as the tilt is increased, and the levels of nulls and sidelobes remain almost unaffected for small tilt angles. The directivity of an array falls with the cosine of the beamtilt—because the array appears shorter when viewed from the position of the tilted beam maximum—and also falls because the grating lobes above the horizontal increase in level as the electrical tilt is increased, especially if the inter-element spacing is electrically large.

2.2.1.5 Passive Intermodulation Products In digital mobile radio networks, base station antennas usually function as both transmitting and receiving antennas, providing diversity on reception and air combining[*] on transmission. Transmissions are usually made on more than one frequency, and for orthogonal frequency-division multiplexing (OFDM) and CDMA systems, a single channel may occupy a wide bandwidth with multiple carriers or signals with noise-like spectra. When multiple-frequency transmissions encounter any conductor or joint with a nonlinear voltage/current relationship, the result will be the generation of *passive intermodulation (PIM) products*. These are spurious signals with frequencies related to those that give rise to them.

For two unmodulated carriers, intermodulation products are generated at frequencies given by:

Second order:	$f_2 - f_1, f_2 + f_1$
Third order:	$2f_2 - f_1, 2f_1 - f_2, 2f_1 + f_2, 2f_2 + f_1$
Fourth order:	$3f_1 - f_2, 2f_1 - 2f_2, \dots$
Fifth order:	$3f_1 - 2f_2, 2f_2 - 2f_1, 4f_1 - f_2, 4f_2 - f_1, \dots$

Some of these frequencies are likely to fall in the receive frequency band, and the situation becomes very complex when up to eight carriers are fed into a single antenna. Severe problems are particularly likely if an antenna operates in both low and high bands, where the second-order product is likely to fall in-band.

Radiated PIMs may cause interference for other spectrum users—for example another network sharing the same base station structure—or they may fall in a receive band served by the same antenna. PIM levels are usually specified for a two-carrier test, with a level of −153 dBc for a carrier level of 2×43 dBm (20 W). Achieving this very low level, which ensures the receiver is not significantly desensitized by locally generated

---

*A term used to indicate that different radio frequency channels associated with the same radio system are transmitted from separate antennas rather than being combined into a single transmitting antenna.

PIMs, is a severe test of the mechanical design and construction of any antenna and is only achieved by rigorous attention to detail at every stage. The practical implications of this requirement are discussed later in this chapter.

2.2.1.6 Grounding for Lightning Protection Base station antennas are often mounted on tall structures, and they are likely to suffer the effects of direct or nearby lightning strikes. To minimize the probability of damage to the antennas or the systems connected to them, careful attention must be paid to bonding of conducting parts of the antenna. Antennas are sometimes protected by a lightning spike mounted to the supporting structure, projecting some distance above the top of the antennas, but sometimes the antennas themselves form the highest point on the structure, in which case they must be fitted with a lightning spike and a solid connection to ground, independent of their connecting cables.

2.2.1.7 Mechanical Design A base station antenna is a complex device that must provide highly reliable service over a period of many years. The achievement of stable operating characteristics despite the effects of wind, rain, and pollution requires careful attention to the selection of materials, the design of joints, and the arrangements for the exclusion of water from critical areas. This requires some careful design judgments because the cost and weight of antennas is tightly constrained. Antenna specifications often cite international test criteria[2,3] typically relating to operation at high and low temperatures, thermal cycling, driving rain, salt mist, and mechanical vibration. Specifications also constrain the maximum permitted lateral wind thrust for an antenna and its mounting hardware, for wind in any azimuth direction at a specified velocity, typically 45 m/s (100 mi/h), but varying with local conditions.

The design of mounting hardware to support an antenna, which may be up to 2.5 m in length and weigh 20 kg, is complicated by the wide variety of mounting poles to be accommodated and the requirement to provide for mechanical tilting of the antenna. The unpredictable velocity and turbulence of wind flow over the roofs of buildings has led to occasional unexpected structural failures of antenna mountings. This has resulted from fatigue failure of the mounting hardware or its points of attachment to the antenna, induced by the coupled resonant behavior of the antenna, its mountings, and the supporting structure. In consequence, some operators now wisely specify that antennas mounted in locations where injury could result from a mounting failure—for example where antennas could fall into a street—are fitted with a secondary method of restraint such as a flexible stainless steel wire rope, firmly anchored to the antenna and the supporting structure.

2.2.1.8 Diversity Systems   When a mobile user moves through the environment, the signal received at a base station antenna fluctuates widely in amplitude because of the interaction of direct and reflected signals from the MS. If another antenna is erected some distance away from the first, the signal fluctuations on the two antennas will be uncorrelated. A diversity system uses two antennas and combines their outputs in some way that exploits this lack of correlation. The result of the use of diversity is that the availability of the received signal is increased. For a given signal availability, this increase could alternatively have been achieved by increasing the transmitted power; so *diversity gain* is defined as the equivalent increase in the power required to achieve some stated reliability.

The multipath nature of mobile radio propagation results in the strong frequency dependence of propagation loss. This is mitigated by the use of slow frequency hopping[†] (SFH) in GSM systems and the use of spread-spectrum techniques such as wideband code division multiple access (W-CDMA) and OFDM. The effect of very short losses of usable signal is reduced by digital signal coding.

Spatial Diversity   Spatial diversity was used on the uplink from the inception of many mobile radio systems, and most readers will be familiar with the sight of pairs of antennas firing in each of the three directions of the sectors supported from a base station. These systems normally use vertically polarized antennas; both antennas are also used for transmitting and the availability of two antennas reduces the loss, complexity, and cost of the required combining arrangements.

Polarization Diversity   The signal radiated by a handset is strongly polarized in the direction of its long axis. Seen from the front or rear of a handset user, the polarization will be dominated by the vertical component, but in the lateral direction a handset is typically held at a large angle to the vertical, between the mouth and the ear of the standing or seated user, typically at least 45°. Because of the typical dominance of the vertically polarized component, the diversity gain obtained from a pair of antennas with horizontal and vertical polarization is exceeded by one receiving linear polarization at ±45° from the vertical, so dual slant-polarized antennas are now in almost universal use for polarization diversity systems. (These are often called *dual-polar* or *cross-polar arrays*.)

The use of polarization diversity is well established in urban areas, where propagation paths are characterized by extensive reflections and scattering.[4] A base station employing polarization diversity antennas

---

[†] *Slow* implies that the hopping period is at least one bit long.

uses only one antenna per sector, and three antennas can be mounted closely together on a single pole. This gives an advantage in cost (one dual-polar antenna replaces two space-diversity antennas), and the lower windload and absence of a head-frame reduce the cost of the supporting structure. The reduced visual profile of a polarization diversity base station reduces the difficulty of obtaining planning consent in urban areas.

Further useful diversity gain can be obtained in some environments by the use of both spatial and polarization diversity; the use of 4-branch diversity can be expected to increase, especially for 3G and later systems. Diversity gain can also be achieved on the downlink;[5] it requires more advanced signal processing techniques than receive-path diversity, but the use of high-state modulation formats—for example for High-Speed Downlink Packet Access (HSDPA)—is making its use necessary.

The future migration of mobile radio systems to higher data rates and the quest for increased spectral efficiency is expected to result in more techniques such as multiple-input, multiple-output (MIMO) and space-time block coding as well as other techniques that further exploit the use of multiple antennas.

Slant linear polarization has the unusual characteristic that the signal transmitted by a +45° transmitting antenna is received without polarization loss on an antenna with −45° polarization. The labeling of ports on a dual-polar base station antenna should therefore be regarded as being purely for identification unless it has been agreed whether the designation refers to transmitted or received signals. Fortunately this has no consequence in operation.

2.2.1.9 Effects of Imperfect Antennas on Diversity Performance

To arrive at appropriate specifications, we need to understand how practical antenna performance defects may impact system performance.

Radiation Pattern Defects The implementation of spatial diversity using identical vertically polarized antennas creates no new problems, as the radiation patterns should match closely in both planes and the gain from both antennas in any specified direction will be almost equal. The only significant risk will be of some elevation pattern mismatch caused by manufacturing tolerances or mismatched mechanical tilts.

A dual-polar antenna creates a much greater challenge in ensuring accurate pattern matching for the two polarizations. A dual-polar base station antenna is an unusual device in that it is a linear array in which the elements are arrayed in a plane at 45° to their own principal plane. This has the consequence that some properties of the azimuth and elevation patterns are dependent on one another in unusual ways. Figure 2.2 shows the two copolar azimuth radiation patterns of

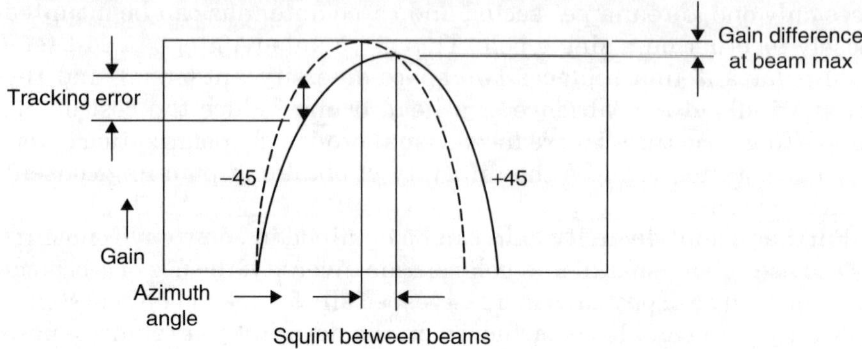

Figure 2.2 Tracking error is the combined result of differential beam squint, peak gain difference, and pattern shape differences. Some contribution to the peak gain difference may arise from a difference in the elevation beamtilt between polarizations.

a dual-polar antenna, one relating to each polarization plane. It will be seen that these exhibit a number of defects:

- The electrical boresight directions of the two patterns are different; the angle between them is referred to as the *squint* angle.

- The two patterns have slightly different gains at beam maximum.

- The slopes on the two sides of each pattern are slightly different

The gains provided by the two patterns at any specified azimuth angle are different by an amount known as the *tracking error.*

The largest practical effect arises from the tracking error; which parameter contributes most is of interest to the designer but has little effect on operation. At the uplink frequency, a large tracking error reduces the available diversity gain; at the downlink frequency, it causes poor handoff between sectors because a mobile making a handoff decision by measuring the received C/I ratio on the BCCH may be assigned a channel on the other polarization (not that occupied by the BCCH), which is lower in mean field strength by the amount of the tracking error at the assigned channel frequency. Handoff between cells is initiated when the signal from the serving cell falls below that of another cell by a margin set by a hysteresis parameter. This margin may be eroded by the tracking error, potentially leading to repeated unwanted handoffs (*ping-ponging*); near the outer edge of the cell—the unexpected change in level between channels caused by tracking error could also lead to a dropped call.

Squint arises from mechanical or electrical asymmetry in one or both arrays (±45°) relative to the mechanical and electrical azimuth axis of the array. A patch element fed from one edge is not a perfectly balanced device and suffers a tendency to squint in its E-plane. In a slant-polar array, this is in the 45° diagonal plane because of the orientation of the elements. In a vertical array, it has the rather unexpected effect of creating an azimuth squint that becomes larger as the beamtilt is increased—an example of the coupling of parameters between planes seen in these arrays. The squint is usually frequency dependent and so is most troublesome when the patch is used in wideband arrays; it can be rectified by using a balanced feed system for the patch, either by driving it in antiphase from opposite edges or by exciting the radiating patch from a balanced structure below it.

In order to recognize the practical limitations of array design, as well as reflecting the significance of potential coverage and handoff defects, the permitted tracking error is usually specified separately between and beyond the nominal –3-dB points of the azimuth pattern. There is probably no need for the related contributory parameters to be separately specified—although they often are.

The elevation radiation patterns in each polarization should match as closely as possible, and because of the high rate of change of signal level with elevation angle below the main beam, the beam tilts should remain equal across the operating frequency band.

Cross-Polar Discrimination   Most dual-polar antenna systems are required to have a high degree of *cross-polar discrimination (XPD)* so each port receives signals only with the designated polarization. It can be shown that the XPD needed to provide effective polarization diversity is not large; this is fortunate because achieving a constant polarization angle over a wide range of azimuth bearings is not easy. A diagonal patch or crossed dipole has a polarization angle of 45° in the boresight direction, but as the angle off boresight increases the polarization angle tends towards 90° (vertical) simply because of the geometrical arrangement. The XPD in the boresight direction is typically around 23 dB, whereas at the edges of a 120° sector it is likely to fall to around 10 dB. The practical result of this is a progressive fall in diversity gain as a distant MS moves off the array axis. To some extent, this behavior matches that of a pair of spatial-diversity antennas where the lateral antenna spacing falls with the cosine of the angle from boresight.

The incoming signal from an MS generally has elliptical polarization with an arbitrary axial ratio and polar angle. Because of this, the diversity behavior of a BS antenna depends on the *orthogonality* of the polarization responses of the two receiving arrays rather than the scalar XPD; the computation of orthogonality requires the measurement of the complex radiation patterns of both arrays for E_V and E_H, i.e., when the antenna is separately illuminated by plane-polarized signals with vertical and horizontal polarization.

Cross-Polar Isolation (Interport Isolation)   We noted previously that with a spatial-diversity antenna system, both antennas are commonly used for the downlink, with half the transmitters connected to each antenna through a hybrid/filter network. In this configuration, there is at least 30 dB isolation between the transmitting antennas so this figure was (and still is) used in the specification of the required intermodulation performance of BS transmitters—unfortunately, achieving 30 dB isolation between the two input ports of a dual-polar antenna is not so easy. Dual-polar antennas usually comprise arrays of dual-polar patches, crossed dipoles, or square dipole arrays, and significant coupling exists between the +45° element of one tier and the −45° elements of the adjacent tiers—another outcome of arraying the elements along their diagonal direction. To avoid the radiation of intermodulation products generated by the PA stage of the transmitters, it is necessary to exceed 30-dB isolation, at least when measured between the transmitter ends of an antenna feed system.

Dual-polar BS antennas provide an example of the fact that the cross-polar isolation (XPI) of a dual-polar antenna is, in general, not related to its boresight XPD.

2.3   The Design of a Practical Base Station Antenna

Engineers unfamiliar with wideband radiating element design should consult the chapters of this book that describe the basic principles of many of the designs mentioned next.

2.3.1   Methods of Construction

The first decision in the design of a base station antenna is the type of construction to be used for both the radiating elements and the feed system. Both may be created using printed circuit techniques or using coaxial cables and fabricated radiating elements. The choice between these methods and various hybrid methods is largely a matter of cost. The optimum design will depend strongly on the cost of labor and materials in the location where manufacture and assembly will take place; where labor is inexpensive a more labor-intensive construction will cost less than a form of minimized labor-input construction—for example, an all-printed circuit design that may have relatively high material and processing costs. There is no optimum design for all occasions. These economic considerations are outside the usual remit of the antenna design engineer, but they will strongly influence the choice of the most economic engineering design for a particular company or manufacturing unit.

2.3.1.1 Radiating Elements   There are many designs for the radiating elements of base station antennas and the discussion here is limited to a brief summary of possible classes of design. Radiating element design has responded to the increasing demand for operation over ever wider bandwidths, but in addition to meeting a stringent electrical specification, a successful design must also be capable of low-cost, high-volume production with consistent electrical performance.

The most commonly used forms of vertically polarized radiating elements are variants of the basic designs shown in Figure 2.3. To provide adequate bandwidth, patch elements (Figure 2.4a) usually take the form of stacked patches in which an upper parasitic radiator is excited by a fed patch lying below it.[6] A stacked patch of this format is easy to integrate into a printed-circuit feed network (corporate feed), so the antenna comprises a printed circuit board (PCB) with the parasitic patches attached to it, typically with push-in plastic spacers.

The reflecting plane behind dipole elements may be flat, curved, bent, or have up-standing flanges along its longitudinal edges. Optimizing the spacing of the element from the reflector, together with the reflector profile, is an important means by which the azimuth beamwidth may be controlled over extended frequency bands. Dipoles are balanced structures and must be driven through some form of balun, the most common forms for BS antennas being the *Pawsey stub* and its printed-circuit derivative the Roberts balun, sometimes called the *hairpin balun* (Figure 2.4).

Microstrip feed networks can be etched on a low-loss laminate with a groundplane on the opposite face, but the cost of suitable materials is high, especially for large low-band antennas. Alternative constructions

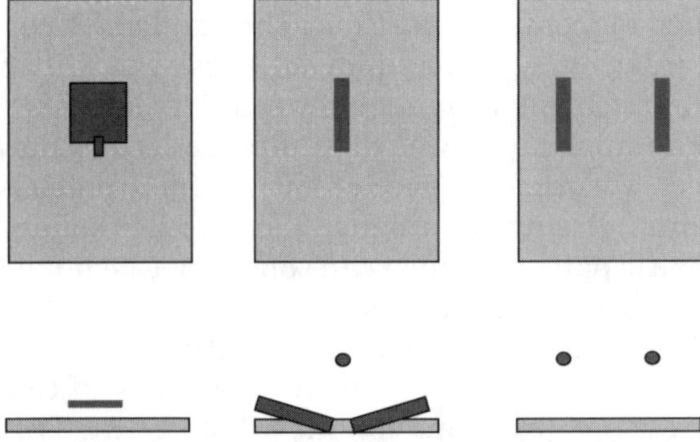

Figure 2.3   Typical designs for a vertically polarized radiating element. The reflecting plane behind the elements may be flat, curved, bent, or have up-standing flanges along its longitudinal edges. Optimization of the spacing of the element from the reflector, together with the reflector profile, is an important means by which the azimuth beamwidth may be controlled over extended frequency bands.

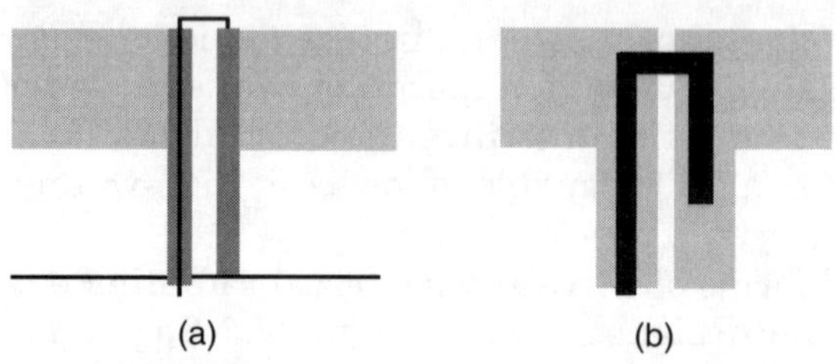

Figure 2.4   Pawsey stub (*a*) and Roberts (hairpin) balun (*b*)

make use of stamped or laser-cut sheet metal conductors or conductors etched on thin low-cost material with the feed system supported on spacers so the effective dielectric is air (Figure 2.5). As a further variation, some constructions use a layer of rigid or semi-rigid foam material to support the transmission line plane above ground.

Figure 2.5   Microstrip line configurations: (*a*) line etched on the same substrate as the ground plane, (*b*) fabricated air-spaced line, (*c*) conductor etched on thin suspended low-cost laminate, and (*d*) line layer separated from ground by low-loss foam

Dual-polar antennas generally use one of three formats of radiating element shown in Figure 2.6, each derived from those just described.

Simple *patch elements* generally have insufficient bandwidth, although a patch with air below it and sufficiently elevated over ground may approach what is needed for some applications. The bandwidth can be increased by stacking parasitically excited patches above the driven patch or by driving a single patch via a capacitively coupled probe.[7]

A *stacked patch element* comprises a lower driven patch, often integrated with a microstrip feed network, with one or more parasitic patches suspended in a parallel plane.[8] It may be designed using either square or circular driven and parasitic patches, which can sometimes be mixed (for example, a round fed patch with a square parasite). In the simplest configuration, the fed patch is excited in two positions mutually at right angles (Figure 2.6*a*), but over a large bandwidth the tendency of this configuration to squint can cause problems for the whole array. This tendency is corrected if for each polarization the lower patch is fed at two opposite points with balanced antiphase voltages, but the configuration is topologically difficult to realize in microstrip format because the feed lines cross one another. Placing the patch in an environment that is itself electrically symmetrical improves matters if the surroundings support the wanted balanced mode but not the unwanted unbalanced

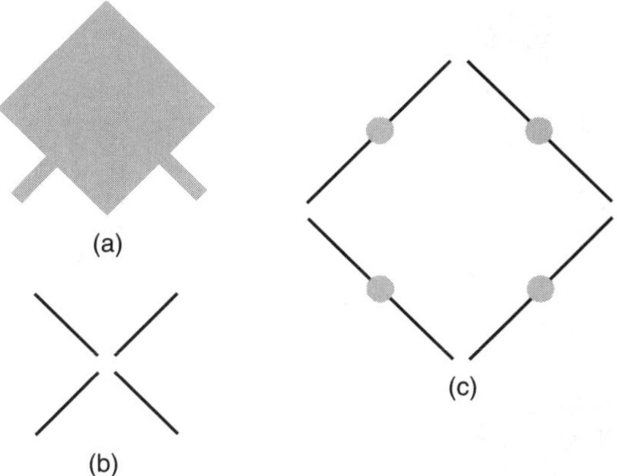

Figure 2.6   Dual-polar radiating elements: (*a*) patch,
(*b*) crossed dipole, and (*c*) square dipole array

mode. A parasitic patch or a ring radiator can be excited by a structure
similar to a crossed dipole, creating a hybrid system.[9,10]

The natural azimuth beamwidth of a dual-polar stacked patch is
around 72°, but it can be reduced to 60° by shaping the reflector in
which it is placed, for example, by bending up its edges. The beamwidth
can also be increased by placing dielectric under the parasite, reducing
the distance between its radiating edges, but this tends to reduce its
impedance bandwidth.

Crossed-dipole elements (Figure 2.6*b*) are used in many designs; when
spaced a quarter wavelength above a reflector, they provide an azimuth
beamwidth of around 90°. To provide stable impedance and pattern
characteristics over wide bandwidths, the individual elements may take
forms approximating bowtie dipoles, pairs of corner-driven squares,
or pairs of rings. The radiating elements may be constructed using
printed circuits, metal castings, or electroplated plastic injection mold-
ings. Crossed dipoles are usually mounted on Pawsey stub or Roberts
baluns, which, as well as supporting the dipoles $\lambda/4$ above the reflecting plane, provide both a high balance ratio and effective impedance
compensation over the necessary extended bandwidths. Parameters
available for optimization include the length, width, and flare angle of
the dipoles and their distance above the reflector. In some designs, the
limbs of the dipoles are bent into a V-shape, sloping backward toward
the ground plane (Figure 2.7). Further parameters are the Z_0 of the
balanced and unbalanced transmission lines of the balun. The effective
shunt capacitance between the inner dipole terminals is an important
impedance optimization parameter because it interacts with the shunt
inductance of the parallel-line stub intrinsic in the balun.

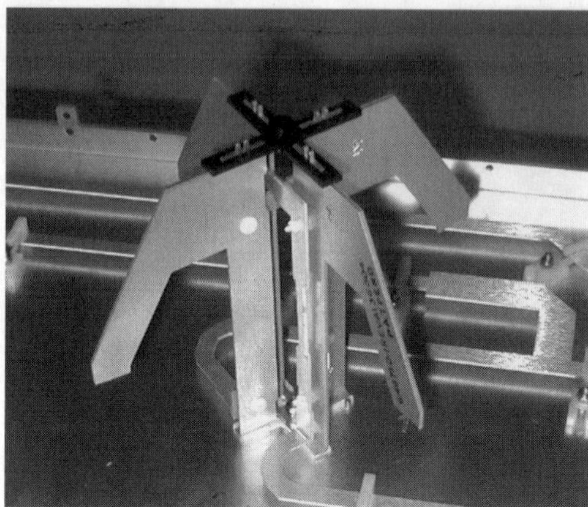

Figure 2.7   Crossed-dipole element pressed and bent from sheet metal with an air-spaced microstrip feed (Photo courtesy of Andrew Corporation)

A Roberts balun provides additional optimization parameters compared with a Pawsey stub because varying the Z_0 of the microstrip line at any point is easy, and the open-circuit $\lambda/4$ line on the unfed side of the balun provides additional impedance compensation. It is possible to arrange for the point at which the feed crosses from one balun leg to the other to be at different heights above the groundplane for the two members of the crossed pair and some very simple and elegant designs have been created in this way.[11]

A *square dipole array*, as shown in Figure 2.6c, provides a further option for a dual slant-polar array with 65° azimuth beamwidth. The radiating currents in the dipoles on the opposite sides of the square are in phase, one pair providing +45° and the other −45° polarization. The individual dipoles can be designed using any of the techniques described previously.

Log-periodic dipole antennas (LPDAs) are sometimes used as array elements. These can be placed with slightly wider interelement spacing than dipoles because the radiation pattern of individual LPDAs has low sidelobes near the vertical axis of the array—the direction in which grating lobes first appear as the element spacing is increased. In a long array (say 8λ or more), there is no significant gain advantage relative to the use of other element forms, although some examples have been produced that have very high F/b ratios. Although individual LPDAs may have wide bandwidths, they are not easy to use effectively in long arrays for applications requiring a bandwidth such as might be needed to cover both low-band and high-band frequencies. Dual-polar LPDAs are bulky, all their dimensions being a substantial fraction of a wavelength at the lowest operating frequency, and the advent of polarization diversity has reduced the use of this element form.

2.3.1.2　General Aspects of Radiating Element Design　The achievement of a well-controlled elevation pattern, stable with frequency, depends on the quality of the impedance match of individual radiating elements. Individual tiers should have a VSWR less than about 1.2:1 across the band of interest. This is a stringent requirement, but unless it is met, errors in the nominal radiating currents in each tier, created by the effect of mismatch at the power dividers in the feed network, will seriously prejudice elevation pattern performance. The reason for this is illustrated in Figure 2.8. If two mismatched loads are connected to a simple branched power divider, then so long as the lengths of transmission line connecting the junction to the loads are equal, the impedances presented at the output ports of the divider will be equal (though mismatched) and equal currents will flow in the loads. Assuming the divider has been well designed, its input VSWR will tend to about the same value as that of the elements. However, if the loads are connected by *unequal* lengths of line, then the impedance plots of the two elements will have rotated round the center of the Smith Chart by different amounts, depending on the electrical distances between the loads and the junction. At the junction, the unequal impedances will cause unequal currents to flow in the two output branches; these will differ from the design values in both amplitude and phase. As the frequency changes, the ratio of the complex load currents will change, so if the load mismatch is large the power division ratio will be strongly dependent on frequency. The same arguments apply to dividers with any number of branches.

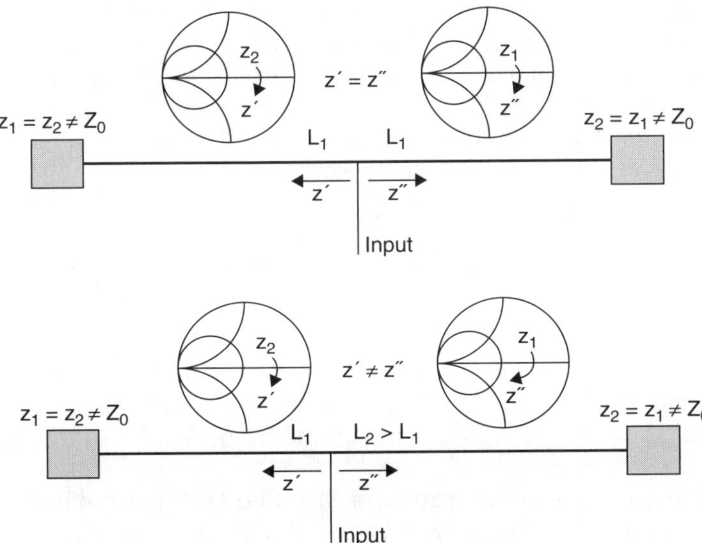

Figure 2.8　Illustration of the reason for the error in power division that occurs if the load impedance is poorly matched

Avoiding these effects requires close impedance matching of individual tiers of elements over the whole frequency band. For a linear array, the azimuth beamwidth and its stability with frequency, the control of squint, and azimuth tracking behavior are all directly determined by the design of the radiating elements. These are all major factors in determining their design, and they are often overlooked in proposals for new broadband element configurations by engineers who have not encountered the practical aspects of BS array design.

Cross-polar discrimination (XPD) is the ratio of the output of an antenna to an incoming signal with its nominal polarization relative to its response to the orthogonal polarization. In the case of ±45° slant linear polarization, we can easily see that for a simple array such as just described, there is an intrinsic problem in maintaining a high XPD over a wide azimuth sector. If we consider an array of dipoles oriented at 45° to the horizontal and placed in front of a vertical reflecting plane, then, in the boresight direction, the polarization will conform with the required 45° plane (although the asymmetry caused by a narrow reflector will limit the XPD). However, as an observer moves away from the array axis, the angle between the plane of polarization and the horizontal gradually increases until at 90° from boresight the plane of polarization is vertical. Unless we design some more elaborate radiating structure, the XPD for signals with ±45° linear polarization will always fall close to 0 dB at ±90°. A practical antenna will have an XPD exceeding 20 dB on boresight, falling to around 10 dB at the −3dB points of the azimuth pattern. Because the incoming polarization of the real signal is an arbitrary ellipse that changes in axial ratio and polarization angle as the source mobile moves through the environment, the precise polarization of the receiving antenna in a particular direction is not really important; what matters is the *orthogonality* of the polarization responses of the two sets of elements. Orthogonality may be computed from the complex radiation patterns of the two arrays as follows[12]:

$$\frac{P}{P_{max}} = \frac{\left| E1_{+45\Upsilon}\, E2_{+45\Upsilon} + E1_{-45\Upsilon}\, E2_{-45\Upsilon} \right|}{\sqrt{E1_{+45\Upsilon}\, E1^{*}_{+45\Upsilon} + E1_{-45\Upsilon}\, E1^{*}_{-45\Upsilon}}\ \sqrt{E2_{+45\Upsilon}\, E2^{*}_{+45\Upsilon} + E2_{-45\Upsilon}\, E2^{*}_{-45\Upsilon}}}$$

where $E1_{+45°}$ is the +45° field component from Port 1; $E2_{+45°}$ is the +45° field component from Port 2; $E1_{-45°}$ is the −45° field component from Port 1; $E2_{-45°}$ is the −45° field component from Port 2; and E^{*} is the complex conjugate of E.

Many arrays are designed to operate over 824–960 MHz (a total bandwidth of 15.3%) or 1710–2170 MHz (24%), and over these bands they must maintain closely constant azimuth patterns, impedance, and cross-polar isolation.

Antenna constructions employing coaxial cables for the feed network generally use cables with solder-flooded copper wire braid. Cables of this type are easy to bend; they stay in place once formed; and they have much lower PIM than cables with conventional wire braid. Some cable-fed array designs make use of printed-circuit power dividers; others are all-cable designs, using cables of different characteristic impedance, sometimes connected in parallel, to form impedance transforming sections. The electrical design of feed networks is discussed below.

In general, the construction methods for the radiating elements and the feed system should be chosen together; particular care must be taken with the interface between them in terms of mechanical integrity, impedance matching, the generation of PIMs, and cost. This interface is often between copper or copper alloys (in the feed network) and aluminum alloy (for the radiating elements). A direct joint between these has a high galvanic contact potential and will rapidly corrode in a warm, humid atmosphere. The availability of reliable low-cost tin plating on aluminum has provided one popular solution, allowing coaxial cables to be soft-soldered directly to aluminum alloy dipoles or patches.

For azimuth beamwidths narrower than about 55°, a tier is often constructed using two radiators. As their lateral spacing is increased to reduce the beamwidth, the level of azimuth sidelobes rises. The achievement of low sidelobes with a 30° beamwidth is difficult and solutions exist in which three elements are used, with their lateral spacing and relative currents chosen to provide a clean azimuth beam.

2.3.2 Array Design

Having chosen the type of radiating elements and feedlines to be used, the next task is the design of the array.

2.3.3 Dimensioning the Array

The gain of a BS antenna is typically many times larger than can be provided by a single radiating element, and it is necessary to array a number of elements to achieve the required gain and pattern characteristics. If the array were uniform and lossless, its directivity would depend only on the azimuth pattern of the elements, the electrical length of the array, and the number of elements used to fill it. In order to minimize cost, we wish to use the minimum necessary number of elements, so we must understand how the vertical spacing between the elements affects array performance. Note that, at this stage, we refer to *directivity* because we have not yet investigated the inevitable losses within the array that will reduce its gain.

A typical relationship between array directivity and element spacing is shown in Figure 2.9, which relates to a uniform broadside array of

n \ d/λ	¼	⅜	½	⅝	¾	⅞	1
4	2.45	3.39	4.29	5.21	6.05	6.84	6.95
5	2.94	4.05	5.30	6.45	7.55	8.59	8.86
6	3.44	4.87	6.29	7.81	9.15	10.37	10.77

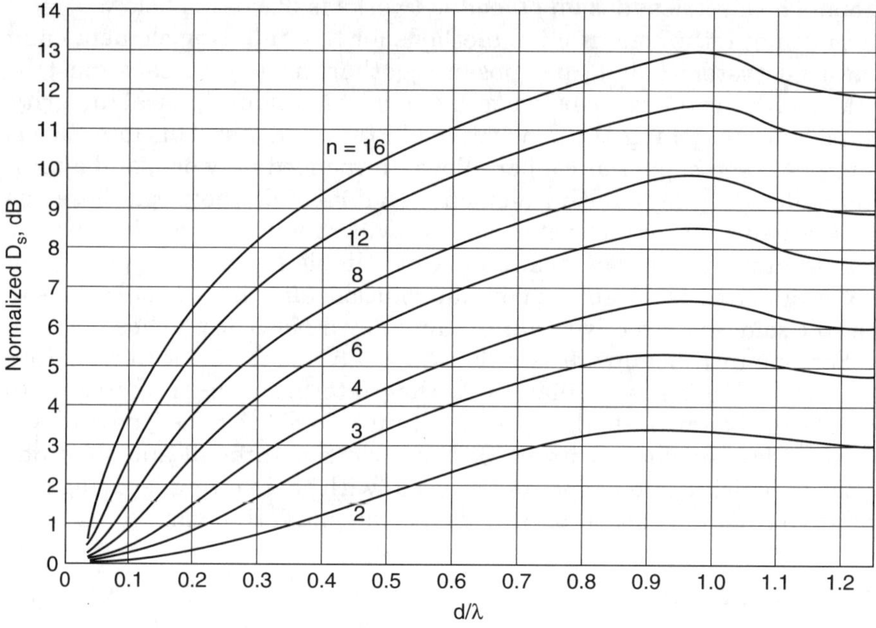

Figure 2.9 Directivity of a broadside array of short dipoles as a function of element spacing, normalized to that of a single element

short dipoles. Directivity increases to a maximum when the spacing is around one wavelength. At a spacing of one wavelength the signals from each element add in phase in three directions: the broadside direction, where we want the main beam, and also along the array axis in both directions. As the spacing exceeds one wavelength, the directions of these *grating lobes* of the array factor move away from the axis, causing the array sidelobe level to increase at a rate determined by the elevation pattern of a single tier. The main beam continues to shrink with increasing spacing but progressively more energy is lost into the grating lobes. To maximize the array directivity over a specified bandwidth, we could choose a spacing that provides equal directivity at the upper- and lower- band edges, but if we add electrical tilt to move the main beam downward, the upper grating lobe will grow in amplitude and the loss of directivity caused by overspacing will increase. We must, therefore, choose the element spacing after computing the directivity/frequency relationship with the maximum required electrical tilt applied. The rate at which the grating lobes grow as the spacing exceeds 1λ depends on the elevation pattern of the individual elements—those with significant radiation close to the array axis must be spaced closer than elements with low radiation in this direction.

To optimize both the technical and the economic parameters of array design, establishing a gain budget is helpful (Table 2.2). Begin with an estimate of the number of elements needed (n) and repeat the process for a different number if the calculated net gain is unsatisfactory.

TABLE 2.2 Gain Budget

Reference (*see below*)	Parameter	dB
a	Directivity of a uniform array of n omnidirectional elements at the chosen spacing	
b	Reduction in directivity at band edges	
c	Beam shaping loss	
d	Allowance for imperfect current distribution	
e	Max/mean power ratio of the azimuth pattern	
f	Directivity of the array	
g	Loss in radiating elements	
h	Attenuation in feed network	
i	Attenuation of cable from connector to primary power divider	
j	Radome loss	
k	Net gain	

$a.$ This value is obtained by integration of the elevation pattern of a collinear array of the chosen radiating elements. For a vertically polarized array an approximation may be made by using the patterns for an array of short vertical dipoles, but for slant-polarized arrays this approximation may lead to underestimation of the grating lobes and a consequent overestimation of the directivity of the real array. The elevation pattern maximum of individual tiers must lie in the horizontal plane. Although the beamtilt of the array will be determined by the array factor, if the individual tiers fire slightly upward, the maximum total directivity will be reduced, especially for arrays with a large electrical downtilt. A slight downward tilt is more acceptable because the element pattern and the array factor are likely to be tilted in the same direction.

b. Because the electrical length of the array increases with frequency, it is often assumed that the elevation-plane directivity of the antenna will also increase with frequency. However, the directivity/frequency relationship is determined by the chosen spacing. The usual specification requirement is to achieve the largest possible *minimum* directivity across the required frequency band. This leads to the choice of the largest acceptable inter-tier spacing consistent with acceptable grating lobes, so the directivity may fall with frequency at the upper band edge. Errors in element currents will tend to be higher at band edges than at band center, and the attenuation of the feed system will increase with rising frequency. To ensure that all these effects are understood and taken into account in the array design, the gain budget should be established for the band-edge frequencies as well as for the center frequency (see also item *e*).

c. A typical array specification will require the first null below the horizon to be filled to at least −18 dB and the two sidelobes above the main beam suppressed to at least −20 dB relative to the main beam. This will result in a beam shaping loss of about 0.3 dB.

d. This factor accounts for the difference between the intended radiating currents, used in the computation of directivity, and those realized in practice. This difference may be as small as 0.2 dB, but is often larger, especially at the band edges.

e. The max/mean ratio of the azimuth pattern can most easily be understood by considering an idealized pattern in which the antenna uniformly illuminates a 90° azimuth quadrant. Compared with the power necessary to provide a given field strength over a complete 360°, the quadrant antenna will require only one quarter of the input power. The ratio of the area of the whole circle to that of the quadrant pattern is 4. In a general case, if we plot the azimuth pattern in terms of relative field, the power in each direction is proportional to the E^2 (i.e., the radius squared) and the area inside the curve is the integral of r^2 over 360°. For any azimuth pattern plotted on a relative field scale, the numerical value of the max/mean power ratio is the ratio of the area of the outer circle ($E = 1$) to that of the azimuth pattern plotted within it. This relationship is true for any number of tiers in the array, and we can obtain its total directivity by multiplying the (numerical) directivities in the azimuth and elevation planes, calculated separately. Any tendency for the azimuth beamwidth to vary with frequency will cause a change in max/mean ratio that will in turn impact the gain/frequency behavior of the array. The highest array gain will be obtained if the azimuth beamwidth is kept close to the specified minimum value at all frequencies in the operating band.

f. This net directivity of the array is the upper limit to the gain that we can achieve, even if the antenna were lossless ($f = a - b - c - d + e$).

g. Losses are inevitable, and will tend to be higher for elements and feed structures etched on low-cost PCB substrates, whereas fabricated metal designs generally have low losses. Unfortunately the estimation of losses by many simulation programs is not accurate and is often optimistic. Element efficiency can be measured by comparing the input power with the integral of the radiated power measured over a sphere in an anechoic chamber. As we have seen, the input VSWR of individual elements should be kept below 1.2:1 for other reasons; at this level the reflection loss is only 1% (0.04 dB) and can be ignored.

h. For an antenna with length 8λ or more, the attenuation of the feed system is considerable, especially in the high band, and the need to control it determines the diameters and types of coaxial cables, and the dimensions and materials of microstrip lines. Losses also cause heating of the feed system of transmitting antennas and limit their maximum power rating, so in a mobile base station antenna both attenuation and heating effects must be considered. Requirements for elevation pattern shaping and the need to control currents to avoid excess shaping loss mean that the feed system must have predictable and repeatable phase characteristics. Because feed networks are usually deployed in a very limited space, cables are formed on assembly with bend radii often close to their permitted minimum limit. Despite this they must accurately maintain their phase lengths; for this reason, cables with solder-dipped braid have become popular because they are reasonably flexible and retain their position and electrical properties after bending. An array design with excessive internal loss will be suboptimal in several ways: for a given gain, it will be longer than it needs to be; it will have a narrower elevation beamwidth than would normally be associated with an antenna of the specified gain; and it will have a larger visual profile, a higher windload, and higher cost.

i. Tower-mounted antennas can be fitted with an external connector at the rear midpoint of the antenna, so the internal cables extend only from the midpoint to the upper and lower ends of the array. Antennas to be pole-mounted are usually fitted with bottom-mounted external connectors, so internal cables must extend from the bottom of the antenna to the midpoint and from there to the ends of the array. This arrangement produces higher internal losses, especially on high band.

j. A radome causes both reflection and absorption of energy radiated by the antenna elements. The extent of these effects depends on the dielectric constant, dielectric loss factor, and thickness of the radome.

Materials used successfully have included glass-reinforced plastic (grp/fiberglass), polystyrene, ABS, ASA, and UPVC. For outdoor installation, the material must be stabilized against ultraviolet deterioration. In the case of grp, the resin used must be chosen with care. A typical resin absorbs a sufficient amount of water to change its permittivity between a dried-out condition under prolonged hot sunshine and equilibrium water content in wet weather. In an experiment by the author, a pultruded polyester resin grp radome 1.2-m long released a wineglass-full of water (125 g) after being dried out following a 24-hour period of exposure to water on its external surfaces. The result of water absorption was a substantial increase in the input VSWR of the antenna.

There are two contributions to the loss caused by a radome. Loss caused by wave propagation through the thickness of the radome is inevitable, but close placement of the radome to the radiating elements results in further loss caused by stored-energy fields, local to the element, intersecting the radome.

k. The net gain is equal to the directivity of the array *(f)* minus the sum of all losses *(g + h + i + j)*.

2.3.3.1 Elevation Pattern Shaping

Because the user specification relates only to the amplitude and not the phase of the signal radiated at different elevation angles, there is no unique set of complex currents that will deliver the specified performance. The antenna designer can choose to constrain a number of variables in order to arrive at a solution that meets the requirements of particular construction methods. A general solution will require both the amplitudes and phases of the radiating currents to be different for each element in the array. Among others, there are classes of solution for which the element currents are equal in magnitude and vary only in phase, solutions in which pairs of adjacent elements have equal complex currents, and solutions in which the required power dividers have a limited number of simple ratios, suitable for construction from coaxial cables.

3.3.3.2 Mutual Impedance

A current flowing in one element of an array induces currents in other elements, particularly in the adjacent elements. The feed impedance of each element is a function of the current that it carries itself and also of the currents in its neighbors. To take a typical example of the third element in a long array:

$$Z_3 = \frac{I_1}{I_3} Z_{13} + \frac{I_2}{I_3} Z_{23} + Z_{33} + \frac{I_4}{I_3} Z_{43} + \frac{I_5}{I_3} Z_{53} + \cdots$$

If the currents in elements 2 and 4 are much larger in magnitude than that in element 3, the input impedance of element 3 may be substantially changed even if the mutual impedance is itself relatively small. If we use ordinary branched power dividers, the change in Z_3 and the corresponding but different changes in other element input impedances will result in an uncertainty in the value of I_3 relative to other element currents. Worse, as the frequency changes the mutual impedances will change, particularly in phase, causing the error in the relative magnitude and phase of I_3 to be frequency dependent. Other array excitations to be avoided include those with large amplitude tapers across the array, causing excessive broadening of the main beam with consequential loss in directivity. We must also avoid any excitation function for which the specified pattern characteristics are very sensitive to the achievement of exact current values.

A wide range of strategies for pattern synthesis are described in the literature, the cited references[13,14] being only a small sample. It is possible to write computer programs using iterative optimization routines and genetic algorithms, the main problem with their use being to specify the constraints described above and to appropriately weight the different features of the solution. Solutions can be tested by applying a Monte Carlo analysis; after, say, 1,000 trials using different sets of random errors, it is possible to list the number of trials that failed because of inadequate directivity, poor sidelobe suppression, inadequate null fill, or any other selected parameter. The use of this method allows a robust solution to be selected from a number of possible candidates and develops a good understanding of the need to control errors in the excitation of a practical array.

For many years the author has used a simple computer-based synthesis technique developed by his former colleague Jun Xiang, which makes use of the superposition of patterns created by superposed excitation functions. In the case of a base station array, we need an asymmetric elevation pattern with adjustments made only to the first lower null and typically three upper sidelobes, so a uniform distribution serves us well as a starting point.

For the purpose of pattern synthesis, we are interested in the region of the pattern within about 20° of the broadside direction, so for elements with limited directivity in the elevation plane (such as dipoles or patches), we can concentrate on the array factor rather than working with the complete elevation pattern.

If we excite the array uniformly, the resulting array factor is a function of the form $E = \sin(nx)/n\sin(x)$, as shown in Figure 2.10. If all currents are cophased, the maximum of this function will be in the broadside direction, but we can apply a linear phase shift across the aperture after the synthesis is complete to move the maximum to the desired angle.

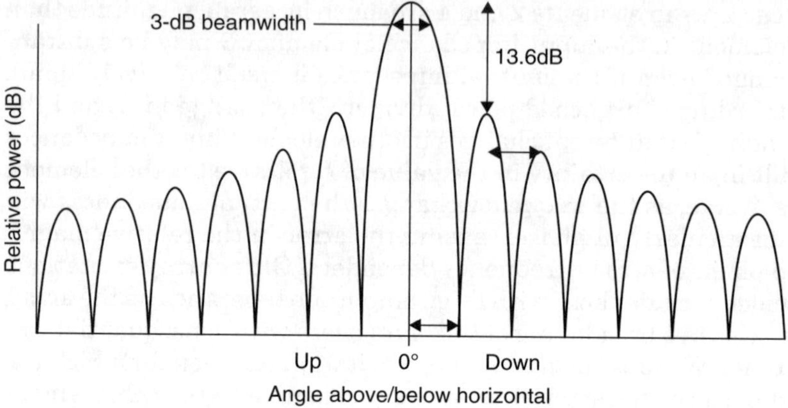

Figure 2.10   Form of the radiation pattern of a uniformly excited broadside array; the horizontal scale depends on the electrical length of the array.

Close-in sidelobes are spaced about one beamwidth apart, but the spacing increases for lobes farther from the broadside direction.

For ease of understanding, the phase reference point in the following discussion is taken to be the center of the array.

1. We begin with a uniform, cophased excitation of the array, producing a pattern of the form shown in Figure 2.10.

2. In order to fill the first null below the horizon, we establish a second excitation of the array, in phase quadrature with the first, with uniform amplitude but with a linear phase shift applied such that the maximum is shifted to the angle of the null we wish to fill. The amplitude of this second excitation is determined by the extent to which the null is to be filled.

3. We apply a third excitation with its maximum directed to the angle of the first upper sidelobe, in phase with the primary excitation and with a magnitude equal to the required reduction in the sidelobe (the first sidelobe of the primary excitation is in antiphase to the main lobe, so applying an in-phase excitation will reduce its magnitude).

4. In the same way, we apply a fourth excitation with its beam maximum directed to the angle of the second upper sidelobe, but in antiphase with the primary excitation (the second sidelobe being in phase with the main lobe of the primary excitation).

5. Some iteration may be needed because the sidelobes of the subsidiary excitations will create small errors in the levels of the null and suppressed sidelobes. If we wish to fill more nulls or suppress more side lobes, we can apply further excitations, but as additional shaping is applied, the directivity of the array may be expected to fall.

At the end of this procedure the currents in each element are given by the vector sum of the currents associated with each excitation that we have applied. To allow for inaccuracy in the practical achievement of the desired current, a margin should be left between the calculated values of nulls and sidelobes and those specified—a margin of 3 dB should generally be adequate. After a suitable array excitation has been produced, it may be modified to suit various practical constraints. It is often convenient to feed adjacent elements with currents of the same amplitude, so for every element pair the mean of the calculated currents can be applied to both. After checking the computed pattern following any adjustments, the required electrical tilt can be arranged by applying a linear phase shift across the array.

To provide a fixed electrical beamtilt as a function of frequency, we must delay the currents in lower elements by a fixed time rather than a fixed phase, as our objective is to tilt the phase-front of the radiated signal away from the normal to the array; the phase angle associated with this delay is directly proportional to frequency. This characteristic is provided by different lengths of transmission line rather than by fixed phase shifters. If we compute the cable length differences required at one frequency (typically we do this at midband), the electrical beamtilt will, in principle, remain constant over the whole band. Errors in the element currents can be regarded as random in phase and with a *circular error probable (CEP)*—the median length of the error vector of, say, 0.5 dB; the CEP can be increased and a Monte Carlo analysis rerun until a significant number of specification infringements occur. This exercise will indicate the precision that must be achieved in the practical control of the element currents. As an example of what can be achieved, Figure 2.11 shows the superimposed measured elevation patterns of a batch of ten 12-element antennas. The degree of consistency of the result is very good above −25 dB.

A wide variety of computer-based tools is used for network coverage planning, and unfortunately many of these have adopted inconsistent conventions regarding the format of the input radiation pattern data. Most antenna manufacturers provide radiation pattern data files on their websites and make them freely available in a selection of the more common commercial program formats.

2.3.3.3 Input Impedance The usual specification for the VSWR at the input of a base station antenna is 1.4:1 or 1.5:1. As well as regarding the antenna as a single unit with an input VSWR specification, we need to consider it as a complex assembly of radiating elements, power dividers, and interconnecting transmission lines, and must consider the control of impedance matching at each stage between the radiating elements and the array input. When designing the intermediate components,

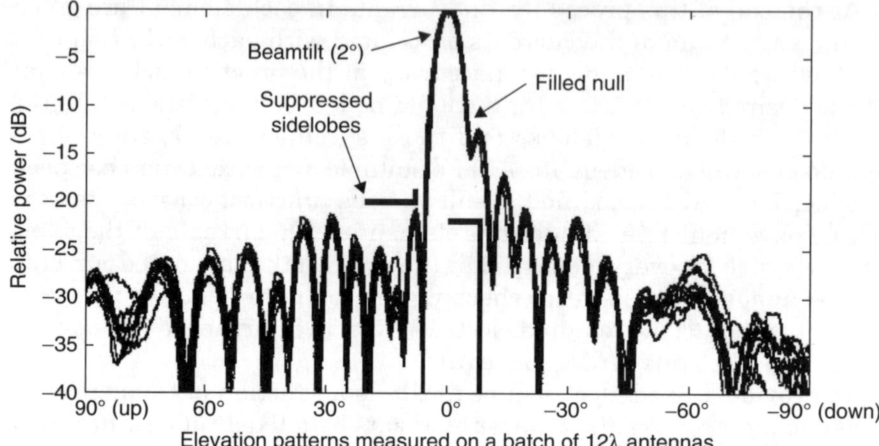

Figure 2.11 Superposition of ten elevation patterns measured at 1800 MHz (Courtesy Jaybeam Wireless)

we will then have a clear understanding of the impedance specification to be met by each one. Because the antenna will operate over a wide frequency band, it is likely that the individual reflections from separate components will add together unfavorably at some frequencies. This is particularly likely for an antenna with 0° electrical tilt, especially if minimal pattern shaping has been applied, because in this case the reflections from various components of the array, like the radiating currents, will be close to being in phase.

An 8-element array is typically constructed as shown in Figure 2.12, where each element is connected through three levels of power divider. Here, the element may have a maximum VSWR of 1.2:1, whereas that of each of the three power dividers should be not greater than about 1.06:1. Considering the large bandwidths involved, these are stringent targets, but their achievement will help control elevation pattern and gain across the operating frequency band(s).

2.3.3.4 Cross-Polar Isolation   For the reasons explained in 2.2.1.8, the cross-polar isolation for a complete dual-polar array is specified

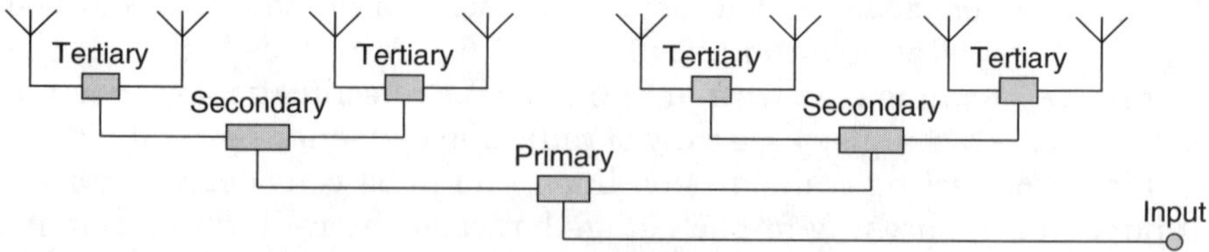

Figure 2.12 Typical feed system for an 8-element array

as >30 dB. The target isolation for an individual tier should be at least this, or perhaps >33 dB, but even if perfect isolation is obtained for a single tier, the mutual impedances between tiers will generally reduce this to around 26 dB, especially for small values of beamtilt. The required isolation can be restored by introducing some intentional compensating coupling. When considering the XPI of a single element pair, coupling may occur because of lack of symmetry in the element or in its environment in the array. A dipole square configuration (Figure 2.6) is symmetrical if the individual dipoles are driven through effective baluns, but a crossed dipole always has some degree of asymmetry caused by the feed method at its center. A square or circular patch or stacked patch is asymmetric unless two balanced feed points are provided in each polarization plane.

Figure 2.13 shows two mechanisms that cause cross-polar coupling in a simple linear array of crossed ±45° elements. Element 2R (solid) is isolated from element 2L, but it couples to elements 1L and 3L. The currents induced in the reflector surface flow in a diagonal direction parallel with the elements, but at the edge of the reflector currents induced by both ±45° elements can only flow parallel with the edge, causing coupling between 2R and 2L.

The mechanisms of element-pair coupling, coupling via edge currents, and intertier coupling all have different path lengths associated with the coupled energy, so they cannot all be compensated in the same way. Coupling between the element pair may best be compensated by some corresponding asymmetry on or near the elements; coupling caused by currents along the reflector edges may require compensation by perturbing the fields at the edges of the reflector, whereas intertier coupling may require compensating coupling between selected adjacent tiers. An effort to offset these separate coupling mechanisms by a single means of compensation is likely to lead to a solution that works well over part of the band but fails elsewhere. The best tactic is to separate the effects, using simulation and experiment, and compensate each mechanism at source.

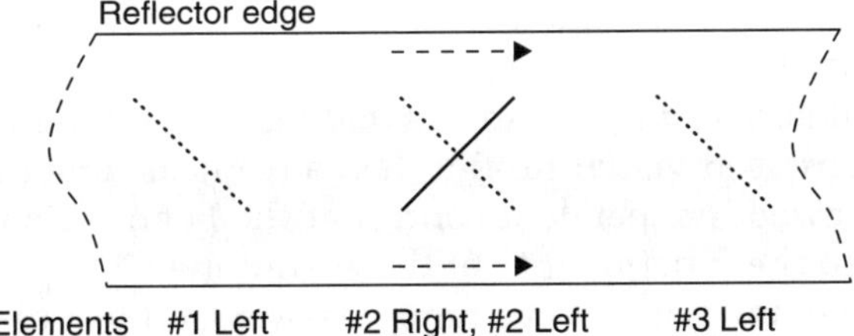

Figure 2.13   Intrinsic cross-polar coupling mechanism

With no steps taken to enhance the isolation, a typical achieved result is of the order of 26 dB for an antenna with a 2° beamtilt and typical beam-shaping, but is lower for an antenna with 0° tilt. The achievement of an isolation of 30 dB or more usually requires additional polarization compensation.

The simplest way to minimize both the unwanted coupling between alternate cross-polarized elements and the asymmetry of the array environment is to place each radiating element in the center of a cavity.[15] This cavity has a side length of a little less than a wavelength, determined by the interelement spacing chosen as described in 2.3.3. Its depth must be sufficient to create the required environmental symmetry, but may also be determined by the target front-to-back ratio (F/b), deeper cavities generally providing higher values. The depth necessary to provide an enhanced F/b ratio usually ensures that the intertier coupling is much reduced and an optimally dimensioned cavity can also reduce the variability of the azimuth beamwidth with frequency.

2.3.4 Multiband and Wideband Arrays

In many countries, an operator commonly has access to two frequency bands for GSM services and an additional band for 3G services. To minimize the costs of antenna systems and tower space rental, as well as to respond to public pressure for less obtrusive antenna systems, operators may use antennas providing service on several frequency bands. The different histories and current environments of operators have created a variety of requirements for these antennas, so there is a correspondingly wide variety of specifications and solutions. Replacing existing single-band antennas with new multiband antennas is always challenging, because the quality of current network coverage must be maintained. Any compromises in performance must be made on the added frequency band, so as a consequence alternative solutions exist for networks approaching dual-band operation from different directions.

In general, low-band GSM operators have added high-band coverage to create additional capacity in dense networks. The use of frequency bands is transparent to users, all of whom now have multiband handsets, and it is not of primary importance that the high-band coverage is contiguous as long as sufficient total capacity is available at any location. Dual-band antennas are attractive for this purpose because they allow extra network capacity to be added without additional antennas, avoiding requirements for added windload and increased visual profile. The advent of remote electrical tilt (RET) has allowed independent control of the beamtilt on each band, greatly adding to the functionality of these antennas.

Where permitted by spectrum availability, many networks that previously had only high-band GSM frequency allocations have added low-band capabilities to enhance their coverage, especially in rural areas. This usually required additional large antennas, but where adequate existing high-band coverage existed, the use of dual-band antennas was not necessary.

The advent of 3G UMTS services added further antenna requirements, and again the optimum solutions depended on what systems and services each network already provided. The preferred network engineering solution is that independent antennas are provided for each band, because this facilitates easy independent optimization, upgrade, and maintenance, but practical constraints have driven the creation of a variety of hybrid solutions. These range from completely separate antennas with independent tower-mounted amplifiers (TMAs) and separate feeder systems, to integrated antennas with common TMAs and single feeder systems to the equipment cubicle. Most solutions fall into one of the following categories.

2.3.4.1 Independent Antennas Mounted Side-by-Side Under a Single Radome These antennas have many arrangements. They generally comprise two arrays of the same overall physical length and are typically comprised of one of the following combinations:

	Array 1	Array 2
a	850 and/or 900 MHz	1800 and/or 1900 MHz
b	1800 and/or 1900 MHz	1900–2170 MHz (UMTS)
c	1700–2170 MHz	1700–2170 MHz

Although both arrays generally have the same physical length, they may have different azimuth beamwidths and polarizations. If equipped with RET facilities, they can also have independent fixed or variable electrical elevation beamtilt. The only constraint on design is that the two arrays have a common physical azimuth pointing direction—though that constraint is being addressed by the introduction of azimuth beamwidth control and beam-steering as described in 2.3.9. Example a is typical of antennas used by a GSM network adding extra coverage on an alternative frequency band. Example b would suit a high-band network adding UMTS coverage, whose requirements for azimuth beamwidth or polarization may differ between bands and the parameters on the lower band must be preserved while 3G capabilities are added. Example c is probably a better solution if the constraints of the older system allow it to be used. The use of broadband arrays enables antennas with a common design to be deployed in 2G and 3G networks, providing a great benefit in terms of logistics. A dual multiband array is less flexible than using separate arrays for each service, although the availability of RET and other remote control techniques is addressing this.

A significant problem encountered with laterally combined arrays is that each array operates in an unsymmetrical environment: the reflector extends farther on one side of each element column than it does on the other, so the azimuth pattern is unsymmetrical and the beam maximum may squint off-axis. The second array suffers a mirror-image squint, so there may be a substantial difference between the boresight direction of the two beams even when the array design is identical for both element columns. A degree of compensation can be obtained if some elements of each array are transposed between columns;[16] although if complete symmetry is to be obtained, the azimuth beamwidth will be reduced. Techniques of this kind must be applied with care because the elevation patterns of a non-collinear array differ as a function of azimuth direction.

Interleaved Arrays   There is approximately one octave separation between the low-band and high-band frequencies, so a high-band array optimally has an interelement spacing that is approximately one half that of a low-band array. As we have seen, this is not a problem if array elements are spaced by slightly less than a wavelength, so, providing we can devise a structure in which high- and low-band elements can be co-located, we can design an entirely satisfactory array in which the radiators for both bands are interleaved as shown in Figure 2.14. The overall dimensions of this interleaved array will be the same as those of a low-band antenna having the same electrical characteristics (gain, azimuth, and elevation beamwidths, F/b ratio, and so on). It is possible to co-locate twice as many elements for the high-band array, so the high-band array potentially has a higher gain than that of the low-band array and would have half the elevation beamwidth. In some cases, a decision is made to use the available aperture for two separate high-band arrays, each having the same number of elements as the low-band array. Ideally the low-band array will cover 824–960 MHz, whereas the high-band array(s) will cover 1710–2170 MHz, although operational requirements may allow some designers to select a subset of the frequency assignments.

Although the azimuth beamwidths of the high- and low-band sections of an interleaved array do not need to have the same azimuth beamwidth,

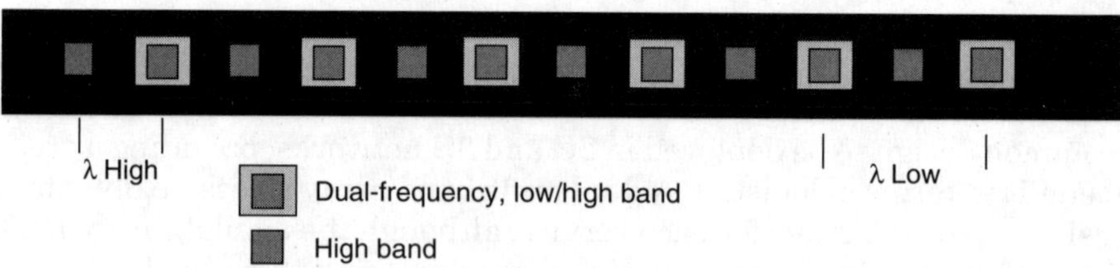

Figure 2.14   Example of an interleaved array

it has become general practice that they do. They also usually have the same ±45° slant linear polarization—antennas where the two bands require different polarizations are usually constructed as side-by-side arrays.

A dual-polar dual-band antenna may contain up to six independent and well-isolated antennas in one housing. Although creating a conceptual design for such an antenna is relatively easy, optimizing it is not. Both band groups have wide bandwidths, so stable performance is not easy to achieve, and many of the parameters available to the design engineer simultaneously influence the performance of both the high-band and low-band arrays.

The requirement for isolation between the cross-polarized ports of a dual-polar antenna is that it exceeds 30 dB. The isolation between the ports for the high- and low-band arrays should also typically have an isolation of 30 dB to avoid unwanted interactions between the transmitters. The close proximity of the elements for each band makes it difficult to provide this isolation directly, so it is common to incorporate a filter circuit in the feedlines where this is needed. The high- and low-band arrays are sometimes combined into only two ports (dual band ±45° ports) by the use of a diplexer—a simple microstrip filter can usually provide the required isolation and power-handling capability.

Wideband Arrays   The reader may wonder why a satisfactory base station antenna design cannot be created by using true broadband radiating elements, such as LPDAs. This is not as easy as it may sound. A dual-polar LPDA with sufficient bandwidth is a quite large device of very considerable mechanical complexity. To avoid grating lobes and consequent loss of gain the required interelement spacing will be around one wavelength at the *high* band, whereas the physical dimensions of each element are of the order of one half-wavelength at the *low* band. Even if the element groups can be compressed by using miniaturization techniques (which usually reduce the available bandwidth and result in lower gain), the very high mutual impedances between the elements of such a close-spaced array create major problems in optimizing elevation pattern performance and input impedance.

Independent Antennas Mounted End-to-End   Two antennas can be mounted end-to-end, one above the other, but for high-gain antennas this leads to very long structures. Antennas mounted end-to-end have occasionally been used to obtain space diversity on a single frequency band, but their small vertical separation results in high signal correlation. (Lateral separation of antennas is more effective than vertical separation because scatterers are predominantly dispersed in the horizontal plane around the MS, not in the vertical plane.)

Compound Arrays   In some instances, integrating a dual high-band array, comprising two separate laterally spaced element columns with a single low-band array, may be required. As an example, such an antenna may provide operation on GSM900, GSM1800, and UMTS2100. In this case, the space is sufficient to place the low-band elements on the center-line of the array with the high-band elements on either side, so no co-located dual-band elements are required.

2.3.4.2 Dual-Band Radiating Elements

Most dual-band radiating elements are conceived as combinations of two separate radiators, with separate feed systems. Combinations include the following:

- A high-band patch mounted over a low-band patch in such a way that the low-band patch forms the groundplane for the high-band patch[17]
- A high-band crossed dipole mounted over a low-band patch[18]
- A high-band crossed dipole surrounded by a 4-square low-band sub-array[19]

The design of these element groups must achieve good control of the equality and stability of the azimuth beamwidth over each operating band, together with adequate cross-polar isolation, despite the presence of conductors in the radiating structure of the "other" band, which act as parasitic polarization-coupling elements. The achievement of a fully symmetrical operating environment for every element is more difficult than it is for a standard single-band array, and it is typical that additional polarization decoupling arrangements will be needed.

2.3.4.3 Operational Considerations

Many mobile network operators now have frequency assignments in two or three bands having selected their BS sites based on the planning of their earlier networks. As new technologies emerge, re-using and adapting their infrastructure to provide support for new networks is a continual challenge. From an operator's point of view, using separate antennas for each system is always preferable, whether or not the antennas operate on different bands, so one network can be optimized, upgraded, or replaced with minimal impact on other networks.

A single multiband antenna provides a solution that is lower in capital, installation, and tower rental costs than an arrangement with separate antennas for each band. The multiband antenna will generally have a lower total weight and windload than separate antennas and will present a much lower visual profile than separate installations for each band/system, especially when considering the complete tower-top equipment including antennas, brackets, and headframes, and the strength, weight, and visibility of the necessary supporting structure. The downside is that a six-port antenna (three bands, two polarizations) is a very complex device and failure in any of its six constituent arrays will require its replacement, potentially putting three networks off the air, possibly for several hours until rigging operations are complete.

2.3.5 Feed Networks

A typical feed network is shown in Figure 2.12. In this example, the input signal is divided in three stages. The primary power divider is a two-way splitter feeding the upper and lower halves of the array; the tertiary dividers each feed a pair of radiating elements; and the secondary dividers have the appropriate number of ports, depending on the number of elements in the whole array. In an array with a number of elements that is not a power of 2 (2, 4, 8, or 16), an additional stage of power division is sometimes used.

2.3.5.1 General Design Considerations The first step in feed network design is to compute the required power division ratio needed at each junction. As noted, it is common for adjacent elements to be assigned currents with equal amplitudes, and this adjustment should be carried out before designing the network as it reduces the number of different power ratios needed.

Power dividers often have only two output ports, especially if cable impedance matching sections are used. If a power divider feeds a different number of elements from each output, then the elements with smaller currents are grouped together so that every power divider has output powers that are as equal as possible; in general, it is easier to ensure a constant ratio from a wideband divider when the output powers are comparable.

The effects of mismatch at junctions can be mitigated by using hybrid or Wilkinson power dividers, but these are expensive and may not be practicable. Not only must the element impedances be well matched, but also any transformation ratio needed between the junction and the transmission line feeding the elements must be achieved in a manner that provides stability over a wide bandwidth. To achieve this, it is common to employ multisection impedance transformers that can be designed initially by using a Smith Chart and then optimized using a suitable simulation program. A high VSWR at the input of the secondary dividers will cause phase errors in the primary divider, resulting in an unstable elevation beamtilt—a problem that must be avoided to prevent troublesome differences in performance between the transmit and receive bands of a frequency-division duplex (FDD) system.

2.3.5.2 Microstrip, Coaxial, and Hybrid Line Systems   Microstrip design is relatively easy because both the Z_0 and the length of any line section can be adjusted at will. Coaxial line dividers must make use of the characteristic impedances of available cables, although cables may be connected in parallel where an alternative impedance is required. Many commercial antenna designs use hybrid techniques to provide design flexibility at the lowest possible cost, using microstrip power dividers connected by coaxial cables. The microstrip components often employ standard microwave PTFE-glass PCB laminates, but alumina and other substrates are also in common use.

The requirements of low PIM, together with the power levels common in high-capacity GSM base stations, has led to the adoption of 7/16-DIN connectors as the almost universal standard connector for base station antennas.

2.3.6 Practical Cost/Performance Issues

The worldwide success of mobile radio services and the prospect of the continuing growth of data- and information-based systems have led to requirements for extremely large numbers of base station antennas. There are usually many ways in which an effective technical solution can be found and antenna design is now strongly determined by the economics of production. Many manufacturers have production facilities in regions with low labor costs, so the emphasis of design has tended to be to minimize material costs while accepting significant mechanical complexity.

There is a significant difference in antenna economics between networks in rural areas, where the network optimization emphasis is on coverage, and networks in urban areas, where network engineering is dominated by the need to maximize capacity. The cost of establishing a base station is high and includes not only the equipment and installation costs but also the costs of planning, backhaul links, site rentals, power, taxes, and maintenance. If the number of base stations required to cover the desired service area can be reduced by using high-cost antennas with the maximum possible gain, this may be a very worthwhile expenditure. In a dense network, the antenna parameter that is of most importance is clean radiation patterns, and the achievement of maximum gain is far less significant, so antennas can be designed using low-cost materials. Optimum network design depends on a wide variety of economic and technical factors, including the local planning régime, labor rates, and site access costs, so there is no single best choice.

2.3.7 Passive Intermodulation Products and Their Avoidance

The advent of multiband networks and antennas, the demand for additional capacity, and the adoption of tower and infrastructure sharing have led to an increasing requirement to reduce PIMs to a level at which they do not impair network performance. The standard test level, -153 dBc relative to two carriers at a $+43$ dBm, is approximately the ratio $(10^{15.3})$ of the distance from the Earth to the sun compared with the thickness of a piece of thin paper. Its achievement is a factor that strongly determines the choice of the materials, finishes, and mechanical design of almost every antenna component. Not only must the PIM limit be achieved consistently in large numbers of production antennas, but antennas must maintain this level of performance for many years while exposed to the corrosive effects of rain, salt, and industrial pollutants as well as the mechanical stresses and vibration caused by wind. Antenna design must be conceived from an early stage in terms of the avoidance of PIMs—there is no easy way in which an inadequate initial design can be modified later in the hope that PIMs can be avoided.

The achievement of low PIM levels depends on following clear rules, both in design and at all stages of production. These include

- Avoid design details requiring any interconductor joint that is not essential to the electrical operation of the antenna. If two conductive mechanical components are in contact, consider how they could be made from a single piece or insulate them so no current can flow across the joint. If current must flow between two large flat surfaces, consider separating them with an insulating membrane and relying on the large capacitance formed between the opposing surfaces. Ensure that every essential conducting joint is tightly compressed and that the force on the joint does not depend on any compressible or creep-prone material. Where possible, reduce the area of contact so the current path is clearly defined and the contact pressure is high.

- Avoid dry solder joints. Well-soldered joints cause few problems, so ensure good access and good lighting so joints can be soldered and inspected easily and also that soldering equipment is right for the job. If lead-free solder is used, the soldering temperature is usually higher than with leaded solder, so these considerations become even more important to success. For printed circuit boards, choose laminate grades that are guaranteed to have low intrinsic PIM levels; these are available from most manufacturers.

- Ensure that printed circuit boards are cleanly etched and well washed after processing. If boards are cut after washing, there is always a risk of new contamination; routing or water-jet cutting are less likely to lead to contamination than laser cutting. Good material badly processed will always fail to deliver the expected result.

- Observe good engineering practice in the selection of metals and finishes at any intermetallic joint. Corrosion, and consequent PIM generation, is driven by the galvanic contact potential between the contacting metals. Protect essential joints from corrosion and wherever possible protect them from contact with water.

- Avoid the use of metals with nonlinear conductivity or contact potentials—nickel is a known PIM source and cannot be tolerated in electroplating or metal alloys. Stainless steel fasteners are frequently used in antenna construction, but they should not form the current path at an interconnection.

- Avoid the use of ferromagnetic materials; the skin depth on these materials is reduced and their RF resistance is correspondingly raised.

- Aluminum is widely used for antenna construction, but it is a very electropositive metal and its surface is always covered by a thin layer of aluminum oxide (alumina). Joints between aluminum components must be provided with corrosion protection and high contact forces.

- Maintain a high level of cleanliness during the handling and storage of components as well as in assembly, testing, and packaging areas.

2.3.7.1 Some Universal Principles Many important design techniques relate to the configuration of joints between conductors, whether these are part of a radiating structure, an internal feed system, or the reflector surface and its supporting hardware. RF currents flow only in a very thin layer on and below a conductor surface; the skin depth at 2 GHz in aluminum is only 1.8 μm. Only the extreme surface layer of the conductors takes part in current conduction, and at every point where conductors touch, the current will pass across the shortest available path. The impedance encountered by the current will change if the surface path geometry changes. At every interconnection, the surface current path must be well defined, maintained independently of production tolerances, and bearing sufficient mechanical pressure to ensure that the path remains stable whatever forces the joint may experience on account of temperature change, vibration, wind load, or other causes. It is essential to remember that in the real world no surfaces are flat—they are always rough and they slope in one direction or another. No surfaces are exactly square, and no conductors are completely rigid. Table 2.3 shows examples that illustrate simple principles that must be followed in every antenna design. These examples are not exhaustive, and they apply whenever the radiating structure and feed system are

TABLE 2.3　Typical Mechanical Details Illustrating Good Design Principles

	Detail	Practical Imperfections	Preferred Detail Example
1	Strip conductors joined to one another or to ground.	Contact pressure at circled point is not defined. Loss, power rating, and PIM are uncertain. Impedance changes if contact point moves.	(a)　(b)　(b) Can also be achieved by bending or pressing strip end(s).
2	Right angle connection between flat conductors.	Any lack of square or deflection of the joint displaces the contact point: loss, PIM, unstable Z.	Move the fixing closer to the bend, or press the flange to define the contact point.
3	Connection through a dielectric layer.	Most thermoplastic materials creep under pressure, so the joint integrity is lost and heating occurs. Brittle material may be fractured if joint is over-tightened.	Add a sleeve over the fastener to remove stress from the dielectric and to form a stable current path.
4	Clamping through a dielectric layer.	The problems are the same as those of the preceding example, but this is worse because the pressure under the fastener is higher for a given joint closure force.	Add a sleeve and a washer to reduce pressure: also observe precautions as in (1).
5	Current path relies on contact through thread.	This is particularly to be avoided. It shares the problems of (3), and the point of current flow between the screw and the upper conductor is undefined. Don't rely on current flowing from a screw into the threaded component.	Defined current path.
6	Spring plug contact is smaller than hole.	These examples are exaggerated to show clearly what is happening. Current path is incorrect: impedance uncertain and current density at contact point is high. Plug will overheat.	Contact is at entry to hole. it is stable and less dependent on spring temper.

chosen. If the reasons for avoiding the details in the left-hand column are understood, a variety of good solutions may often be found. This kind of improvement in design details is not costly and the rewards are of value both to the antenna manufacturer and the user: There will be fewer failures in post-production VSWR and PIM tests, together with improved stability in performance and higher levels of lifetime reliability.

2.3.8　Use of Computer Simulation

The increasing power and affordability of computing resources has led to the development of a range of electromagnetic simulation programs that can be used to accelerate the design of base station antennas.

Although these tools are valuable aids to a competent antenna design engineer, their use is no substitute for adequate understanding of the processes at work in the physical antenna. All the available tools are far better at analysis than at synthesis—given a complete description of an antenna structure, the tools will compute its electrical behavior with considerable accuracy. Optimization routines are usually provided, but these are comparatively unintelligent processes that allow the designer to explore a range of input parameters and compute the corresponding performance of the model, seeking an optimum within a defined search area.

Simulation is powerful in predicting the radiation patterns that will be obtained from a specified structure and its use can significantly shorten the time needed to develop a radiating element and reflector structure with the needed characteristics (VSWR, beamwidths, F/b ratio, and polarization). The element currents required to provide a specified elevation pattern can be obtained either by an interactive or a self-optimizing routine. Having decided on the required radiating currents, simulation is almost essential for the rapid optimization of array feed networks following a first-pass design from basic principles.

One of the most useful aspects of simulation is the way in which the design engineer can examine fields and currents in the antenna. This knowledge provides a key to understanding how the antenna operates, and this in turn helps in the understanding of the performance deficiencies that usually beset a new design. Useful insights are often obtained by comparing the operation of the computer model with that of a physical prototype.

There is a constantly changing point of balance between computer simulation and practical experiment. Dramatic increases in the speed and capabilities of physical measurement systems have provided a much better insight into the performance of antenna prototypes, but neither impressive measurement systems nor the latest simulation tools can replace the competent design engineer who can engage in the creative process of new design and can understand what needs to be done when problems arise.

2.3.9 Arrays with Remotely Controlled Electrical Parameters

Electrical beamtilt has been used since the inception of mobile radio systems as a means of controlling the range of communication from a base station while maintaining near-in coverage in shadowed regions and inside buildings. Its application to reduce the interference to and from users outside the intended coverage area of a mobile radio cell, and so to improve frequency re-use, was described in 1981 by Lee.[20]

The advent of 2G systems with their digital signal formats led to the universal adoption of higher levels of control of antenna elevation sidelobes and the general use of beamtilt to optimize network capacity. When new networks commence operation, they are generally coverage-limited, so antennas are installed with small beamtilts, typically 2° for an antenna with a 5° elevation beamwidth (Figure 2.15). As the networks became denser, it became necessary to increase the mechanical tilt of existing antennas or to replace them with antennas with larger electrical tilts. This process was labor intensive and required manufacturers and network operators to maintain inventories of antennas with a variety of electrical tilts.

The advent of 3G networks provided a further incentive to optimize beamtilts to a higher level of precision than was previously necessary. In a CDMA network, all users share a common frequency and their signals are identified and separated by code sequences assigned to each user. When users are located between cells, or between the sectors of the same cell, their signals are received by more than one cell (or sector), and they are combined together by processes known as *soft* (or *softer*) *handoff.* When a mobile is in soft handoff between cells, it enjoys improved BER

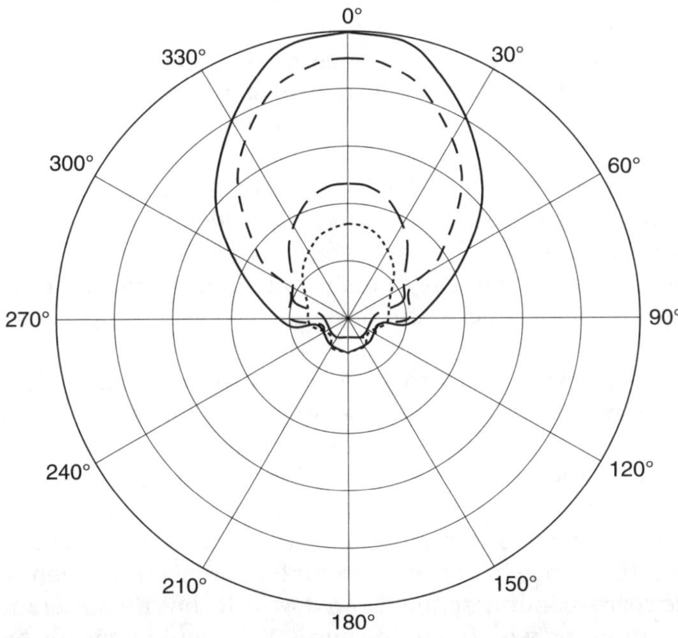

Figure 2.15 Constant field strength azimuth footprint for a 65° antenna with electrical tilts of 0° (outermost), 2°, 5°, and 8°, assuming well-controlled elevation patterns and flat terrain

performance but it consumes resources at both base stations and also in the backhaul link joining them. System optimization requires careful control of coverage overlap, and the adjustment of electrical tilt is a powerful tool to enable this. In a situation in which one cell is overloaded with traffic but an adjacent cell has spare capacity, the use of adjustable tilt can allow the overloaded station to reduce its footprint (and therefore its traffic load) while, at the same time, the adjacent station can expand its footprint by reducing its antenna tilt angle. The ability to remotely control the beamtilt of base station antennas[21] is a powerful technique that can be extended to optimize network capacity by including real-time adaptation to traffic patterns.[22]

2.3.9.1 General Design Issues

The usual method for producing electrical tilt is to apply a linear phase shift across the array aperture. To maintain the levels of sidelobes and null fill, no change must occur in the amplitudes of the currents, and their relative phases must remain unchanged apart from the added linear incremental shifts.

When designing phase shifters, we could consider the use of PIN-diodes or varactor diodes or the use of more mechanical methods such as dielectrically loaded lines, trombone lines, or tapped lines. The severe constraints on PIM generation and the relatively high power ratings required currently rule out the use of electronic phase shifters for BS antennas, so mechanical techniques must be used. Variable transmission-line structures must avoid the use of sliding mechanical contacts, so electrical connections usually rely on capacitance and not on physical contact.

Lines with variable velocity ratio, loaded by sliding dielectric slugs, appear attractive because no conductor contact is needed, but designing a phase shifter with a large range of adjustable phase and low VSWR over a wide frequency band is not easy. Some sliding dielectric arrangements are in use,[23] but many successful designs have used tapped-line phase shifters.[24]

Figure 2.16 shows an 8λ array with a linear phase shift introduced, retarding the phase of successive elements by a variable phase $\varphi°$. To drive an 8-element array, this method requires four phase shifters delivering a phase shift in the range $0°-\varphi°$, together with two delivering $0°-2\varphi°$ and one delivering $0°-4\varphi°$. One simplification that is generally adopted, at the cost of some loss in gain and in the control of sidelobe levels, is to use a fixed phase between immediately adjacent elements, providing a phase corresponding to the mean downtilt that the antenna will provide; for an antenna with $0°-10°$ downtilt, this would typically be set at $s(\sin5°)$ where s is the interelement spacing in electrical degrees at band center. The arrangement now requires only three phase shifters, two providing $0°-2\varphi°$ and the other $0°-4\varphi°$.

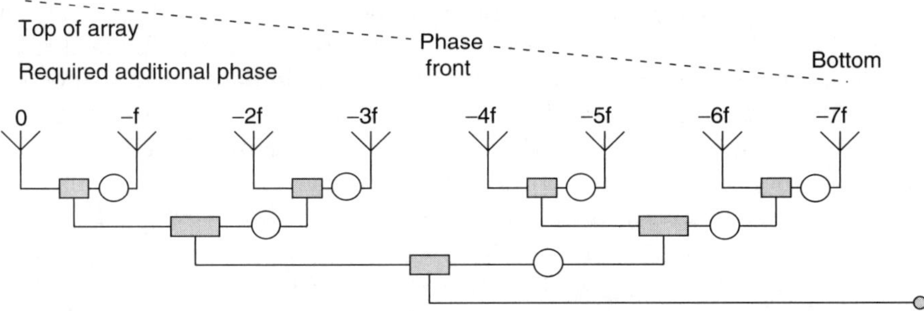

Figure 2.16   An 8λ array with linear variable phase shift

In Figure 2.17 adjacent elements are provided with a common fixed phase difference, and variable phase is applied between the pairs of elements. The phase shifters are designed as tapped transmission lines—this design method has the advantage that a movement of the tap position over a length of line $\psi°$ long provides a relative phase advance between outputs of $+\psi°$ when the tap is at one end of its travel and of $-\psi°$ when at the other end, a relative total of $2\psi°$. If the tapped lines are arranged in circular arcs, the maximum phase shift for a given angular movement of the input shaft can be adjusted by choosing the appropriate radius. Because the phases of the elements at each end of the array move symmetrically about zero, an additional pair of elements with fixed (0°) phase can be introduced at the center of the array. This allows a 10λ array to be driven with only two phase shifters, which can be nested concentrically to provide a compact mechanical construction. An identical arrangement is needed for each polarization, and in a multiband antenna, separate feed networks and variable phase shifters are required for each frequency band.

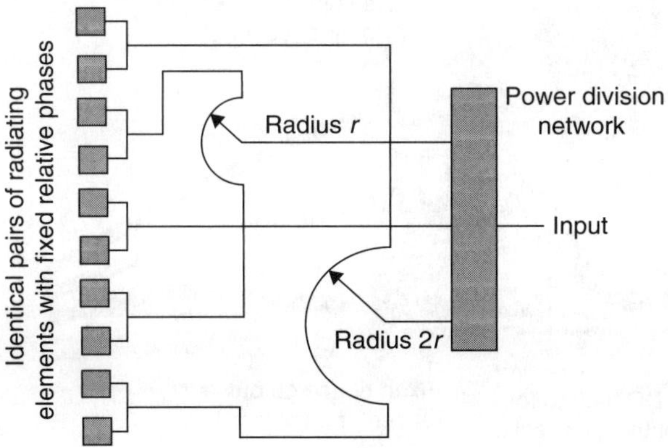

Figure 2.17   Feed network for a 10λ array using two tapped-line phase shifters

A 10λ array with downtilt adjustable between 0° and 10° requires a total phase shift across the array of 562°, so the outer arc phase shifter needs a length of about 280°. If the radiating elements are connected to the phase shifters using lines of equal length, the available tilt will be ±5°, so to provide a beamtilt of 0°–10° an intrinsic 5° tilt must be provided by the fixed lines in the feed network.

There is no single optimum design—each variant provides different technical and economic trade-offs. An array with uniform phase shift between every element at all tilt angles requires more phase shifters but can provide better-shaped elevation patterns and higher gain at extreme tilt angles. A design with concentric phase shifters will be mechanically less complex but may provide less control of the current amplitudes in the inner and outer element pairs—and so may have less ability to shape the elevation pattern. The use of subarrays of two or three elements simplifies the design and reduces cost, but at some compromise to performance over the range of supported tilt angles.

A microstrip design for a variable phase shifter is shown in Figure 2.18. A successful design must provide a low-input VSWR over the whole operating band and for any selected value of phase shift.

A dual-polar 10λ antenna covering one frequency band (including wideband antennas covering 1710–2170 MHz) requires two sets of phase shifters, one for each polarization. Both phase shifters are usually operated from a single drive, and the beamtilts on both polarizations must

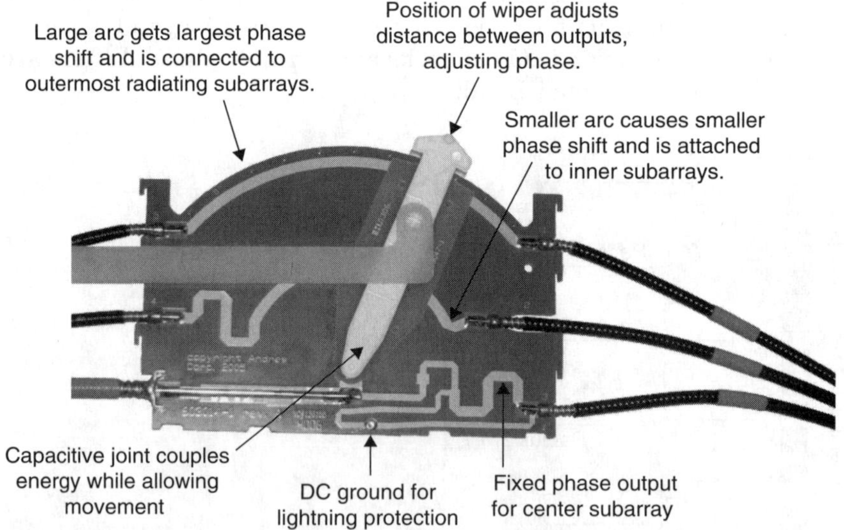

Figure 2.18 Microstrip phase shifter with two concentric arcs (Photo courtesy of Andrew Corporation)

track accurately. The phase shifters are connected to elements at the top and bottom of the array through cables of similar length, so the natural location for the phase shifters is close to the center of the array.

Many dual-band antennas with multiple input ports are used on more than one radio network, perhaps employing different air interface standards, and to facilitate separate optimization on each band, these antennas are almost invariably provided with RET control. Their design combines the engineering challenges referred to in the context of dual-band antennas with those of the control of mutual impedances and broadband polarization decoupling that are needed to provide satisfactory RET performance. For the same application reason, remote azimuth steering (RAS) and remote azimuth beamwidth control (RAB) will also be implemented on these complex antennas.

Direct current (DC) motors or stepper motors are commonly used to provide remote operation of the phase shifters; they are typically located close to the bottom of the antenna to allow access for replacement in case of failure. Early RET antenna designs were fitted with external motor drives, but with the increasing acceptance of remote control techniques, the complete drive system will be installed within the profile of the array (some manufacturers are already providing this arrangement). Some means of manual adjustment is usually provided for use in case of motor failure. A mechanical indicator to show the current tilt setting provides confidence if there is any doubt that the remotely set parameter has not been achieved.

2.3.9.2 Remote Control Interface and Protocol Following early proposals for antennas with RET it became clear that the development of this technology would progress much more quickly if there were a standard interface for the control system. Proprietary systems would present problems if an antenna manufacturer discontinued a product line or if a network operator wished to change its supplier for commercial reasons. Antenna manufacturers, network operators, and infrastructure vendors formed the Antenna Interface Standards Group (AISG) in 2002, and this body, in conjunction with The 3rd Generation Partnership Project (3GPP), has developed interface standards whose objective is to provide interoperability among different vendors' equipment. The standards define a three-layer protocol stack.[25] On the PHY layer, the standards[26,27] define alternative connection means between tower-top equipment and the controller. These are an RS485 bus and a system using a low-frequency modulated subcarrier introduced into the coaxial connection between the base station and the tower-top equipment. Among other parameters, the standards define relevant supply voltages and currents, subcarrier characteristics, connectors, and pin-out details. The data link layer[28] is based on a subset of High-level Data

Link Control (HDLC); the system uses a single primary (control) device, and all other devices (including RETs and TMAs) are secondary devices that respond to the primary device. The application layer[29] defines a set of elementary procedures (a command set) that provides all the required functionality for the connected secondary devices.

The trend toward the increasing complexity of tower-top equipment and the application of RET techniques to other radio systems has led to the continuing development of AISG standards, and the availability of a low-cost control bus on the tower is stimulating the development of other equipment, including various forms of system monitors that make use of a remote data connection from the tower top.

2.3.9.3   Azimuth Pattern Control   For many years, the prospect of implementing "smart" antennas has provided a continuing incentive to researchers. It has been demonstrated that such systems can deliver great benefits in terms of system capacity and range of communication, but their implementation in current mobile radio systems has proved difficult and expensive. There is a large gap between what can be provided by a simple "dumb" antenna and a smart antenna providing adaptive beamforming, C/N optimization, and capacity maximization. RET techniques may be seen as a first step in filling that gap. A further important step is the provision of remote control of the azimuth beam-width[30,31,32] and the azimuth beam direction.[33,34] Antennas using these relatively low-cost techniques will operate with any radio air interface, any modulation technique, at any power level, and with any number of RF carriers. They also provide symmetrical characteristics on transmit and receive bands. Sometimes described as "semi-smart," these methods provide relatively slow changes in antenna characteristics, typically taking seconds or tens of seconds to change the beam shape or position, but they are fast enough to allow adaption to match most changes in the pattern of network traffic demand.[35]

2.3.10   Antennas for TD-SCDMA Systems

Air-interface protocols that employ time-division duplex (TDD) transmission use the same transmission frequency for the up- and downlinks and are able to make generally more effective use of "smart" antenna techniques than are protocols that use frequency-division duplex (FDD). The use of a single frequency allows valid channel state information (CSI) to be obtained rapidly for transmission in both directions. In the case of an FDD system, especially one with large frequency spacing between the up- and downlinks, obtaining up-to-date CSI at the base station for the downlink is challenging; rapid updating requires significant channel capacity and consequent loss of user capacity.

An example of a TDD system is the Chinese Time Division-Synchronous Code Division Multiple Access (TD-SCDMA) standard, and this already employs adaptive beamforming techniques using both omnidirectional and directional antenna arrays.

An example of a directional smart antenna is shown in Figure 2.19. This uses the same methods for the control of elevation patterns as more conventional "dumb" antennas, but the whole antenna array is formed from multiple subarrays of the more conventional type. The multiple subarrays, each a column of elements, are closely matched in performance, and sampling facilities are provided in each column to allow the whole array to be calibrated. As with a conventional antenna, it is possible to achieve diversity using dual-polar elements. A high-band, four-column array is typically around 320 mm wide and allows beam-steering to ±60° from boresight. The amount by which the beam can be steered off boresight is limited by the appearance of potentially high azimuth sidelobe levels as the steering angle increases. To optimize

Figure 2.19 Example of a directional beam-forming antenna for a TD-SCDMA system (Photo courtesy of Comba Telecom)

the received C/I ratio, beam-steering is augmented by the use of null-forming methods by which unwanted signals are suppressed and interference radiated toward the sources of interfering signals is reduced at the same time. In these systems, beams can be switched, and radiation patterns adapted on a user-by-user, burst-by-burst basis. Currently, this adaptivity is available only in the azimuth patterns, the elevation patterns being those of a conventional fixed array. The application of adaptive beam-shaping in the elevation plane adds considerable cost and complexity and may not be justifiable, but RET techniques could easily be applied.

The use of a single-carrier TDD system greatly reduces the specification requirements for PIM levels, so some lower-cost methods, materials, and components can be used without penalty; for example, premium PCB materials are not needed and Type-N connectors are entirely adequate for this application.

2.3.11 Measurement Techniques for Base Station Antennas

Base station antennas present a number of measurement challenges because of their closely specified radiation patterns and PIM performance, as well as the very large volumes of product for which conforming performance must be assured.

2.3.11.1 Radiation Pattern Measurements

Far-field measurements will generally be made within the service area of one or more local networks. For this reason the rotating antenna under test (AUT) should always receive the signal radiated by a fixed illuminating antenna that should be aimed in a direction in which it is least likely to cause interference to any network user.

Accurate far-field measurements depend on the availability of a clear and open test range. Because we are interested in the levels of nulls and sidelobes at levels around −20 dB relative to the main-beam maximum, the traditional criterion for the necessary range length is inadequate, and a range length of at least $4d^2/\lambda$ is needed. (At a range of $2d^2/\lambda$, there is a parabolic phase error of 45° at the ends of the array relative to the center—if you doubt that this is too short, try applying that phase error to a set of typical array currents.) The aperture occupied by the AUT should be probed to establish that the amplitude and phase of the illuminating signal are constant to acceptable limits. Time gating methods can be used to offset the effects of site reflections, but implementing these tends to slow down the measurement process. The azimuth pattern of a base station antenna must be measured at the elevation angle equal to the nominal beamtilt. This requires the use of an azimuth-over-elevation mount for the AUT or a high mounting location for the AUT with the illuminating antenna at a lower elevation. For slant-polar antennas, the AUT must be illuminated in turn with plane waves polarized at ±45°.

Some facilities are equipped with very large anechoic chambers, but if valid far-field measurements are to be obtained, their use requires the application of correction techniques. These may include the installation of movable or multiple source antennas and the recording of complex field values for later computation. Such systems can be considered as a class of very large near-field ranges.

Far-field measurements are fast to perform, and data for a large number of frequencies can be gathered in a single rotation of the AUT. It is also easy to conduct a swept gain/frequency measurement that provides confidence that there are no frequencies at which the gain of the antenna is unexpectedly lower than elsewhere. The main drawbacks are the need for land for the facility, the cost of handling antennas, and problems caused by inclement weather.

When measuring the radiation patterns of a dual-polar array, it is important to appreciate that the radiation in the forward direction has the opposite polarization sense to that of the rearward radiation (seen from behind, the radiating element appears to have a mirror image orientation compared with when it is seen from the front). This means that to assess the F/b ratio, we must compare the forward radiation of one polarization with the maximum rearward radiation of the other polarization (or both of them to be certain of capturing the largest value). Sometimes a total rearward power is computed by adding both polarization contributions power-wise.

Near-field measurements are usually carried out in a screened anechoic chamber; they have the advantage of ease of accessibility and use and being unaffected by weather conditions. The field radiated by the AUT is probed by a small horn or open-ended waveguide. Because the main area of interest in the radiation pattern of a BS antenna is within a few degrees of the horizontal plane, a cylindrical near-field system can be used rather than a full three-dimensional system—although high-angle grating lobes will not be seen. The main constraint of a near-field system is that gain measurements are only available at those frequencies at which a pattern measurement is made, so no frequency-swept gain measurement is possible.

Some near-field ranges are provided with multiple illuminating horns, allowing electronic switching of the source antenna rather than much slower physical manipulation. A system of this type provides fast measurements allowing more frequencies to be measured in a given time, but the security of a continuous gain/frequency measurement is still missing.

With some near-field systems, measuring two components of the E-field (E_V and E_H) may be necessary in order to synthesize a pattern with ±45° polarization; with other systems, illuminating the AUT directly with the required polarization may be possible, so it may take only half the time to measure a single pattern—often this is all that is needed to compare results during the development process. A wideband illuminating horn allows measurements on both low and high bands to take place during one measurement process. The use of a dual-polar horn with electronic switching is an effective way to accelerate measurements.

An ideal arrangement is a combination of both types of range as there will always be some need for the measurements that can only be performed on the far-field range, whereas the ease of use and accessibility of the near-field range is a huge bonus, especially in the winter!

2.3.11.2 Gain Measurements The internal complexity of a base station antenna and the long electrical length of the transmission lines connecting its components make it possible that, in some regions, the gain-frequency curve has minima affecting relatively small frequency ranges. This possibility makes continuous frequency-swept gain measurements important, at least during product development and until confidence is established that a product has stable and substantially frequency-independent characteristics.

Comparing the gain of a base station antenna with a standard gain horn is prone to errors caused by the very different ground reflections that affect measurements with antennas with such different beamwidths (say 6° and 90° in the vertical and horizontal planes for the BS array and around 20° in both planes for the standard gain horn). A better procedure is to use the three-antenna method[36] to calibrate a base station antenna as a secondary standard. By this method, site reflections affecting the (secondary) gain standard and the AUT are similar in magnitude, so more accurate results can be obtained.

2.3.11.3 VSWR and Cross-Polar Isolation Measurements VSWR and XPI measurements are normally made using swept-frequency network analyzers. In development, these will usually be vector network analyzers (VNAs). In production, vector or scalar analyzers may be used, and it is common for the test gear to be controlled by a computer that, following a scan of a barcode on the AUT, sets up the correct frequency bands and test limits, prompts the test technician to make the appropriate connections for each measurement, records the result to an archive, and prints a test certificate.

2.3.12 Array Optimization and Fault Diagnosis

Once a prototype is constructed, it is often found that its performance does not fully meet expectations. A modern dual-polar multiband antenna is a complex structure, and determining what has gone wrong when performance expectations are not met is not easy. Finding the source of the problems requires some clear thinking and careful experimental techniques.

2.3.12.1 Single Radiating Elements and Element Groups

Measurements on a single element or element group will usually include the measurement of azimuth patterns, input impedance, polarization, and XPI. The most troublesome parameter to contain over a wide bandwidth is the azimuth beamwidth, and this is best optimized by a mixture of computer simulation and experiment. Simulation is good for rapid optimization of simple mechanical parameters, for example, the spacing between the radiator and the reflector or the reflector width and the depth of a flange on the front edge of the reflector. When the simulation doesn't produce the desired result, it may be useful to carry out some practical experiments to provide new ideas about what can be done and then to optimize the idea using the simulator.

It is usually worth checking the magnitude of the mutual impedance between adjacent array elements to make sure the driven impedance of the elements will remain within expected bounds when the phases and amplitudes of currents in neighboring elements are changed as a result of beam-shaping or electrical tilt. In a RET antenna, the effect of mutual impedance will be to modify the complex currents in the array elements in unintended ways when the beamtilt changes, causing unwanted variations in the elevation pattern. In an array covering a single band group, mutual impedance can be controlled with separating fences between the elements, but this solution is not easy to apply to a dual-band antenna where adjacent elements are of very different dimensions. For this reason, the elevation pattern performance for dual-band interleaved arrays tends to be less well controlled than for separate low-band and high-band arrays.

A high value of measured XPI for a single tier does not ensure that the performance of the complete array will be adequate, but if the isolation of the components of a single tier is significantly less than the required 30 dB for the complete array, obtaining 30 dB from the complete array will be difficult. It is always hard to compensate a local effect by a compensating coupling at some point electrically far away; such compensation tends to work over some relatively narrow band where the phase relationships between the error vector and the compensating coupling are correctly phased.

In a dual-band array, the coupling between high- and low-band element groups can give rise to high-band radiating currents flowing in low-band elements, and this causes frequency-dependent ripples or other deformities in the high-band patterns of the array. The amplitude of low-band currents flowing in high-band elements is usually small and is seldom troublesome. In an array in which inputs at both bands are combined using a diplexer, excessive coupling at the element may interact with the behavior of the diplexer to cause unwanted ripples in the input VSWR of the high-band array.

2.3.12.2 Feed Networks and Arrays Before building a prototype array, ensure the feed network provides the expected complex currents at its output ports and is well matched when all the output ports are correctly terminated. Although the network may have been simulated before it was constructed, coaxial feed systems are likely to show some differences in the stray reactances associated with cable junctions. Inaccuracy in modeling of interline coupling in microstrip networks may require some impedance correction on a physical model, particularly if line spacings are small to economize on the use of expensive laminate materials.

A simple method for the diagnosis of feed network problems is to sample the outputs of different elements of an array using a balanced loop. As the loop is moved to couple to successive elements, the measured ratio of the currents in the sampled elements should be constant in amplitude, with a phase difference having the correct nominal value at midband and a value at other frequencies proportional to f/f_0. This measurement is easy to carry out using a VNA, and the result can be displayed in polar or Cartesian form. A corresponding measurement can be carried out on the feed network itself, terminated with matched loads, and the difference between the two results is a clear indication of the effects of element mismatch (including mutual coupling). The periodicity of ripples in the response give an immediate indication of the distance separating the sources of interacting reflections, allowing the matching to be improved to reduce the excursions from the wanted values. The results become confusing if the ports of the VNA are not themselves well-matched (as is common), so matched attenuators should be inserted in series at the VNA ports.

The feed network of a RET antenna can be measured in this way to provide confirmation that there are no unwanted interactions within the network and that the output currents are correct for the required range of tilt angles. As an extension of this technique, it is possible to construct an anechoic "box," lined with absorber, with arrangements for a probe to be scanned along the array to measure the relative complex element currents. These sampled currents can be used to compute the elevation pattern of the array, often with surprising accuracy. Measurements of this kind are sometimes used to carry out routine testing of production antennas; they provide rapid confirmation that connections have been correctly made and that the array has no serious problems.

If the feed network appears to provide the correct output currents when measured on the bench with matched loads, yet the complete array shows variability of element currents with frequency, the cause is almost certainly mismatch at the input of the elements. If the elements (or element groups comprising each tier) are known to match correctly when measured separately, then attention must be given to the mutual impedance between them. The mutual isolation between tiers is easy to measure, and the driven impedance of a tier in the presence of its immediate neighbors can be measured using multiple directional couplers or a multiport VNA. The coupling to neighboring elements at 2λ spacing or more can generally be neglected.

2.3.12.3 Passive Intermodulation Products

PIM is usually measured with the AUT placed in an anechoic box with a typical clearance between the antenna and the absorber of around 1 m. The supporting frame or slide on which the antenna is placed during measurement should be made of wood, GRP, or other insulating material. While a workable measurement system can be made using two signal sources, two power amplifiers, a triplex filter, and a spectrum analyzer, a complete integrated system is available for this task, allowing fast and flexible measurement down to a level of around −160 dB for two input signals at +43 dBm. Most BS antenna manufacturers measure PIM on a routine basis on 100% of production. As noted, PIM is likely to be generated at improperly compressed conductor joints, and the general practice is to measure PIM while the antenna is vibrated, sometimes by a motor-driven device, but often using a rubber-faced hammer or a walking stick!

The field intensity close to the antenna created during a PIM measurement is likely to exceed safe limits,[37] and access to the antenna by personnel should be restricted while power is applied by a physical bar or walk-on pressure pad.

The diagnosis of PIM problems in production is not easy. Consistent failure suggests inadequate design and specification, while occasional failing of batches of product may point to a defective batch of material. A sudden increase in failures may indicate a change in a material, a process, or a member of the production team. Isolated failures may relate to a production error (wrong part, loose screw, bad solder joint, or contamination). As with VSWR performance, the use of statistical process control (SPC) is a powerful method for spotting trends and understanding the most likely reasons for failure. Practical measured results will always display statistical variability, but by understanding the scatter of results and the mean values obtained on each product over a period of time it should be possible to relate changes in performance to factors related to design, procurement, and manufacture. Without some history of results, pinning down the reasons for changes is very difficult.

2.3.13 RADHAZ

The possibility that the radio emissions from mobile radio base stations may present a long-term hazard to health has been of concern to radio engineers and to members of the public for many years. Because of the distance between a typical base station antenna and members of the general public, the levels of electromagnetic field (or power flux density) the public encounters are typically three orders of magnitude below the levels recommended by the International Committee on Non-ionizing Radiation Protection (ICNIRP) and most national authorities. In some countries, however, there is considerable public anxiety about this matter and large numbers of base stations are moved every year because of public concern. The public perception of the electromagnetic radiation hazard (RADHAZ) issue relating to base stations is probably that the larger the number of antennas and the greater the visual profile of an installation, the higher the possible risk, so more compact systems are likely to draw less criticism on this matter.

In general the levels of power flux density created by a handset in use next to the head are much closer to the ICNIRP limits, and the total dose of electromagnetic energy received by most individuals is heavily dominated by their handset use.

Riggers and engineers who work on BS systems should be made aware that in close proximity to an antenna they may be exposed to field intensities that exceed recommended limits, and when working on radio towers, all personnel should be required to wear calibrated field hazard monitors. The transmitter(s) connected to any antenna on which they intend to work should always be turned off before accessing the antennas; if they must pass close to the front of any operating antenna in order to reach their working location, the connected transmitter should also be turned off to allow them to pass safely.

Where base station antennas are mounted on short towers or on masts erected on roofs that can be accessed by building residents or other members of the public, it is good practice to install barriers to identify any area in which high power flux densities may be encountered. Small compact base stations built into street furniture usually operate at low power, and there is no area in which fields exceed safe limits for public exposure.

Everyone working with base station antennas, their design, testing, planning, installation, and maintenance should be aware of issues relating to RADHAZ. They should take responsibility for their own safety and that of their colleagues and the general public. Although most research findings indicate that there is probably little ground for concern at field intensities below the recommended limits, the engineering community should remain fully informed about the results of current research, both in order to reassure members of the public and to respond in an effective and timely manner if future findings require any changes in current practice.

2.3.14 Visual Appearance and Planning Issues

For the engineering reasons discussed in this chapter, base station antennas are large devices and have a substantial visible profile. The provision of an extended high-capacity mobile radio network requires very large numbers of base stations, and in many countries each region is served by several competing network operators. Whether in historic cities, residential areas, or rural and wilderness landscapes, the erection of base stations is frequently a matter of controversy, the antennas and their associated supporting structures attracting most criticism. Network operators must usually obtain planning consent for proposed base stations from local authorities, and both to increase their probability of successful planning applications and to be seen as organizations responsive to the feelings of local communities, they have developed a number of solutions to soften the impact of their installations on the visual environment.

Figure 2.20*a* shows a typical three-sector cell site with pairs of space diversity antennas. The large-diameter monopole is needed to provide sufficient stability for the high-gain microwave back-haul antennas, which in this example operate on 18, 23, or 38 GHz. The presence of the microwave antennas and LNAs adds to the cluttered appearance of the installation; the headframe and safety rails add to the visual impact; and the wind forces on all these components must be taken into account when dimensioning the monopole. Large-diameter cables cannot be bent easily—in this example no attempt has been made to hide them and some smaller cables have been left hanging untidily.

The example in Figure 2.20*b* is clearly an improvement. The use of dual-polar antennas has removed the need for a large headframe and the cable installation has been well managed. The monopole is still heavy for a 10-m structure, and various components appear to be standard fitments that were not really needed for this installation. The access ladder and the rail for the fall-arrest system are seen to the left of the monopole.

Dual-polar antennas as so small—at least for the high bands—that a variety of low-profile solutions has been created. Figure 2.21 shows typical Streetworks installations, both free-standing and wrapped around

(a)   (b)

Figure 2.20 (*a*) Space diversity and (*b*) polarization diversity high-band antenna systems mounted on monopoles typically 15-m high (Photos by the author)

a street lighting column. In this second case, the cabinet housing the base station equipment is no more conspicuous that the control box for the traffic light (lower right). Urban sites can usually be connected by underground data lines so there is no requirement for a rigid structure to support a narrow-beam microwave antenna.

Base station antennas have successfully been integrated into advertising signs and their incorporation in a functional flagpole can allow them to be erected with little controversy in many locations where a utilitarian engineering design would be completely unacceptable. Disguise is also possible using dielectric panels that simulate building features. Simple measures like avoiding the projection of antennas above the roof line are very effective—of course, this impacts radio planning, but there are many instances in which this impact can be accommodated. In rural areas, measures such as avoiding placing structures on the skyline, choosing sites near the edge of woodland, or simply painting structures in harmonious colors can limit the impact of base stations. The use of artificial "trees" is well known; although unsympathetic and obtrusive examples can be seen in some locations, the most successful examples are probably simulated coniferous trees located close to coniferous woodlands.

(a)

(b)

Figure 2.21 These base stations quickly become part of the street scene and are hardly noticed by the average passerby. (Photos by the author)

Glass-reinforced plastics (grp/Fiberglass) structures can be used to create visual profiles for base station antenna installations on old and modern buildings. Figure 2.22*a* shows a grp imitation chimney during tests to check the effect of the enclosure on the radiation patterns of the antennas housed inside. In Figure 2.22*b* the antenna and a large amount of electronic equipment is housed in a cylinder that forms an architectural feature. Figure 2.22*c* shows the installation of a "tree."

The Scottish Government's website[38] points out that public works of art have been commissioned that incorporate antennas or complete base stations. They can enhance the landscape and strengthen the identity of a place. Possible locations for public art are in squares and plazas, alongside major transport routes, at transport intersections, or close to important vistas.

The challenge of minimizing the visual impact of base stations varies in different countries according to local architectural styles and the nature of the rural scenery. In some cities, excellent coverage has been provided with minimal impact on the urban scene, whereas other cities have been less active in encouraging network operators to achieve the highest standards of visual engineering. While some of these solutions are very expensive, other measures are relatively inexpensive and should always be considered.

(a)

(b)

(c)

Figure 2.22   Disguised base station antenna solutions: (*a*) A grp structure surrounding the antenna (Photo courtesy of Jaybeam Wireless); (*b*) an integrated three-sector BS antenna with TMAs presented as an architectural feature (Photo courtesy of Powerwave, Inc.); and (*c*) a BS disguised as a tree being lifted into position (Photo courtesy of Kitting Telecom)

2.3.14.1 Site and Antenna Sharing

The sharing of base station facilities exists at several levels. Networks may share sites, but use separate antenna structures; they may erect antennas on a common structure or share antennas. Radio access network (RAN) sharing between operators is becoming common in some countries. A mobile virtual network operator (MVNO) shares an entire physical network belonging to another

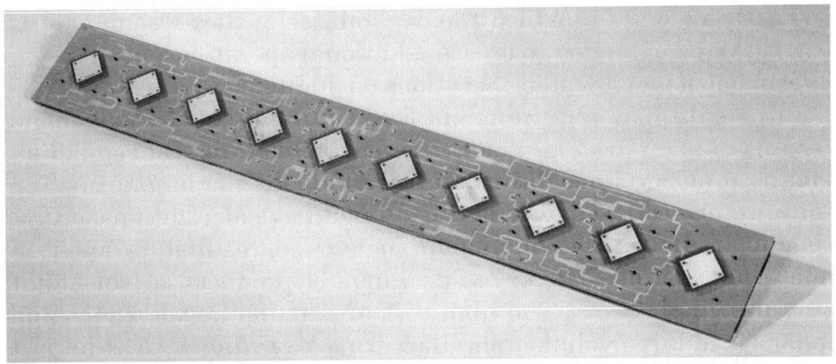

Figure 2.23 Microstrip feed network for a multi-user antenna with independently controlled beamtilts (Photo courtesy of Quintel Ltd.)

organization, which may simply be a hardware facility provider rather than a service operator. Sharing creates some constraints on the ways in which the sharing organizations can design and optimize their networks, but it has an important contribution to make both to network economics and also to minimizing the visual impact of mobile infrastructure. Figure 2.23 shows an example of a dual-polar antenna with a microstrip feed network and stacked-patch elements that can provide individually controlled beamtilts to multiple network users.[39]

In some countries, the radio and TV broadcasting industry has a clear distinction between a small number of companies who operate the hardware infrastructure; they competitively sell capacity to a much larger number of companies who are responsible for the content. This pattern of division reduces the total cost of operation while maintaining a wide variety of choice for the consumer and may be one that mobile radio systems follow increasingly in the future.

2.3.15 Future Directions

Network operators will continue to be under pressure to reduce costs, while at the same time to provide increased capacity, better coverage, and higher data rates to support new user services. The costs of new equipment roll-out on new frequency bands are very high, and the greatest challenge in many parts of the world will be to wring the highest possible levels of optimization from the existing networks. The wide adoption of RET techniques and the advent of remote azimuth steering (RAS) and remote azimuth beamwidth control (RAB) provide new tools for the dynamic optimization of network performance. The network software to integrate these technologies with traffic dynamics is still under development but will potentially have a major effect on network operation. Once this layer of software becomes available, the antenna industry will likely devise new ways in which the capabilities of these "semi-smart" techniques can be extended to create additional operational flexibility.

The licensing of mobile radio operations on additional frequency bands is likely to create new requirements for multiband and/or wideband antennas. This is an area in which the changing policies of regulators may have a significant impact on antenna design and manufacturing. The very large number of combinations of frequencies, polarizations, beam-widths, and other parameters has, for some time, resulted in antenna manufacturers requiring a very wide range of products, a trend that has persisted for 20 years. The high cost of antenna development and the economics of large volume manufacturing have had a great impact on the structure of the antenna industry worldwide and are some of the drivers that have encouraged the advent of wideband RET antennas as well as the rationalization of the industry in recent years.

Moves to provide higher user data rates are stimulating intensive research on multiple antenna techniques. Planning constraints as well as hardware costs and the public's mistrust of large antenna installations will limit the possibility of simply adding more hardware to existing installations, but operators can respond to these new requirements by sharing their antenna estate, pooling their resources (tower space, planning consents, and other hardware), and using them in a more technically advantageous manner.

It seems unlikely that what many would regard as true "smart" antennas will be introduced in any quantity into networks using the existing GSM/UMTS air interfaces,[35] but practice in East Asia shows that they may play a wider and more significant role if future air interfaces are designed to exploit their advantages while accommodating their limitations. Technical developments such as the introduction of transmit diversity and MIMO raise the performance bar of standard antennas combined with semi-smart capabilities. In both cases what is being done is effectively to combine intelligent signal processing with conventional antenna elements, and the fact that there are many different ways in which some near-optimum solution may be found by combining these techniques in various ways comes as no surprise.

There is currently much interest in possible applications for antennas using artificial materials such as electromagnetic bandgap (EBG) structures and frequency selective surfaces (FSSs), and there may be some possibilities for their use in base station antenna design. However, requirements for wide bandwidth and dual-polarized operation, together with constraints on cost and efficiency, may limit their possible uses for mainstream applications.

The use of optical fiber to carry RF signals is well-known, and this technique could be extended to the development of antennas in which beamforming is performed in fiber, the signals being converted to RF at each radiating element. This would continue the existing trend of moving electronics to the tower top and, among other advantages, could provide a great improvement in the power efficiency of base stations.

References

Note: Where patents are cited they should be regarded only as examples for reference. The citations are not always to the primary documents (which may not be in English), and those that are cited as applications may not have been granted. In some cases, the referenced documents may appear to conflict with one another and no opinion is implied or intended as to which may take precedence or claim priority.

1. W. C-Y. Lee, *Mobile Communications Design Fundamentals,* 2nd Ed, New York: J Wiley, 1993.
2. "Environmental conditions and environmental test for telecommunications equipment; Parts 1–4: Classification of environmental conditions: Stationary use at non-weatherprotected locations," *ETS300-019-1-4,* European Telecommunications Standards Institute: www.etsi.org.
3. "Guidance for the correlation and transformation of environmental condition classes of IEC 60721-3 to the environmental tests of IEC 60068 – Stationary use at non-weatherprotected locations," *IEC/TR 60721-4-4:* www.iec.ch.
4. B. S. Collins, "Polarization diversity antennas for compact base stations," *Microwave Journal,* vol. 43, no. 1 (January 2000): 76–88.
5. S. M. Alamouti, "A simple transmit diversity technique for wireless communications," *IEEE J. Select Areas of Comms.,* vol. 16, no. 8 (October 1998): 1451–1458.
6. D. M. Pozar and D. H. Schaubert, *Microstrip antennas: the analysis and design of microstrip antennas and arrays,* IEEE Press, 1995.
7. K-M Luk et al, "Wideband patch antenna with L-shape probe," US Patent 6,593,887.
8. R. B. Waterhouse, *Microstrip Patch Antennas: A Designer's Guide*, Springer, 2003.
9. RYMSA, European Patent EP 1,879,256 A1.
10. Argus Technologies, PCT Patent Application WO2006/135956 A1.
11. CSA Ltd., British Patent GB 2,424,765.
12. N. Cummings, "Methods of polarization synthesis," in *Antenna Engineering Handbook,* 4th Ed. J. Volakis, New York: McGraw-Hill, 2007.
13. R. C. Hansen, "Array pattern control and synthesis," *Proc IEEE,* vol. 80, no. 1 (January 1992).
14. R. S. Elliott, "Design of line-source antennas for sum patterns with side lobes of individually arbitrary heights," *IEEE Trans. AP.,* vol. AP-24, (January 1976): 76–83.
15. Kathrein-Werke KG, US Patent 6,930,651.
16. Kathrein-Werke KG, US Patent Application US 2004/0178964.
17. CSA Ltd., PTC Patent Application WO99/59223.
18. CSA Ltd., PTC Patent Application WO9959223A2.
19. Kathrein-Werke KG, US Patent 7,079,083.
20. W. C-Y Lee, "Cellular mobile radiotelephone system using tilted antenna radiation patterns," US Patent 4,249,181.
21. Deltec Telesystems International Ltd., US Patent 6,198,458.
22. I. Siomina, P. Varbrand, and D. Yuan, "Automated optimization of service coverage and base station antenna configuration in UMTS networks," *IEEE Wireless Comms,* vol. 13, no. 6, (December 2006).
23. Teillet et al, US Patent Application US2005/0057417.
24. Andrew Corporation, US Patent Application US2008/0024355.

25. "UTRAN I_{uant} interface: General aspects and principles," Technical Specification TS 25.460, 3rd Generation Partnership Project: www.3gpp.org.
26. "Control interface for antenna line devices," Standard AISGv2.0, Antenna Interface Standards Group: www.aisg.org.uk.
27. "UTRAN I_{uant} interface: Layer 1," Technical Specification TS 25.461, 3rd Generation Partnership Project: www.3gpp.org.
28. "UTRAN I_{uant} interface: Signalling transport," Technical Specification TS 25.462, 3rd Generation Partnership Project: www.3gpp.org.
29. "UTRAN I_{uant} interface: Remote electrical tilting (RET): Antennas application part (RETAP) signalling," Technical Specification TS 25.463, 3rd Generation Partnership Project: www.3gpp.org.
30. Jaybeam Ltd, PCT Patent Application WO2007/141281.
31. Izzat et al, US Patent Application US2004/0160361.
32. "Remote azimuth beamwidth extension to the control interface for antenna line devices," Extension Standard AISG-ES-RAB, Antenna Interface Standards Group: www.aisg.org.uk.
33. Powerwave Inc., PTC Patent Application WO2007/136333.
34. "Remote azimuth steering extension to the control interface for antenna line devices," Extension Standard AISG-ES-RAS, Antenna Interface Standards Group: www.aisg.org.uk.
35. C. Parini et al, "Semi-smart antennas," UK: Ofcom, 2006: www.ofcom.org.uk/research/technology/research/emer_tech/smart/finalb.pdf.
36. "IEEE standard test procedures for antennas," ANSI/IEEE Std. 149–1979, December 1979.
37. International Commission on Non-Ionizing Radiation Protection, "Guidelines for limiting exposure in time-varying electric, magnetic and electromagnetic fields (up to 300GHz)," *Health Physics*, vol. 74 (1998): 494–522. (Refer to your national regulations for regulatory information.)
38. Many good planning ideas and examples of disguised antennas can be found at http://www.scotland.gov.uk/Publications/2001/09/pan62/pan62-.
39. Quintel Technology Ltd, Patent Application PCT WO2006/008452 A1.

Antennas for Mobile Communications: CDMA, GSM, and WCDMA

Ka-Leung Lau and Kwai-Man Luk

City University of Hong Kong

3.1 Introduction

The operating frequencies for 2G and 3G mobile communications fall into the 820–960 MHz and the 1710–2170 MHz frequency bands. In these lower microwave frequency ranges, antenna size is an issue for both base station antennas and handset antennas. In general, we prefer base station antennas that have a low profile and a streamlined structure. Otherwise, our living environment would be crowded with the huge number of antennas for different systems and operators. Also for aesthetic reasons, we prefer that a handset antenna be embedded in the mobile phone cover.

Designing base station antennas is more challenging and interesting than designing handset antennas, since more stringent requirements are imposed on the performance of different base station antennas for indoor and outdoor coverage.

3.1.1 Requirements for Indoor Base Station Antennas

In indoor environments, such as shopping malls, parking garages, or airports, integrated networks serving all mobile phone operators are commonly installed. For this implementation, very wideband or multiband antennas are required. The gains of these antennas need not be too high. In wall-mounted cases, the radiation pattern should be unidirectional and should have a horizontal beamwidth of at least 90° to ensure wide angle coverage. This can be realized by combining two separate dipole arrays operated at different frequencies. A typical 7-dBi gain dual-band indoor panel antenna consisting of two two-element dipole arrays, one for GSM900 and the other for GSM1800, is schematically shown in Figure 3.1. The two arrays are connected through a specially designed dual-band feed network.

In ceiling-mounted cases, the antenna should have a conical radiation pattern in the horizontal plane. This antenna can be implemented using broadband electric dipole technology. Two typical broadband indoor ceiling-mounted antennas for indoor coverage are shown in Figure 3.2. The first one is based on a dual-band monopole design whereas the second one is basically a biconical antenna.

3.1.2 Requirements for Outdoor Base Station Antennas

Highly sophisticated base station antenna arrays are in demand for outdoor coverage in mobile communications. The various required performance characteristics of this antenna are not easy to achieve. Different wideband arrays with a variety of beamwidth (60°/90°/105°), gain (10 to 20 dBi), and downtilt (0° to 20°) options must be made available to cover various deployment scenarios in urban, suburban, and rural environments. The antennas are designed with null fill capability for

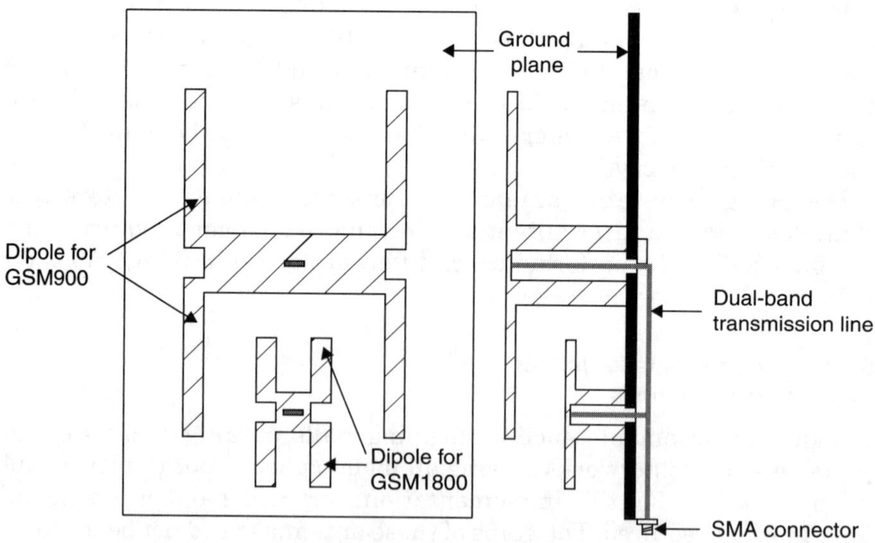

Figure 3.1   A dual-band wall-mounted patch antenna

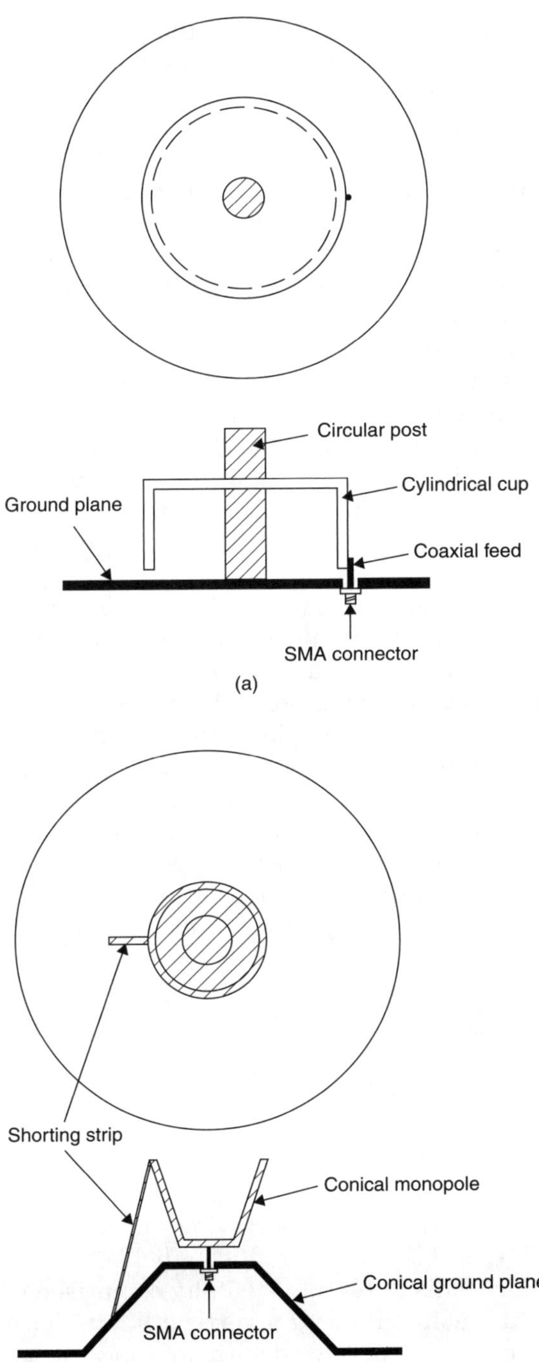

Figure 3.2 Broadband ceiling-mounted antennas

mitigating the coverage issue for subscribers closer to the base station antenna. The antennas should also provide upper sidelobe suppression and high front-to-back ratio for minimizing interference with neighboring cells. To reduce interference among different mobile phone channels, the intermodulation distortion (IMD) must be less than −103 dBm.

In the past, a space diversity technology with one transmitting array and two receiving arrays was usually realized. Nowadays, a dual-polarized antenna with two independently operating slanted arrays, one at +45° degree and the other at −45° polarization, is implemented predominantly. For good diversity performance, the decoupling between the two arrays is required to be at least 30 dB.

For reducing installation effort and costs, antenna arrays are needed that are able to operate in all mobile communication bands, which can considerably reduce the number of antennas required by network operators. To provide more flexibility in network design, the gain in each frequency band can be different from other bands. Again, decoupling between output ports for different frequency bands is a critical issue in the design process.

Commercially available base station antenna arrays are based on dipole antenna technology. Although different brand names are used, such as the vector dipole,[1] butterfly dipole,[2] and directed dipole,[3] they all use a reflector to produce a directional radiation pattern from electric dipoles. Other companies employ aperture-coupled patch antennas[4] as basic elements to develop their outdoor base station antenna series. In general, the performances of these antennas is excellent.

In the following section, we present five novel designs of base station antennas for outdoor and indoor coverage, all based on L-probe fed patch antenna technology.[5] We hope these provide practicing engineers with alternative design solutions and new insights when designing their base station antennas.

3.2   Case Studies

3.2.1   Case 1: An Eight-Element-Shaped Beam Antenna Array

In this section, the design of a linearly polarized, shaped beam antenna array using a genetic algorithm (GA) is presented. Having a base station antenna with null fill to avoid possible areas of reducing field strength in the service area is desirable.[6] This antenna is usually comprised of a number of identical elements mounted along a vertical line to form an antenna array. If these elements are arranged with appropriate element spacing and excited with currents of appropriate amplitudes and phases, a characteristic-shaped radiation pattern in the vertical plane can be obtained. Indeed, the current amplitudes and phases can be controlled by adjusting the power ratios at the output ports of the power divider and the lengths of the connecting cables.

3.2.1.1 Array Geometry The geometry of the proposed array is shown in Figure 3.3.[7] It is an eight-element L-probe-fed patch antenna array. Each array element consists of a rectangular patch that is supported by plastic posts above the ground plane and is proximity-coupled by an L-shaped probe located under its edge. In order to reduce the cross-polarization of the array, which is mainly caused by the vertical portions of the L-shaped probes, the vertical portions of adjacent probes are located at opposite edges and are excited in anti-phases. Normally, a wide-band eight-way power divider will be used to feed these probes. The design of this divider will not be discussed here. We will spend more time describing the optimization of the current amplitudes and phases characteristics and the element spacing for achieving a shaped-beam radiation pattern through the use of GA. The antenna's performance, including the S-parameters, radiation patterns, and gain, was simulated by the MoM solver IE3D with GA capability.

3.2.1.2 Genetic Algorithm Employing the moment method to evaluate this array requires long computation time and numerous memories. Global searching methods for optimization are not recommended. To reduce the computation time, the choice of boundaries for a chromosome involved in the genetic algorithm becomes critical. Here the optimization starts with the current amplitudes and phases obtained by the Orchard-Elliott synthesis procedure. Those chromosomes from this

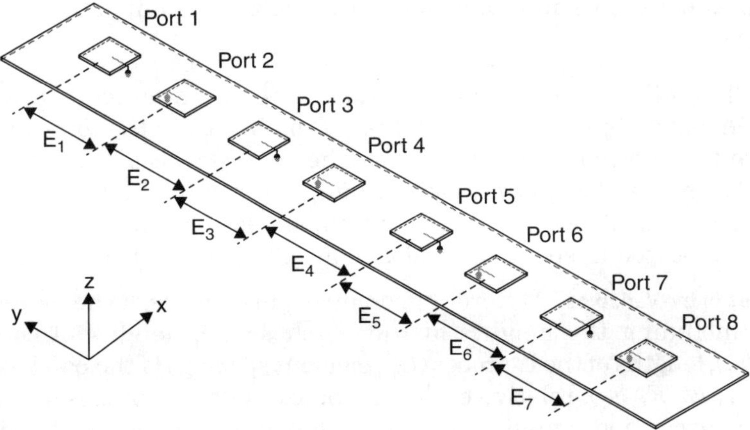

Figure 3.3 Geometry of the eight-element shaped beam antenna array[7]

calculation are varied within a preset range and the element spacing is assumed to be equal to $0.8\lambda_o$. After the boundaries for the current amplitudes and phases are determined, we need to find the boundaries for the element spacings. By considering the physical sizes of the elements and neglecting the mutual coupling effect between the elements, we assume that the lower and upper boundaries for the element spacing are $0.75\lambda_o$ and $0.85\lambda_o$, respectively. Next we need to define the objective functions in the genetic algorithm.

Objective Functions   There are several criteria for the radiation pattern to be synthesized. The main lobe has to be synthesized with the absolute maximum value along a specified direction. Also, the power levels of the nulls have to be specified to form a nearly $\mathrm{Cosec}^2\theta$ pattern. Moreover, the peak gain of the main beam has to be maximized. Other than the radiation pattern, a requirement is also placed on the input return loss of each array element. The objective functions of the genetic algorithm are listed here:

- Minimum gain > 15 dBi (from 1.71 to 2.17 GHz)
- Upper sidelobes < –25 dB (from 1.71 to 2.17 GHz)
- S_{ii} < –10 dB, where ii = 11, 22...88 (from 1.71 to 2.17 GHz)
- Peak angle of main lobe = 0° (from 1.71 to 2.17 GHz)
- $\mathrm{Cosec}^2\theta$ pattern (from 1.71 to 2.17 GHz)

  1. Lower first null and lower first lobe ~ –10 dB

  2. Lower second null and lower second lobe ~ –15 dB

  3. Lower third null and lower third lobe ~ –20 dB

  4. Lower fourth null and lower fourth lobe ~ –25 dB

- The ripples between a null and the nearest lobe are limited to 1 dB difference

Practically realizing a cosecant radiation pattern is a difficult task. By using GA, eight objective functions are set to synthesize the cosecant patterns for the eight-element array case. The power levels of the lower first null, lower first lobe, lower second null, lower second lobe, lower third null, lower third lobe, lower fourth null, and lower fourth lobe are included and located in specified angles to form the cosecant pattern.

Boundaries of the Variables   Different geometric parameters of the array elements, including the heights of the patches (h_p), heights of the L-probes (h_{lp}), lengths of the L-probes (L_p), element spacings of the patches ($E_1, E_2, E_3, E_4, E_5, E_6, E_7$), relative excitation phases of the L-probes ($a^\circ{}_n$), and relative excitation amplitudes of the L-probes ($|a_n|$) are adjusted in the optimization process, where $n = 1, 2, ..., 8$. Impedance matching can be optimized by varying the dimensions of the L-probes and the patches, which have no effect on the far-field radiation patterns. The chromosomes are described in the first illustration and the boundaries of those parameters are shown the second illustration.

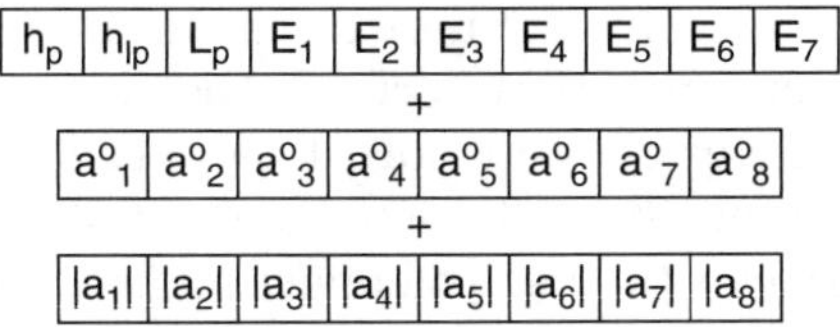

h_p	h_{lp}	L_p	E_1–E_7	a^o_n	a_n
0.8–$1.2\lambda_{0-}$	$0.8 - 1.1\lambda_0$	$0.2 - 0.25\lambda_0$	$0.6 - 0.9\lambda_0$	$0 \leq a^o_n \leq 360°$	$-1 \leq a_n \leq 1$
Phase (a^o_1)			$0°$		
Phase (a^o_2)			$-20° \leq a^o_2 \leq -10°$		
Phase (a^o_3)			$-18° \leq a^o_3 \leq -8°$		
Phase (a^o_4)			$-30° \leq a^o_4 \leq -20°$		
Phase (a^o_5)			$-56° \leq a^o_5 \leq -46°$		
Phase (a^o_6)			$3° \leq a^o_6 \leq 13°$		
Phase (a^o_7)			$10° \leq a^o_7 \leq 20°$		
Phase (a^o_8)			$-2° \leq a^o_8 \leq 8°$		
Amplitude (a_1)			$0.3 \leq a_1 \leq 0.6$		
Amplitude (a_2)			$-0.8 \leq a_2 \leq -0.5$		
Amplitude (a_3)			$0.6 \leq a_3 \leq 0.9$		
Amplitude (a_4)			$-1 \leq a_4 \leq -1$		
Amplitude (a_5)			$0.4 \leq a_5 \leq 0.7$		
Amplitude (a_6)			$-0.5 \leq a_6 \leq -0.2$		
Amplitude (a_7)			$0.2 \leq a_7 \leq 0.5$		
Amplitude (a_8)			$-0.6 \leq a_8 \leq -0.3$		

For efficient computation, the step sizes for those parameters can be different. Each of them induces a quantization error. For the L-probes, the relative excitation phases (a^o_n) and relative excitation amplitudes ($|a_n|$) induce a quantization error of 1° and 0.01, respectively. The other parameters (h_p, h_{lp}, L_p and E_1—E_7) have a quantization error of approximately $0.01\lambda_0$.

GA Parameter Settings   The GA parameters are set as follows:

- The number of chromosomes: 40
- The number of generations: 100
- Crossover rate: 0.8
- Mutation rate: 0.05

Simulation Results   After 100 generations, all criteria were achieved in most frequency points, and thus the program was terminated. A set of nondominant solutions was found. Two sets of parameters for the optimum simulation results, which were obtained by the Orchard-Elliott method with and without the GA optimization, are shown in the following illustration. Actually, the eight-element antenna array with parameters obtained by the GA optimization performs better. Its simulated input return loss against frequency at port 1 to port 8 is illustrated in Figure 3.4. It is clearly seen that all ports have satisfied the return loss requirement (<–10 dB) over the desired operating bandwidth (1.71–2.17 GHz).

	Without GA optimization	With GA optimization
The height of the patches (h_p)	N/A	$0.187\lambda_0$
The height of the L-probe (h_{lp})	N/A	$0.1\lambda_0$
The length of the L-shaped probe (L_p)	N/A	$0.063\lambda_0$
Element spacing (E_1)	$0.8\lambda_0$	$0.83\lambda_0$
Element spacing (E_2)	$0.8\lambda_0$	$0.77\lambda_0$
Element spacing (E_3)	$0.8\lambda_0$	$0.7\lambda_0$
Element spacing (E_4)	$0.8\lambda_0$	$0.85\lambda_0$
Element spacing (E_5)	$0.8\lambda_0$	$0.75\lambda_0$
Element spacing (E_6)	$0.8\lambda_0$	$0.8\lambda_0$
Element spacing (E_7)	$0.8\lambda_0$	$0.85\lambda_0$
Phase (a^o_1)	0	0
Phase (a^o_2)	$-15°$	$-14°$
Phase (a^o_3)	$-13°$	$-13.4°$
Phase (a^o_4)	$-25°$	$-22.5°$
Phase (a^o_5)	$-51°$	$-46°$
Phase (a^o_6)	$8°$	$11.2°$
Phase (a^o_7)	$15°$	$16°$
Phase (a^o_8)	$3°$	$6.7°$
Amplitude (a_1)	0.47	0.49
Amplitude (a_2)	−0.62	−0.63
Amplitude (a_3)	0.7	0.71
Amplitude (a_4)	−1	−1
Amplitude (a_5)	0.5	0.52
Amplitude (a_6)	−0.38	−0.39
Amplitude (a_7)	0.38	0.39
Amplitude (a_8)	−0.47	−0.47

The radiation patterns in the elevation plane are shown in Figure 3.5. It can be observed that the beam peak is located at 0° over the operating frequency band. Referring to the tables shown in Figure 3.5, the sidelobe

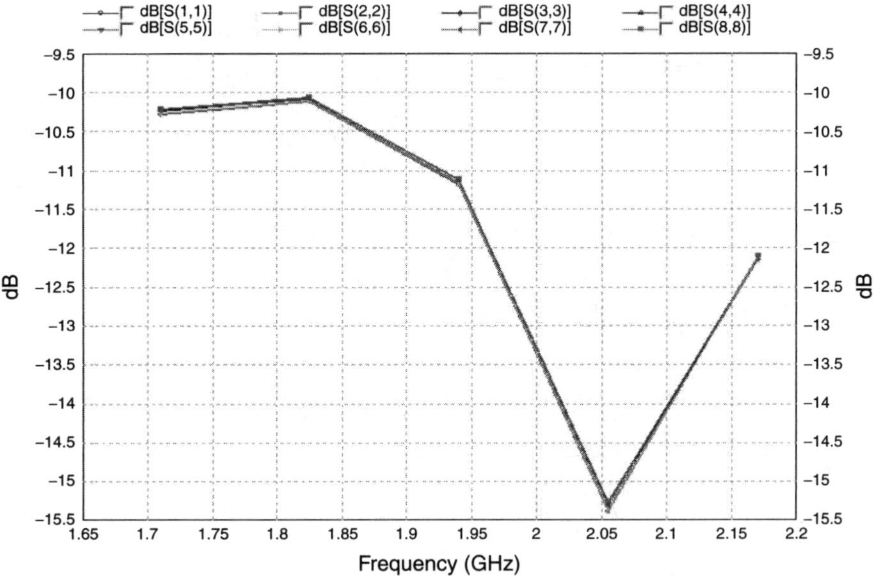

Figure 3.4 Simulated input return loss against frequency of the eight-element shaped beam antenna array at port 1 to port 8[7]

levels are close to the specified values with respect to the objective functions, and most of the power levels of the nulls and lobes fulfill the requirements. The upper sidelobes can be suppressed to −20 dB beyond

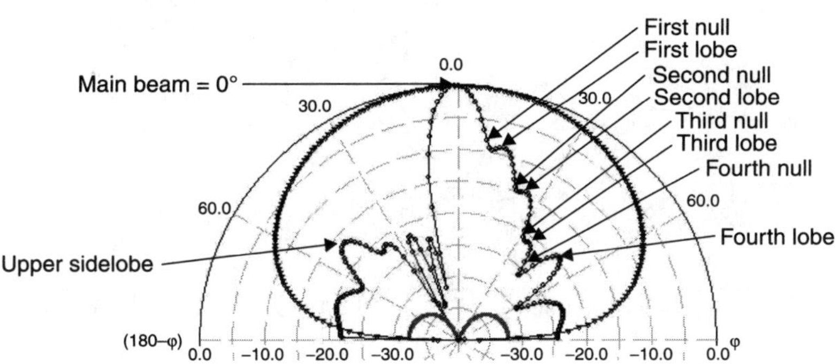

First null	First lobe	Second null	Second lobe	Third null	Third lobe	Fourth null	Fourth lobe
−9.7dB	−9.1dB	−14.9dB	−14.6dB	−21.4dB	−21.2dB	−26.2dB	−19.7dB

3-dB beamwidth in horizontal plane = 80.3°				3-dB beamwidth in vertical plane = 9.8°			

(a)

Figure 3.5 Simulated radiation patterns of the eight-element shaped beam antenna array at (*a*) 1.71 GHz, (*b*) 1.83 GHz, (*c*) 1.94 GHz, (*d*) 2.06 GHz, and (*e*) 2.17 GHz[7]

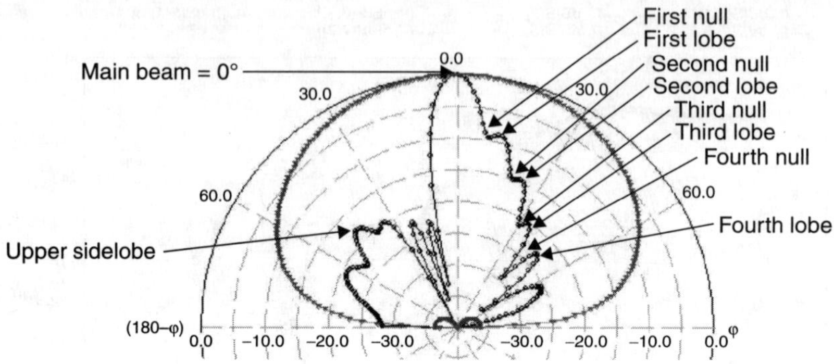

First null	First lobe	Second null	Second lobe	Third null	Third lobe	Fourth null	Fourth lobe
−9.6dB	−9dB	−15.1dB	−14.8dB	−21.3dB	−20.3dB	−29.7dB	−23.8dB

3-dB beamwidth in horizontal plane = 78.6°	3-dB beamwidth in vertical plane = 9.2°

(b)

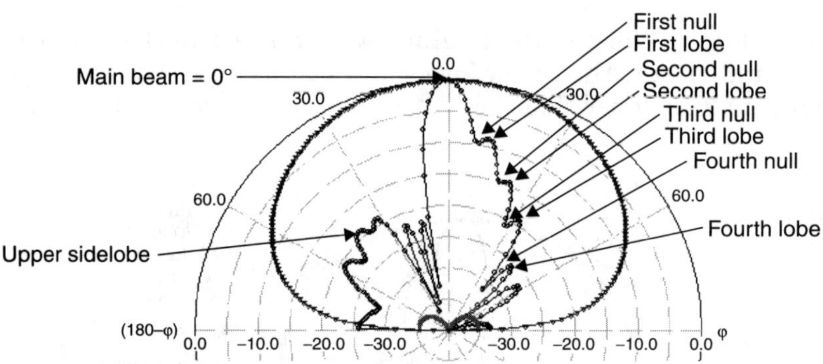

First null	First lobe	Second null	Second lobe	Third null	Third lobe	Fourth null	Fourth lobe
−9.5dB	−9.2dB	−15.2dB	−14.6dB	−21.0dB	−19.3dB	−31.4dB	−26.0dB

3-dB beamwidth in horizontal plane = 76.7°	3-dB beamwidth in vertical plane = 8.6°

(c)

Figure 3.5 Simulated radiation patterns of the eight-element shaped beam antenna array at (a) 1.71 GHz, (b) 1.83 GHz, (c) 1.94 GHz, (d) 2.06 GHz, and (e) 2.17 GHz[7] (*Continued*)

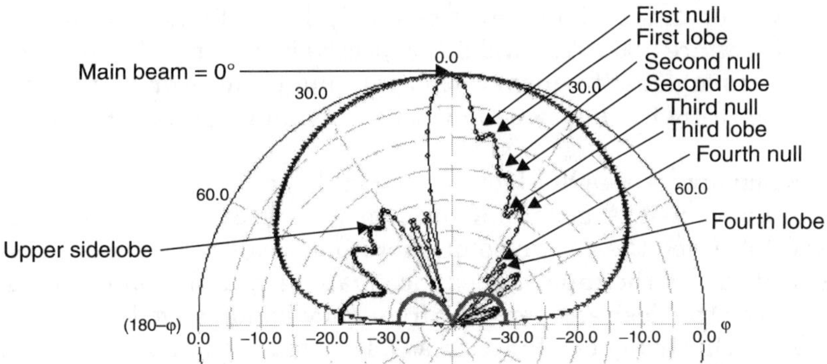

First null	First lobe	Second null	Second lobe	Third null	Third lobe	Fourth null	Fourth lobe
−9.8dB	−9.2dB	−14.8dB	−14.7dB	−20.7dB	−18.5dB	−31.0dB	−27.5dB

3-dB beamwidth in horizontal plane = 74.6°	3-dB beamwidth in vertical plane = 8.1°

(d)

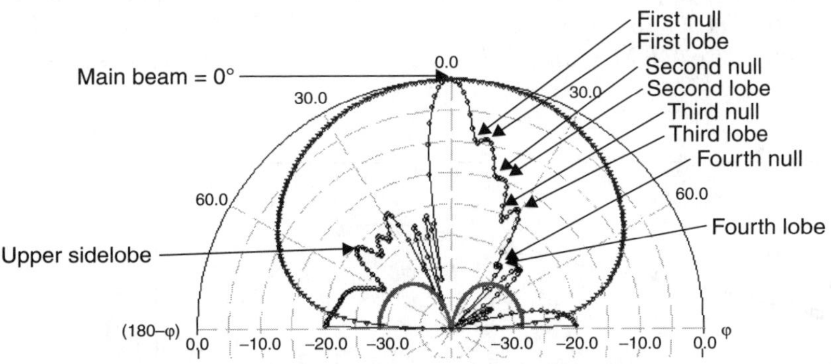

First null	First lobe	Second null	Second lobe	Third null	Third lobe	Fourth null	Fourth lobe
−10.0dB	−9.2dB	−15.0dB	−15.0dB	−20.9dB	−18.2dB	−28.0dB	−17.2dB

3-dB beamwidth in horizontal plane = 72.3°	3-dB beamwidth in vertical plane = 7.6°

(e)

Figure 3.5 Simulated radiation patterns of the eight-element shaped beam antenna array at (*a*) 1.71 GHz, (*b*) 1.83 GHz, (*c*) 1.94 GHz, (*d*) 2.06 GHz, and (*e*) 2.17 GHz.[7] (*Continued*)

the peak value, and the half-power beamwidths in the horizontal and vertical planes are about 76° and 8°, respectively. Figure 3.6 shows the simulated gain against frequency. This antenna yields over 15 dBi gain over the operating frequency band with a maximum gain of 16.9 dBi.

3.2.1.3 Summary An eight-element shaped beam antenna array was designed and tested. A genetic algorithm with IE3D simulation was employed to search the design solution automatically. The array satisfied the criteria on the input impedance matching, radiation patterns, and gain over the desired operating frequency band ranged from 1.71 to 2.17 GHz. This array can function as a base station antenna for both CDMA and GSM wireless communication systems. In this study, the IE3D solver was utilized to simulate the antenna structure, which we used to change the dimensions of the structure manually after each simulation so as to achieve the desired results.

3.2.2 Case 2: A 90° Linearly Polarized Antenna Array

In this section, we present a linearly polarized antenna array with a novel backlobe suppression technique.[7] The antenna has an impedance bandwidth of 18% for a SWR < 1.5, ranging from 0.81 to 0.97 GHz. It can serve the CDMA and GSM wireless communication systems.

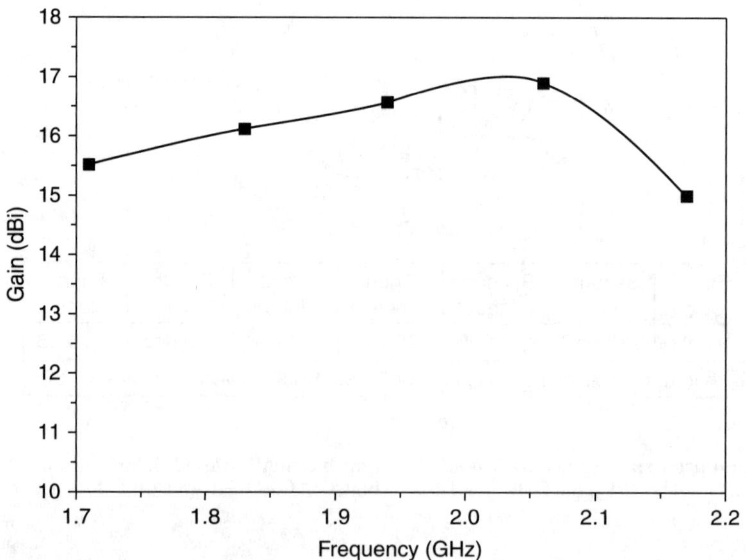

Figure 3.6 Simulated gain against frequency of the eight-element shaped beam antenna array[7]

The antenna has over a 90° horizontal beamwidth and low backlobe radiations in both planes, which make it very suitable for use in sectors with large angles.

3.2.2.1 Array Geometry   The geometry of the array is shown in Figure 3.7. It is a two-element L-probe-fed stacked patch antenna array. Each array element consists of two rectangular patches in a stacked configuration.

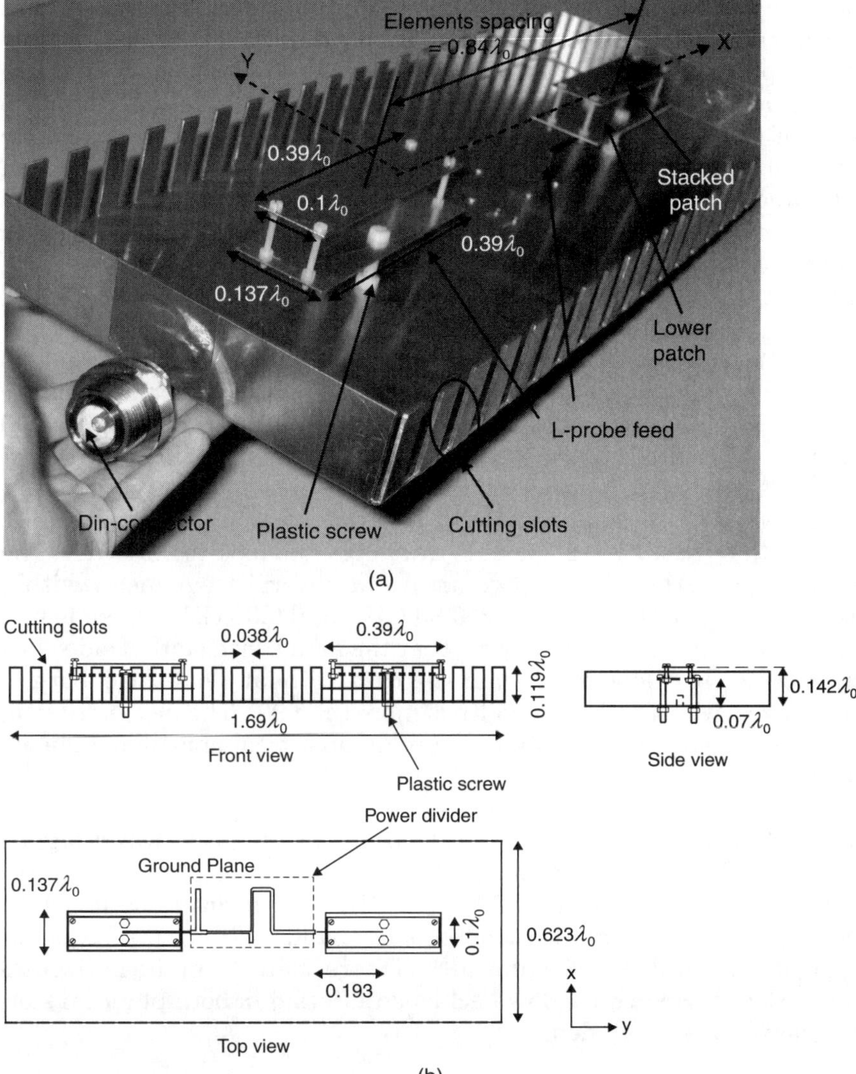

Figure 3.7   Geometry of the 90° linearly polarized antenna array: (a) prototype and (b) different views[8] (© 2005 IEEE)

They are supported by plastic posts above the ground plane. The lower patch is proximity-coupled by an L-shaped probe located under its edge. In order to reduce the cross-polarization of the array, which is mainly contributed by the vertical portions of the two L-shaped probes, the horizontal portions of these probes are located at the opposite edges with out-of-phase excitations. Thus, a wideband two-way power divider with an output phase difference of 180° is needed to feed these probes. The short-circuited stub of the power divider is used to improve impedance matching and provide a dc ground to the antenna. This divider is fabricated on a microwave substrate with a dielectric constant of 2.65. Each L-probe is made from a copper strip with a width of 4.13 mm and a thickness of 1 mm. Four vertical sidewalls are mounted on the edges of the ground plane for the purpose of reducing the backlobe level. To further enhance their effects, a novel backlobe suppression technique is employed, which is implemented by cutting slots on the two sidewalls parallel to the axis of the array periodically. The antenna dimensions are given in Figure 3.7, where λ_0 is the free space wavelength at 0.89 GHz.

3.2.2.2 Measurement Results   In order to evaluate the effect of the two slotted sidewalls on the standing wave ratio and backlobe level of the array, an antenna array without slotted sidewalls (shown in Figure 3.8) was also fabricated and tested. Its standing wave ratio against frequency is shown in Figure 3.9. Clearly, the antenna array with slotted sidewalls located parallel to the array axis has a wider impedance bandwidth of 18% for a SWR < 1.5, ranging from 0.81 to 0.97 GHz. The radiation patterns of these two arrays at several frequencies within the operating band—0.824 GHz, 0.89 GHz, and 0.96 GHz are shown in Figures 3.10 and 3.11. We observe that the array with slotted sidewalls has a better backlobe level than the case without slots. Also, both of them have a wide 3-dB beamwidth of more than 90°. The measured gain of the array is about 9.5 dBi, the gain against frequency is not shown for brevity.

3.2.2.3 Summary   A 90° linearly polarized antenna array was designed and tested. Measured results demonstrate that it has a standing wave ratio of less than 1.5 over 0.82 to 0.96 GHz, which covers both CDMA and GSM mobile communication frequency bands. Within this frequency range, it has a wide 3-dB beamwidth of more than 90° in the horizontal plane. Also, it can achieve low backlobe radiation in both planes if slotted sidewalls are included.

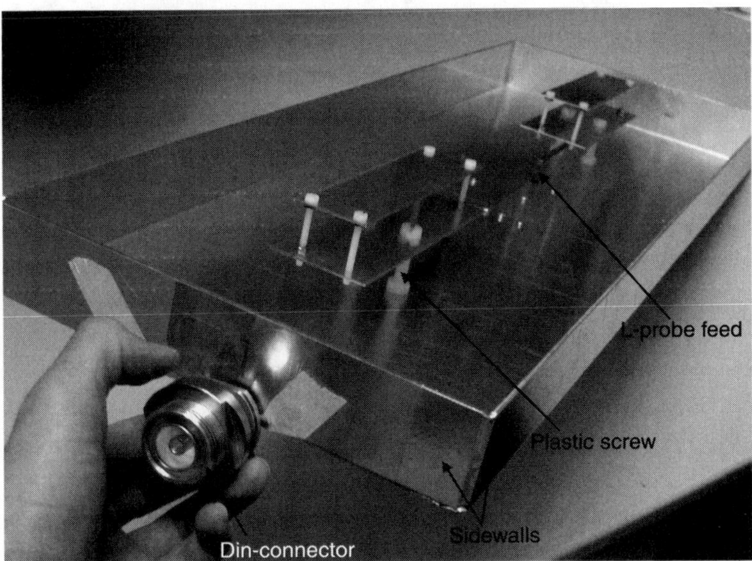

Figure 3.8 Geometry of the 90° linearly polarized antenna array (without slotted sidewalls)[8] (© 2005 IEEE)

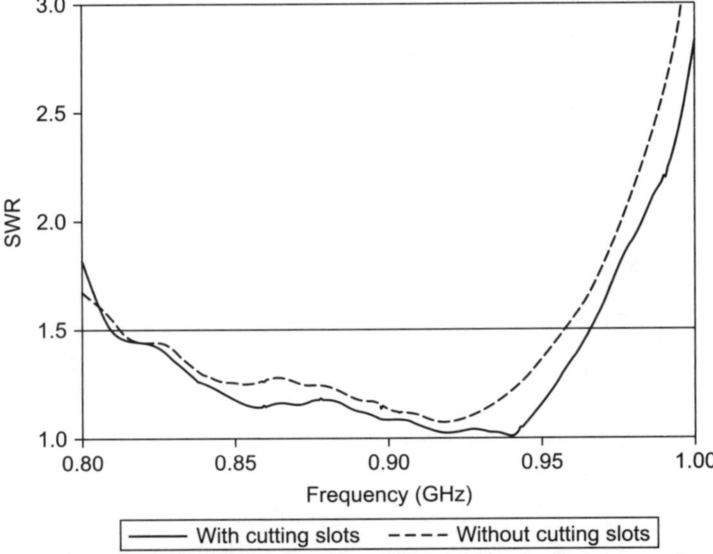

Figure 3.9 Measured standing wave ratio of the 90° linearly polarized antenna array[8] (© 2005 IEEE)

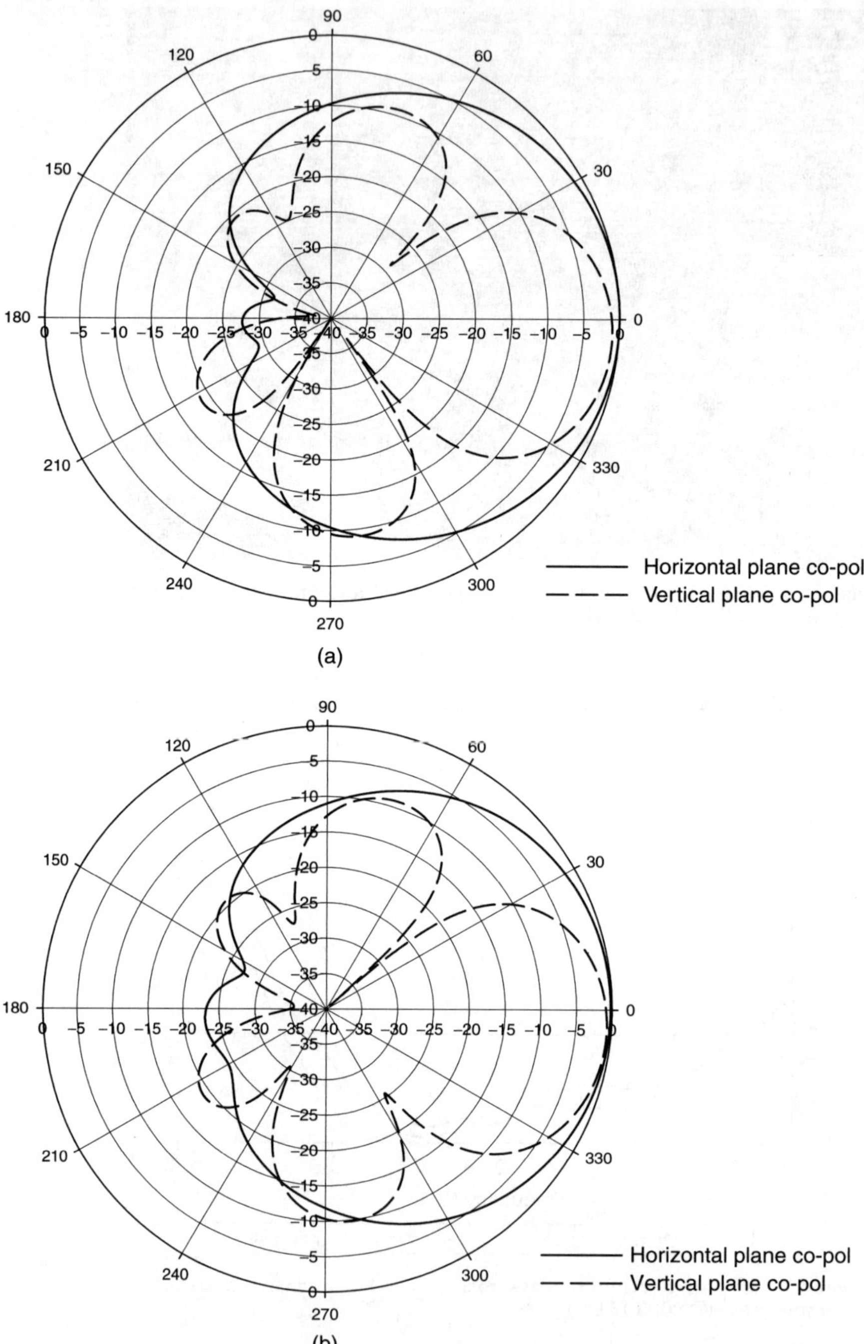

Figure 3.10   Measured radiation patterns of the 90° linearly polarized antenna array with slotted sidewalls at (*a*) 0.824 GHz, (*b*) 0.892 GHz, and (*c*) 0.96 GHz[8] (© 2005 IEEE)

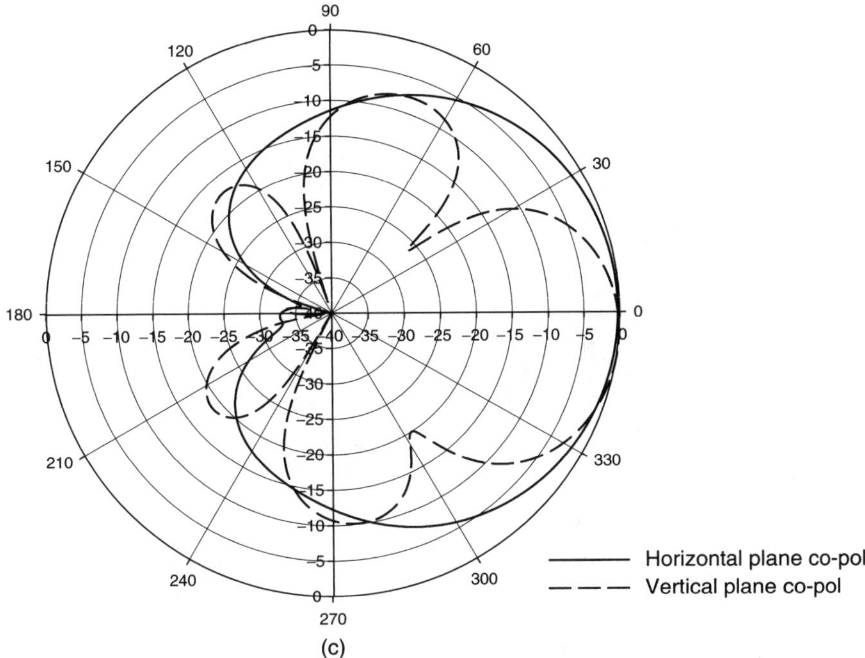

Figure 3.10 Measured radiation patterns of the 90° linearly polarized antenna array with slotted sidewalls at (*a*) 0.824 GHz, (*b*) 0.892 GHz, and (*c*) 0.96 GHz[8] (© 2005 IEEE) (*Continued*)

3.2.3 Case 3: A Dual-Band Dual-Polarized Array

The performance of a dual-band dual-polarized patch antenna array is presented here. This array is designed to operate in two separate frequency bands—820–960 MHz and 1710–2170 MHz—which cover the operating frequencies of most mobile communication systems, including CDMA, GSM, PCS, and UMTS. For this array to function effectively, an isolation of more than 30 dB and a cross-polarization of less than −15 dB are required over the operating frequency bands.[9]

3.2.3.1 Array Geometry The geometry of the array is shown in Figure 3.12. It consists of six elements with two larger patches for the lower band and four pairs of smaller patches for the upper band. In order to maintain equal element spacing in terms of wavelength in both bands, two pairs of smaller patches and two larger patches are arranged appropriately. Each patch element is excited by two L-probes orthogonally from the side for the ± 45° polarizations. The antenna dimensions are detailed following Figure 3.12.

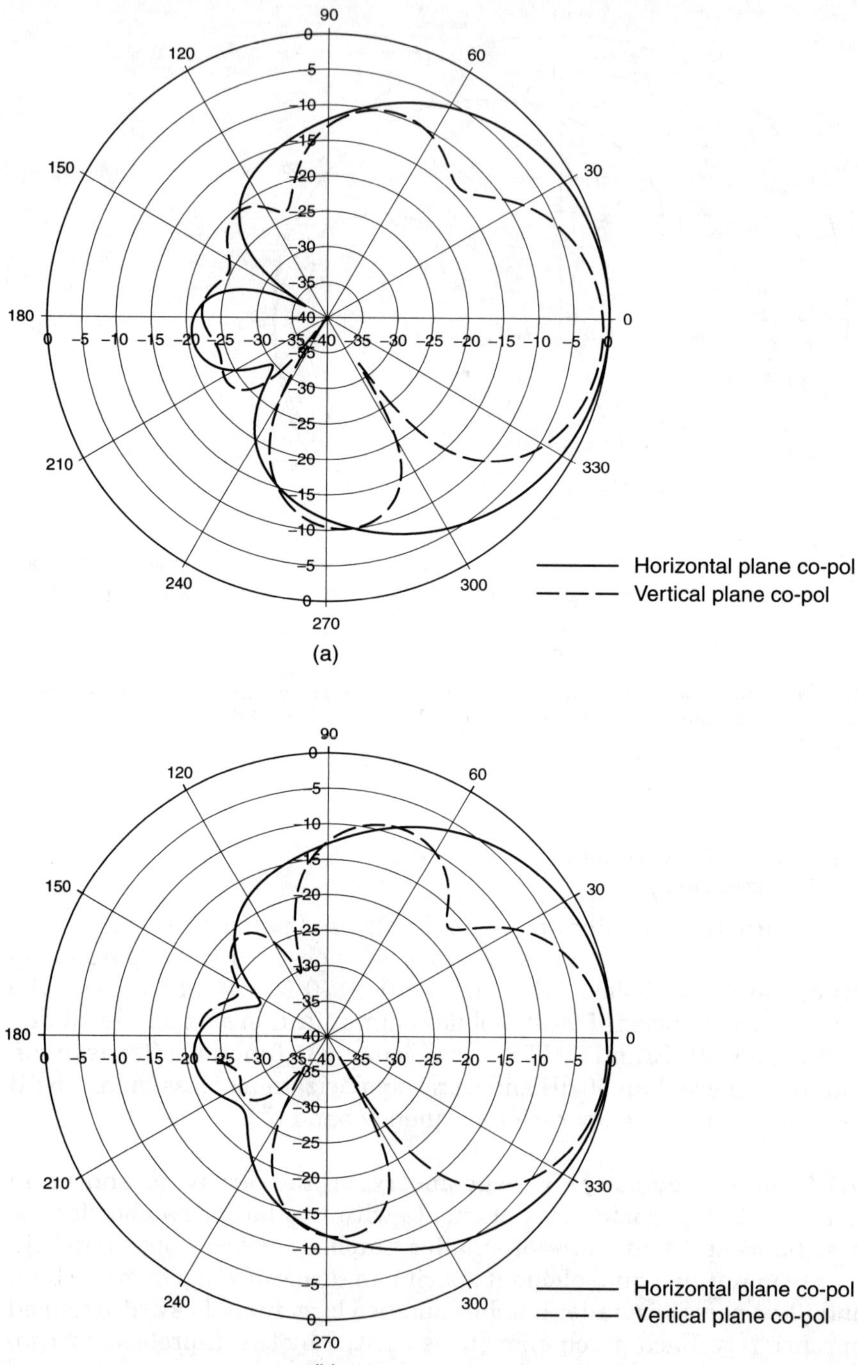

Figure 3.11 Measured radiation patterns of the 90° linearly polarized antenna array without slotted sidewalls at (*a*) 0.824 GHz, (*b*) 0.892 GHz, and (*c*) 0.96 GHz[8] (© 2005 IEEE)

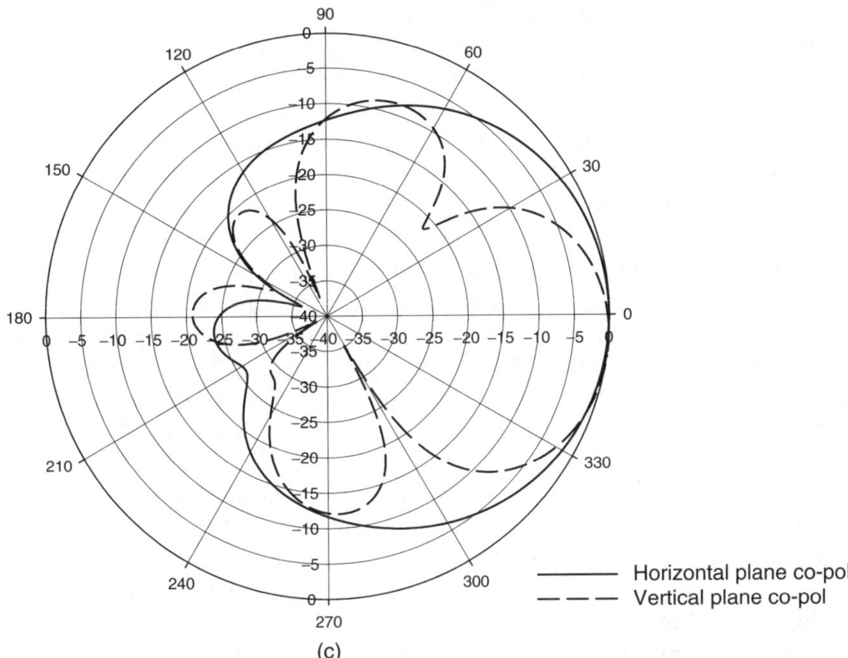

Figure 3.11 Measured radiation patterns of the 90° linearly polarized antenna array without slotted sidewalls at (*a*) 0.824 GHz, (*b*) 0.892 GHz, and (*c*) 0.96 GHz[8] (© 2005 IEEE) (*Continued*)

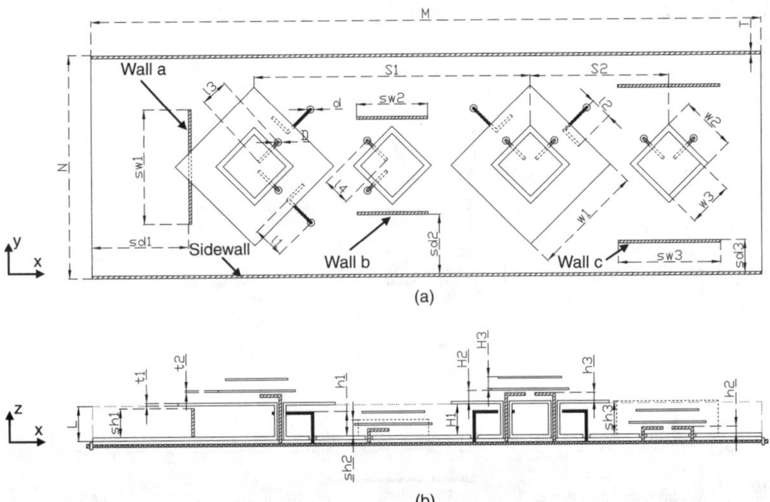

Dimensions: M = 752, N = 244, W1 = 126, W2 = 62, W3 = 50.5, l1 = 38, l2 = 15, l3 = 29, l4 = 31, S1 = 310, S2 = 155, d = 2, D = 4.6, T = 2, t1 = 2, t2 = 1, h1 = 27.5, h2 = 10.5, h3 = 11, H1 = 38, H2 = 14, H3 = 14, L = 35, sw1 = 130, sd1 = 114, sh1 = 34, sw2 = 80, sd2 = 74, sh2 = 15, sw3 = 116, sd3 = 38, sh3 = 29 (mm)

Figure 3.12 Geometry of the dual-band dual-polarized patch antenna array: (*a*) top view and (*b*) side view[9] (© 2005 IEEE)

3.2.3.2 Isolation Enhancement

Probe-fed antennas have the weakness of a higher cross-polarization level due to strong radiation from the vertical probes. In the dual-polarization case, these radiations will also increase the coupling between the two input ports. To solve this issue, two methods were employed to suppress the input port coupling of the array with *feed networks phase cancellation* and *adding auxiliary metallic sidewalls*. The design of the feeding networks is shown in Figure 3.13.

In order to realize the first method, the feed probes are either excited with in-phase (0°) or anti-phase (180°) signals. The phases are controlled by using different lengths of feed cables connecting to the power dividers. For the lower band, probes 1 and 2 are anti-phased, whereas probes 3 and 4 are in-phased. Thus, the coupling from probes 1 and 2 to probe 3 can be partially cancelled. Probe 4 is in the same situation. On the other hand, the coupling from probes 3 and 4 to probe 2 is partially counteracted by the coupling on probe 2 for the lower band (port 1). Therefore, the input port isolation for the lower frequency band can be enhanced. The same effect occurs in the upper frequency band. As the structure of the array is not ideally symmetrical, however, the cancellation effect is limited. Consequently, isolation enhancement cannot be achieved over a wide range of frequencies.

The vertical metallic sidewalls are usually used to reduce the back radiation of the antenna array. They can also be used to change the coupling between the feed probes by multipath reflections. The second method is implemented by placing two long sidewalls and several auxiliary sidewalls strategically on the ground plane of the array as depicted in

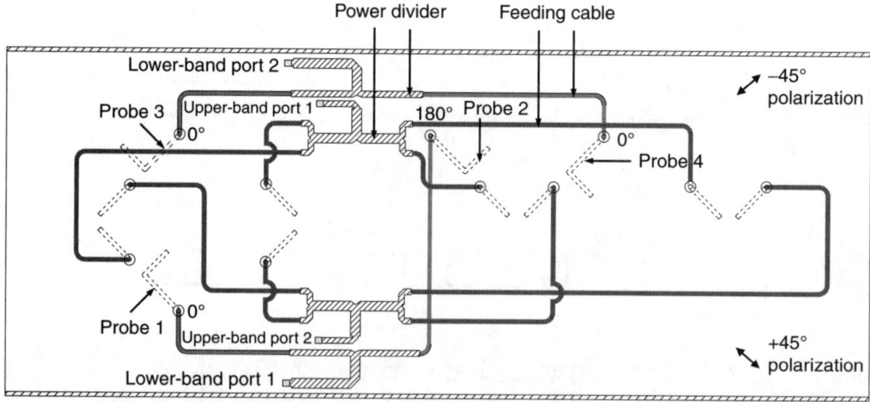

Figure 3.13 Feeding networks of the dual-band dual-polarized patch antenna array[10]

Figure 3.12. This arrangement can enhance the cancellation effect of the first method. For instance, "wall a" mainly affects the coupling between probes 1 and 3 so as to increase the total input port isolation between lower frequency bands (ports 1 and 2). Similarly, the auxiliary "wall b" and "wall c" are mainly used for the upper frequency band isolation enhancement. By optimizing the dimensions and positions of these walls, high input port isolation can be achieved over the frequency band.

3.2.3.3 Measurement Results Because the size of the array is quite large in terms of wavelength, simulating its performance with a commercial simulation package is difficult. Therefore, only measurement results are presented here. Figures 3.14a and 3.14b show the measured return loss and input port coupling of the array in the lower and upper frequency bands, respectively. It is observed that, for RL < −14 dB, the bandwidth is 22% (0.81–1.01 GHz) for port 1 and 19% (0.81–0.98 GHz) for port 2 in the lower frequency band. Over the frequency range of 0.82 to 0.96 GHz, the input port coupling is less than −30 dB. For the upper frequency band, the bandwidth is 40% (1.48–2.23 GHz) for port 1 and 42% (1.53–2.34 GHz) for port 2. The input port coupling is less than −30 dB over the frequency range of 1.68 to 2.2 GHz. In the same figures, it can also been seen that the peak gain is around 10.5 dBi in the lower band and 14 dBi in the upper band.

Figures 3.15 and 3.16 depict the measured far-field radiation patterns of the two input ports, corresponding to the ± 45° polarizations, at the center frequency of the lower frequency band (0.9 GHz) and the upper frequency band (2.0 GHz), respectively. We define the x-z plane as the vertical plane and the y-z plane as the horizontal plane. For the lower frequency band (port 1), the 3-dB beamwidths are 30° and 61° in the vertical and horizontal planes, respectively. The corresponding values for the lower band (port 2) are 30° and 66°. In both planes, the cross-polarization level is less than −17 dB in the broadside direction. Also, the backlobe level is less than −20 dB. For the upper frequency band (port 1), the 3-dB beamwidths are 12° and 67.5° in the vertical and horizontal planes, respectively. The corresponding values for the upper frequency band (port 2) are 13° and 75°. In both planes, the cross-polarization level is less than −19 dB in the broadside direction. Moreover, the backlobe level is less than −25 dB.

3.2.3.4 Summary A dual-band dual-polarized patch antenna array was designed and implemented successfully. The measured results demonstrate that it has an input return loss of less than −14 dB, input port isolation of more than 30 dB, and a cross-polarization level of less than −15 dB over the 0.82 to 0.96-GHz and 1.71 to 2.17-GHz bands. This array is suitable for CDMA, GSM, PCS, and UMTS bands simultaneously.

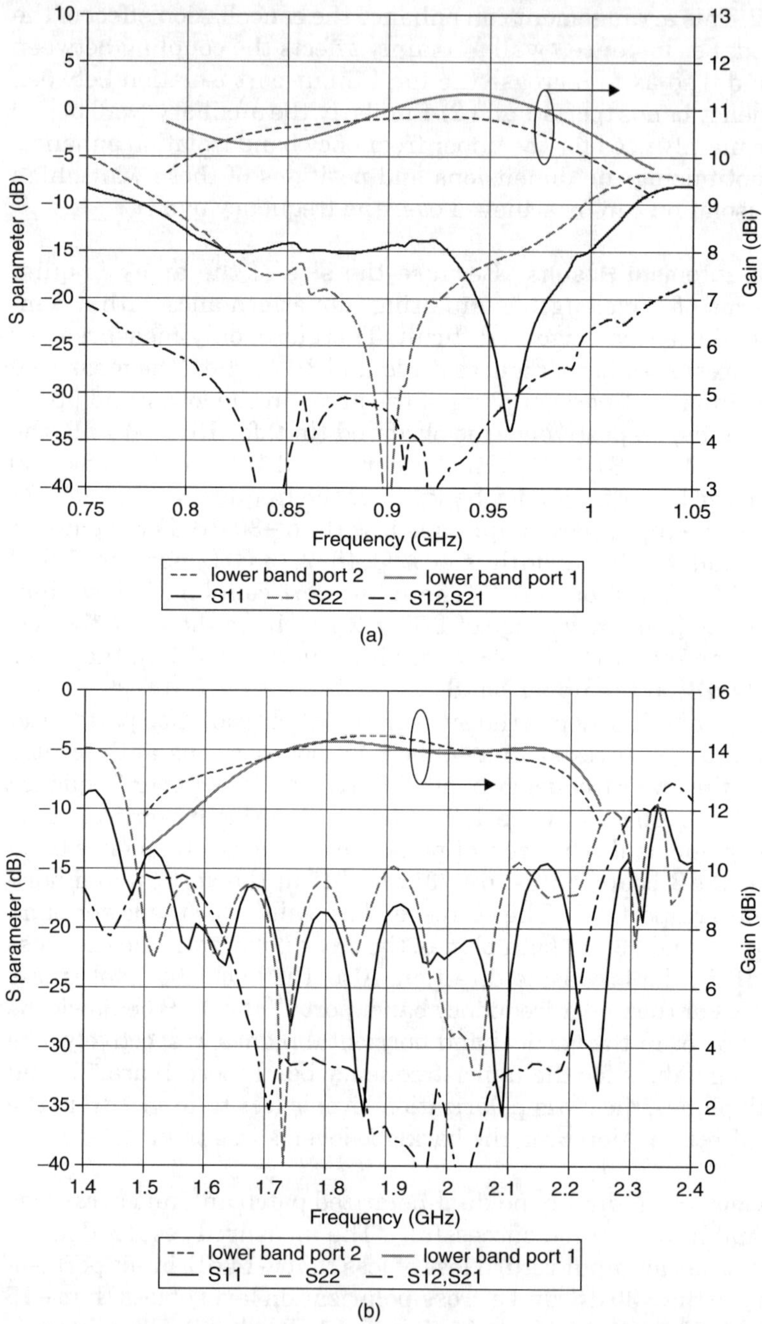

Figure 3.14   Measured S parameter and gain of the dual-band dual-polarized patch antenna array: (*a*) lower band and (*b*) upper band[9] (© 2005 IEEE)

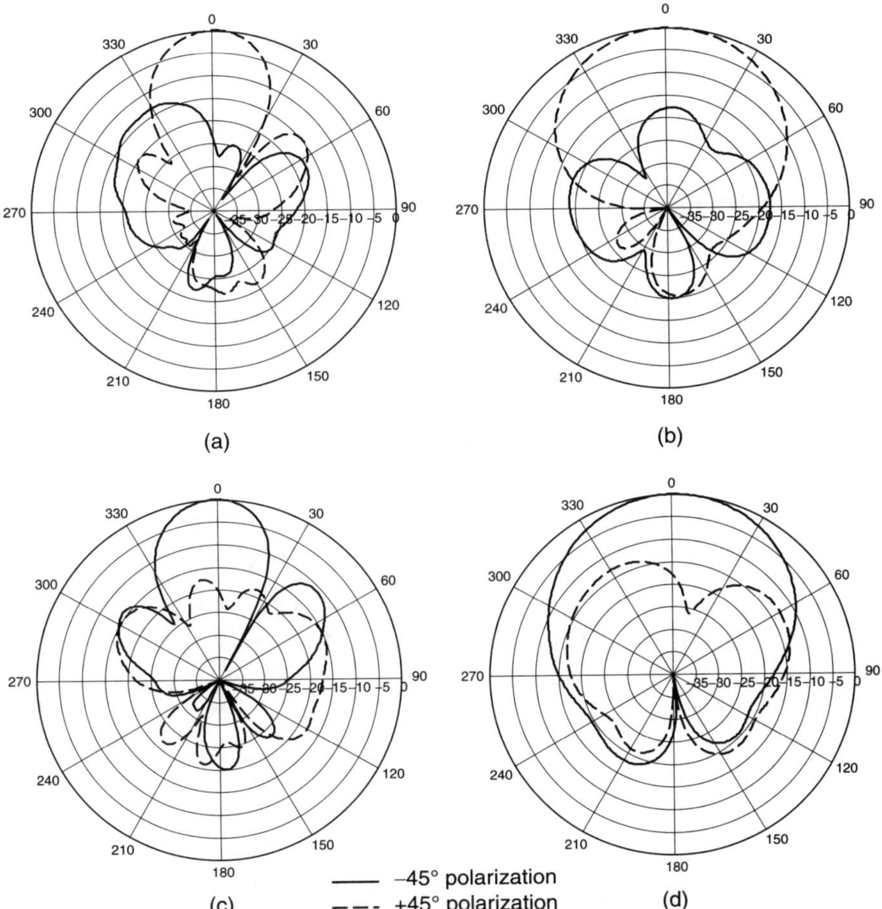

Figure 3.15 Measured radiation patterns of the dual-band dual-polarized patch antenna array at 0.9 GHz: (*a*) *x-z* plane, port 1; (*b*) *y-z* plane, port 1; (*c*) *x-z* plane, port 2; and (*d*) *y-z* plane, port 2[9] (© 2005 IEEE)

3.2.4 Case 4: A Broadband Monopolar Antenna for Indoor Coverage

In this section, a broadband monopolar antenna is presented. It has a wide impedance bandwidth of 62% for a SWR < 1.5, ranging from 1.48 to 2.81 GHz. It possesses nearly omnidirectional radiation patterns over the bandwidth. Moreover, it has a moderate peak gain of 6.1 dBi. Due to these features, it is very suitable for the modern multiband indoor mobile communication systems.

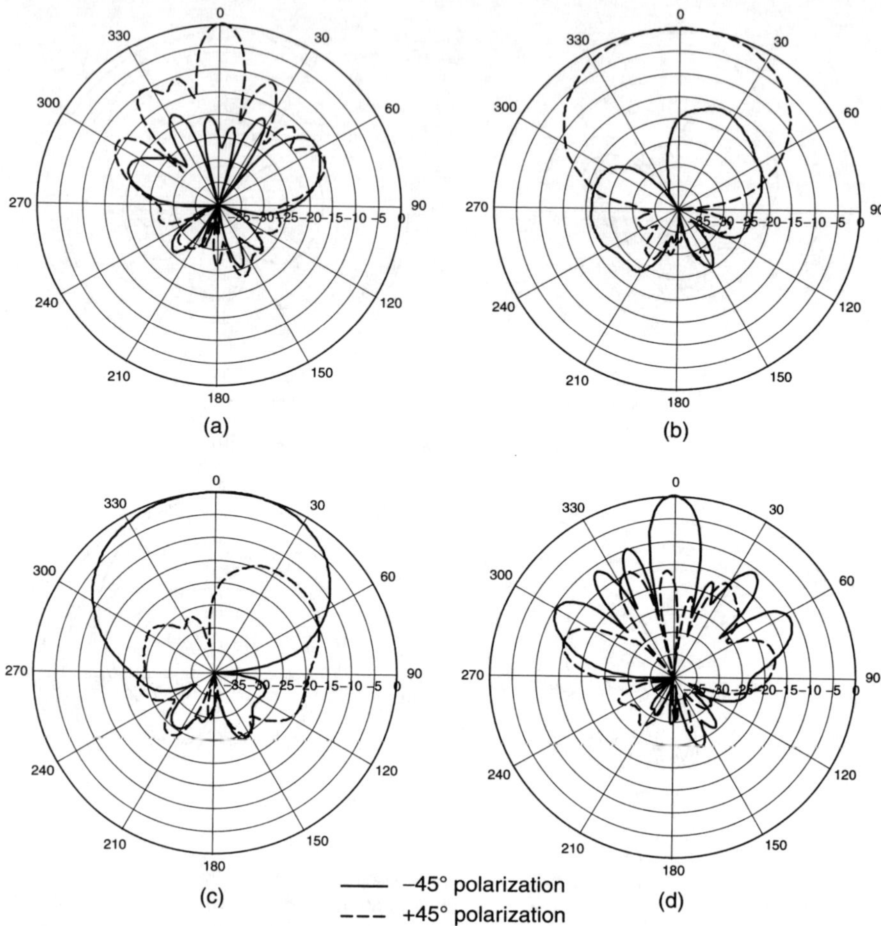

Figure 3.16   Measured radiation patterns of the dual-band dual-polarized patch antenna array at 2.0 GHz: (*a*) *x-z* plane, port 1; (*b*) *y-z* plane, port 1; (*c*) *x-z* plane, port 2; and (*d*) *y-z* plane, port 2[9] (© 2005 IEEE)

3.2.4.1 Antenna Geometry The geometry of the antenna is shown in Figure 3.17. This antenna is mounted in the middle of a circular ground plane. It consists of a circular patch supported by four identical shorted trapezoidal plates above the ground plane. This patch is proximity-coupled by a disc-loaded wire/probe under its center. The antenna dimensions are detailed at the bottom of Figure 3.17.

3.2.4.2 Simulation and Measurement Results Figure 3.18 shows the standing wave ratio and gain against frequency. The term *resonance* is defined as the frequency achieving a local maximum in the input

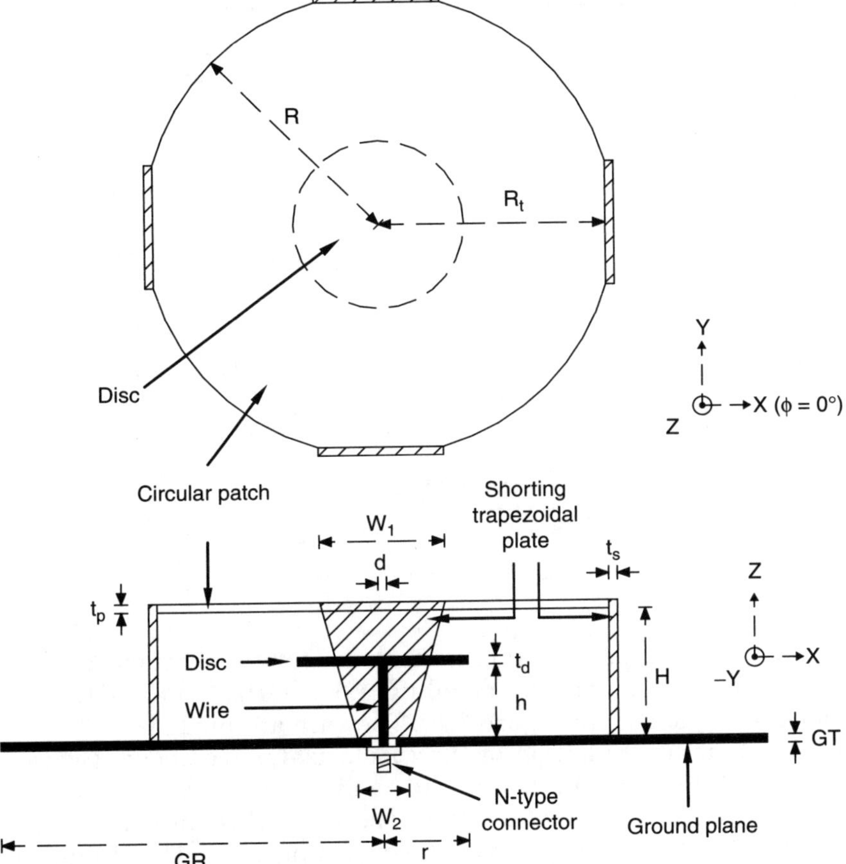

Dimensions: $H = 15$, $R = 39$, $R_t = 38.2$, $W_1 = 15.6$, $W_2 = 2$, $h = 11.3$, $r = 5.5$, $d = 2.9$, $t_p = 0.5$, $t_s = 0.5$, $t_d = 1.5$, $GT = 2$, $GR = 91$ (mm)

Figure 3.17 Geometry of the monopolar antenna

resistance curve of the antenna. This definition is somewhat arbitrary and is selected from a few alternative definitions. In the simulated SWR curve, three local minima can be found. They are produced by the two resonances of the antenna, which are located at 1.56 GHz and 2.01 GHz. Since the separation of these resonances is reduced in measurement, the measured SWR curve exhibits only two local minima. As these minima are close to each other, wide impedance bandwidths (SWR < 1.5) of 62% (1.48–2.81 GHz) and 65% (1.43–2.81 GHz) are achieved in measurement and simulation, respectively. Also observed in Figure 3.18, the measured and simulated peak gains are 6.1 dBi and 5.8 dBi, respectively.

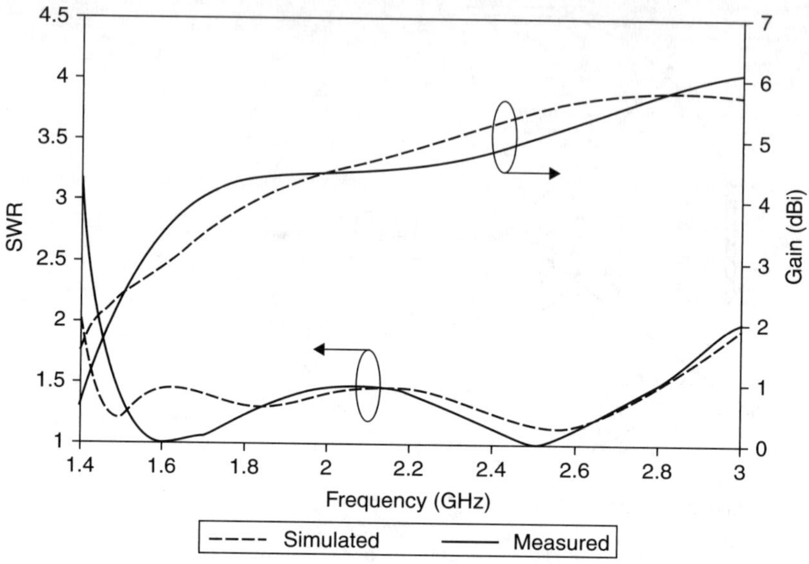

Figure 3.18   Standing wave ratio and gain of the monopolar antenna

The elevation plane radiation patterns at several frequencies over the impedance bandwidths, including 1.545 GHz, 2.145 GHz, and 2.745 GHz, are shown in Figure 3.19. They are attained at $\phi = 0°$. The angle of maximum radiation of the copolarization component varies from $\theta = \pm 32°$ to $\theta = \pm 45°$, theoretically and experimentally. Also, deep nulls appeared at $\theta = 0°$ and $\theta = 180°$. Therefore, this antenna has off-broadside elevation patterns over the operating band. The simulated cross-polarization components (dashed dot lines) cannot be observed because they are very small in the ideal case. The azimuth plane radiation patterns at these frequencies are shown in Figure 3.20. For the measured and simulated copolarization components, the ripple levels are 1.96 dB and 0.29 dB at 1.545 GHz, 1.61 dB and 0.94 dB at 2.145 GHz, and 5.05 dB and 2.46 dB at 2.745 GHz, respectively. Hence, this antenna also has nearly omnidirectional azimuth patterns over the operating band. In fact, unlike in the simulation, it is very difficult to ensure that each shorted plate of this antenna is perpendicular to the patch and the ground plane in the experiment because detecting a $\pm$ 1–2° deviation from 90° (perpendicular) is arduous. This small deviation is, however, large enough to add 1–2-dB ripple to the measurement even though no ripple is observed in the simulation. Consequently, the difference between the measured and simulated ripple levels at these frequencies can be considered as less than 1 dB, so they are not significant. As a result, the proposed antenna has moderate peak gain and nearly omnidirectional radiation patterns over the wide operating band.

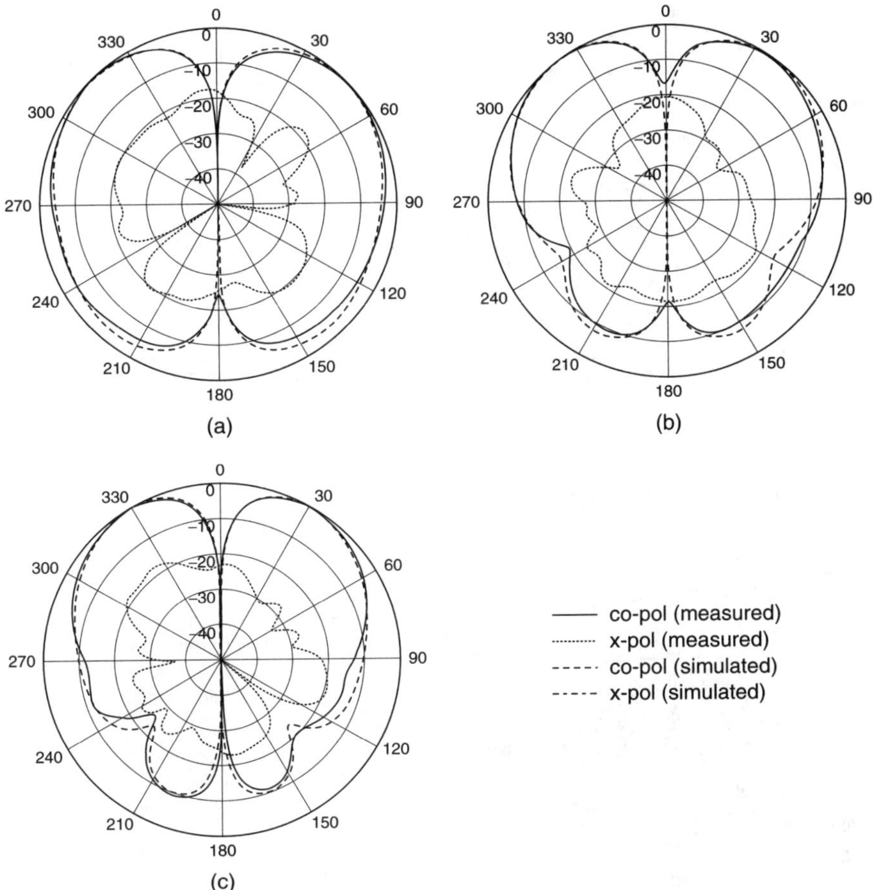

Figure 3.19 Elevation plane radiation patterns ($\phi = 0°$) of the monopolar antenna at (a) 1.545 GHz, (b) 2.145 GHz, and (c) 2.745 GHz

To summarize, the antenna performance determined by the measurement and simulation are in close agreement.

3.2.4.3 Summary A broadband monopolar antenna was designed and implemented successfully. Measured results demonstrate that it has a wide impedance bandwidth (SWR < 1.5) from 1.48 to 2.81 GHz, which is equal to 62%. Within this frequency band, it has a nearly omnidirectional radiation pattern in the azimuth plane and off-broadside radiation pattern in the elevation plane. Moreover, it has other advantages such as low profile, small lateral size, and moderate peak gain. This antenna is suitable for an indoor base station network that serves several wireless communication systems, including PCS, UMTS, Bluetooth, and WLAN, in an integration manner.

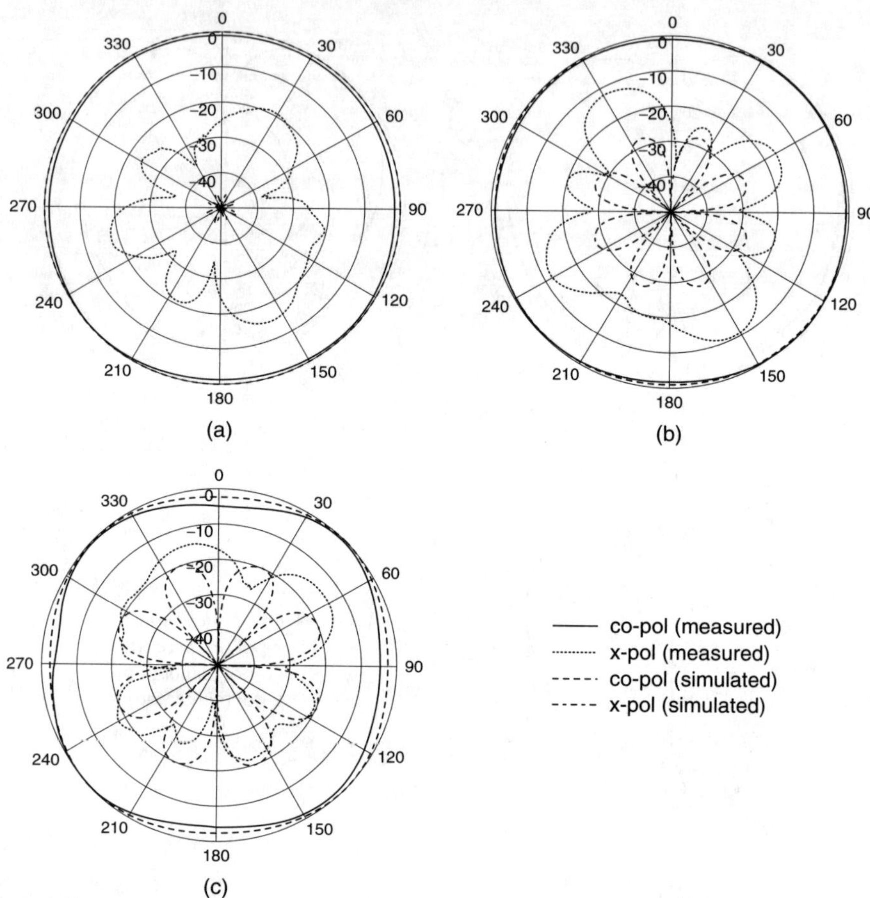

Figure 3.20 Azimuth plane radiation patterns ($\phi = 0°$) of the monopolar antenna at (a) 1.545 GHz, (b) 2.145 GHz, and (c) 2.745 GHz

3.2.5 Case 5: A Single-Feed Dual-Band Patch Antenna for Indoor Networks

A single-feed dual-band patch antenna is presented in this section.[10] Similar to the antenna described in Section 3.5.3, this antenna can also operate in two separate frequency bands across 820–960 MHz and 1710–2170 MHz, which are the frequencies covering most existing mobile systems. In order to reduce the interference between the two frequency bands, two low-pass filters are incorporated into the antenna.

3.2.5.1 Antenna Geometry The geometry of the antenna is shown in Figure 3.21. This antenna consists of two square patches of different sizes. The smaller patch is mounted on the top of the larger patch in

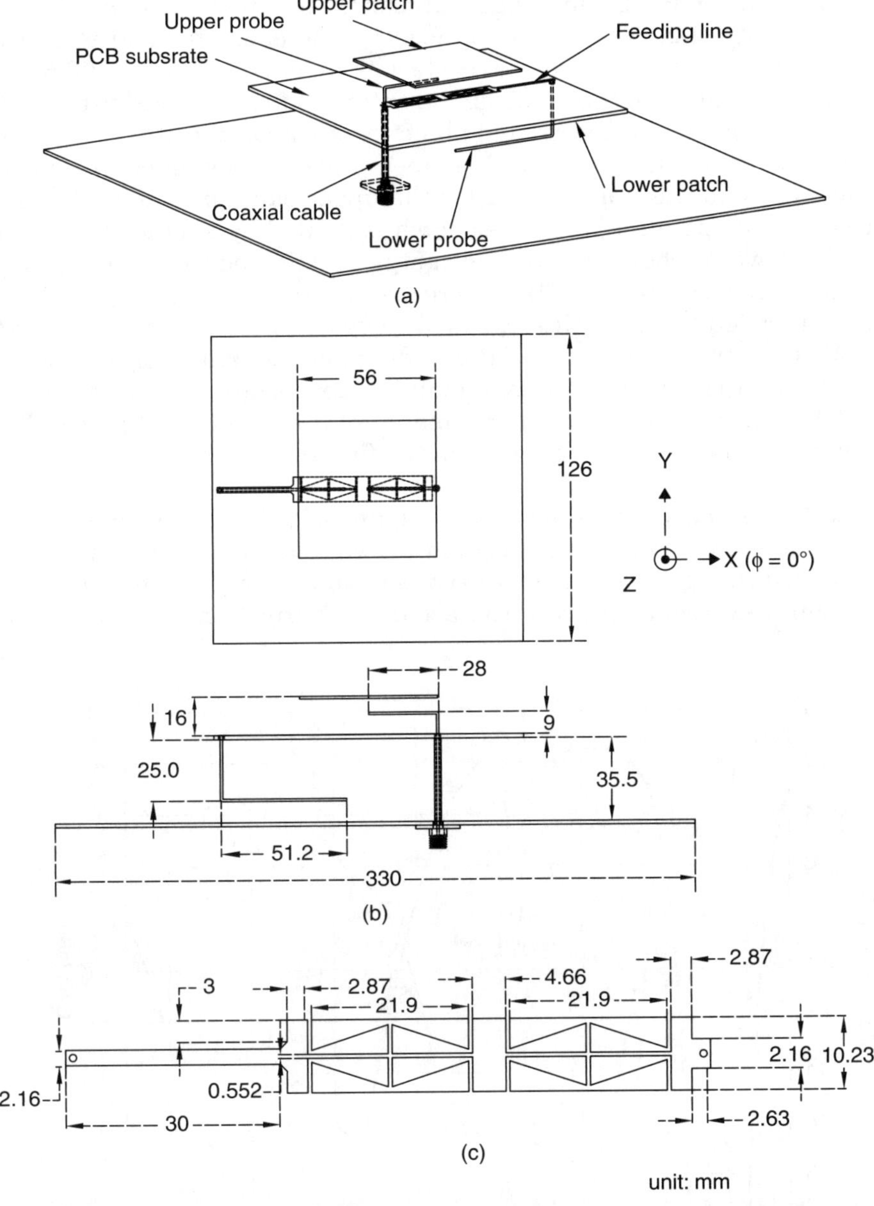

Figure 3.21　Geometry of the single-feed dual-band patch antenna (*a*) 3D view, (*b*) top view and side view, and (*c*) cascaded CMRC[10]

a stacked configuration. To a certain extent, the larger patch functions as the ground plane for the smaller patch. The upper patch is made of copper whereas the lower patch is etched on the lower side of a double-layer PCB, which has a thickness of 1.5 mm and a relative permittivity of 2.65. A microstrip feeding line is printed on the upper side of the same PCB. A coaxial cable with an outer diameter of 3.6mm is also introduced to this antenna. Its outer conductor is used to connect the lower patch and the ground plane together. On the other hand, its inner conductor is bent to an L-shaped probe (upper probe) to couple the energy to the upper patch for the upper-band operation. For the lower-band operation, the lower patch is excited by another L-shaped probe (lower probe) under it. The energy is delivered from one L-probe to the other one through the microstrip feeding line. In Figure 3.21, two cascaded low-pass filters, called *compact microstrip resonant cell (CMRC)*[11], are integrated into the feeding line. Actually, they can effectively suppress the excitation of higher-order modes of the lower-band patch, which can affect the performance of the upper-band patch. The antenna dimensions are detailed in the figure.

3.2.5.2 Simulation and Measurement Figure 3.22 shows the input return loss and gain against frequency in the two frequency bands. It can be observed that good agreement between simulation and measurement is achieved, resulting from the accurate fabrication technique and

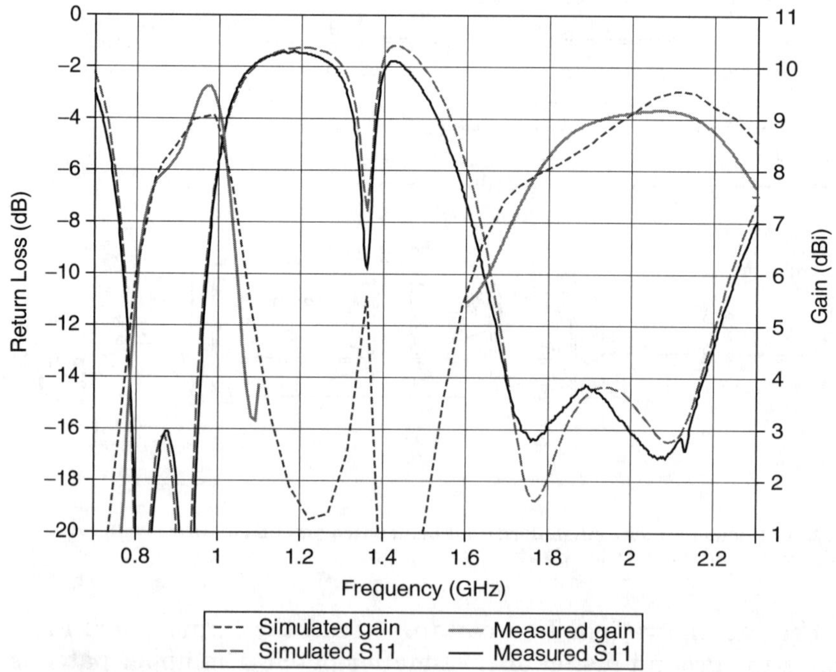

Figure 3.22 Input return loss and gain of the single-feed dual-band patch antenna[10]

the simple antenna structure. Indeed, the impedance bandwidths (RL < −14 dB) obtained by simulation and measurement are identical, which are 21% (0.782–0.965 GHz) in the lower band and 25.3% (1.69–2.18 GHz) in the upper band. Also, in Figure 3.22 the simulated and measured peak gains are around 8 to 9 dBi in both bands.

Figure 3.23 depicts the x-z plane radiation patterns at the center frequency of the lower band (0.89 GHz) and the upper band (1.94 GHz), respectively. The y-z plane radiation patterns at these frequencies are illustrated in Figure 3.24. The simulation and measurement are in good agreement. All the patterns are symmetrical with respect to the broadside direction. In both frequencies, the cross-polarization level in the x-z plane is very low (< −30 dB). However, the cross-polarization is around −12 dB in the y-z plane, which may be due to the vertical currents flowing on the probes and the cable. The backlobe level in both planes is less than −16 dB in the lower band. In the upper band, it is close to −20 dB because of the electrically large ground plane.

3.2.5.3 Summary

A single-feed dual-band patch antenna was designed and tested. Measured results demonstrate that this antenna has a standing wave ratio of less than 1.5 across 0.82 to 0.96 GHz and 1.71 to 2.17 GHz. Within these two frequency bands, good broadside radiation patterns are achieved in the x-z plane and y-z plane. This antenna is suitable for the base station that serves several mobile communication systems together, including CDMA, GSM, PCS, and UMTS.

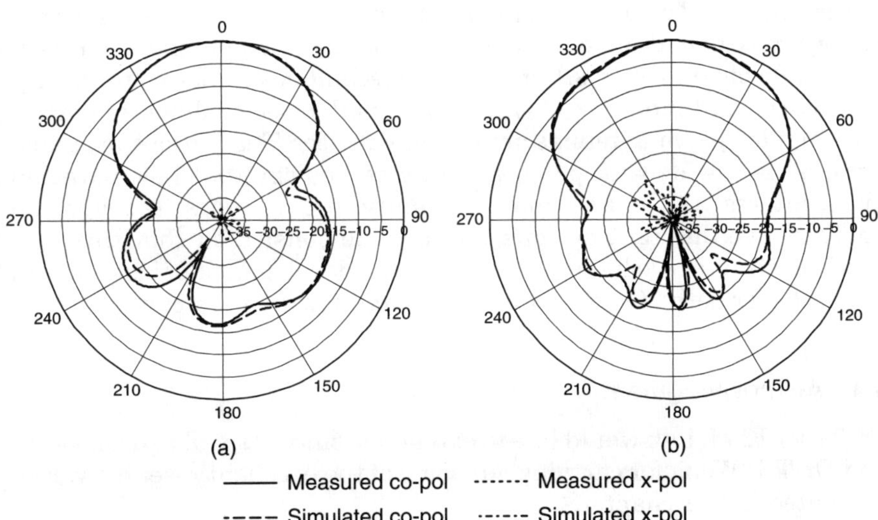

Figure 3.23 x-z plane radiation patterns of the single-feed dual-band patch antenna at (a) 0.89 GHz and (b) 1.94 GHz[10]

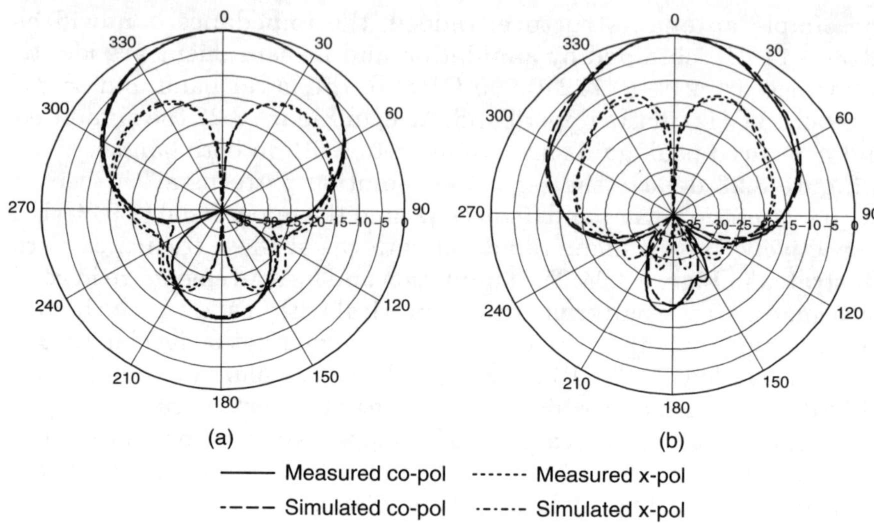

Figure 3.24 y-z plane radiation patterns of the single-feed dual-band patch antenna at (a) 0.89 GHz and (b) 1.94 GHz[10]

3.3 Conclusion

Design of base station antennas for mobile communication systems is very challenging. The required characteristics for bandwidth, gain, pattern shape, beamwidth, and intermodulation level are not easy to achieve. In this chapter, we have presented the commercial requirements for the performance of both indoor and outdoor base station antennas for various mobile phone systems. Conventional techniques employed to develop base station antennas are also reviewed. Both directed dipole and aperture-coupled patch antennas are popular choices. The L-probe fed patch antenna is a wideband patch antenna technology that has attracted much interest in the antenna community in past years. Through the study of five different designs, we have demonstrated that this new antenna structure is highly suitable for the development of base station antennas for both 2G and 3G mobile communication systems.

3.4 Acknowledgment

Professor K. M. Luk would like to extend his sincere thanks to Dr. Pei Li and Dr. T. P. Wong for carrying out some of the excellent research works presented in this chapter.

References

1. M. Gottl, M. Hubner, F. Micheel, and M. Burs, "Antenna, in particular a mobile radio antenna," US Patent, no. US7079083B2, July 18, 2006.
2. P. S. Carter and P. Jefferson, "Short wave antenna," US Patent 2,175,253, October 10, 1939.
3. K. Le, L. Meyer, and P. Bisiulers, "Directed dipole antenna," US Patent, pub no. WO/2005/122331, December 22, 2005.
4. Y. T. Lo, D. D. Harrison, D. Solomon, G. A. Deschamps, and F. R. Ore, "Study of microstrip antennas, microstrip phase arrays, and microstrip feed networks," Rome Air Development Center, Tech. Rep. TR-77-406, October 21, 1977.
5. K. M. Luk, C. L. Mak, Y. L. Chow, and K. F. Lee, "Broadband microstrip patch antenna," *Electronics Letters* (July 1998): 1442–1443.
6. K. Y. Hui and K. M. Luk, "Design of wideband base station antenna arrays for CDMA800 and GSM900 systems," *Microwave and Optical Technology Letters,* vol. 39, no. 5 (December 2003): 406–409.
7. T. P. Wong, "Design of L-probe coupled patch antennas and arrays," Ph.D. thesis, City University of Hong Kong, 2006.
8. T. P. Wong and K. M. Luk, "A wide bandwidth and wide beamwidth CDMA/GSM base station antenna array with low backlobe radiation," *IEEE Transactions on Vehicular Technology,* vol. 54, no. 3 (May 2005): 903–909.
9. P. Li and K. M. Luk, "Novel wideband dual-frequency patch antenna array for dual-polarization operation," *Proceedings of the 2005 Asia-Pacific Microwave Conference, Suzhou,* vol. 1 (December 2005): 4.
10. P. Li, "Novel wideband dual-frequency L-probe fed patch antenna and array," Ph.D. thesis, City University of Hong Kong, 2006.
11. Q. Xue, Y. F. Liu, K. M. Shum, and C. H. Chan, "A study of compact microstrip resonant cells with applications in active circuits," *Microwave and Optical Technology Letters,* vol. 37, no. 2 (2003): 158–162.

Advanced Antennas for Radio Base Stations

Anders Derneryd and Martin Johansson

Ericsson AB Ericsson Research

The rapid growth in the number of mobile communications users, as well as in the data rates required by more demanding applications such as mobile Internet and streaming services, has generated a demand for new and efficient ways of increasing capacity in cellular networks. At the same time, the need to offer communication services for areas with low traffic density, for example, in rural regions and developing markets with initial focus on voice services, has made wide-area coverage solutions essential. Solutions to either or both of these goals include increased frequency band allocation, frequency-hopping techniques, micro cells, underlay/overlay, and advanced antennas. There is an increasing interest in deploying advanced antenna techniques at the radio base station because these offer the potential to exploit the spatial domain and also because many of the other available solutions either have been fully utilized or are considered impractical or cost-inefficient.

The concept of advanced antennas for radio base stations is not well defined but, in general, can be taken to mean any antenna or strongly antenna-related solution more sophisticated than that of a conventional three-sector base station. Such solutions include higher order sectorization, higher order receive diversity, and in-air combining, as well as more elaborate techniques such as adaptive antennas. The latter means that automatic adaptation to the environment is performed on a relatively short time basis during transmission and/or reception. Examples of adaptation techniques are dynamic sectorization and electrical beamtilt and antennas with beamforming capabilities on a user basis as well as multistream MIMO (Multiple-Input Multiple-Output) systems. Several other frequently used terms for these antennas include advanced antenna systems, adaptive antenna base stations, smart antennas, smart base station antennas, MIMO antennas, and multi-antenna systems.

In this chapter, antenna technologies ranging from a conventional three-sector antenna system to a fully coherent adaptive antenna system for radio base stations are discussed, with a focus on macro base station antennas for second generation (GSM) and non-MIMO third-generation (UMTS) systems.

4.1 Benefits of Advanced Antennas

In both second- and third-generation mobile systems, interference is the main limiting factor with respect to achieving high traffic capacity. In second-generation systems, a standard procedure to increase capacity is cell splitting. Cell splitting typically means that new base stations are introduced in the network, and the area served by each base station is reduced. The maximum capacity per base station (site or cell) is basically constant in a well-planned mobile cellular network. By adding more base stations, i.e., cells, for serving a given area, the total system capacity per unit area is increased.

Advanced antennas have been suggested and deployed as an efficient means to meet the rapidly increasing traffic volume.[1–8] The move toward advanced antennas is driven by the shortcomings of the standard method, the cell split, which has become more and more difficult to implement in urban areas, both due to lack of locations that are suitable from a system perspective (propagation, transmission, and so on) and also due to aesthetical reasons (zoning regulations) and site cost aspects.

Advanced antennas in the setting discussed here offer a means for increasing system capacity or coverage while making use of existing sites and without affecting the user device requirements. Increased capacity is obtained mainly by reducing interference in the network using improved spatial filtering. This means that directive beams, narrower in azimuth or elevation than in the conventional three-sector system, are used for communicating with a user. Examples of capacity solutions are antenna beamtilt, higher order sectorization, multibeam array antennas, and steered-beam array antennas. The beamwidth in multibeam and steered-beam array antennas is narrower than the cell or the sector, i.e., only part of the cell where users are located is covered. This requires information regarding the user's location within the cell, either explicitly in terms of actual location of the user device or implicitly in terms of which base station antenna beam arrangement is most beneficial. In a second-generation system, like GSM, the improved spatial filtering allows for tighter frequency reuse giving more channels and higher capacity per sector. In a third-generation system, like WCDMA, the improved spatial filtering allows more traffic per sector and frequency.

Coverage solutions exist for uplink only, downlink only, as well as for both up- and downlink. Extended downlink or transmit coverage is obtained either by increased antenna gain or higher power amplification or a combination thereof. The transmit coverage is characterized by the Effective Isotropically Radiated Power (EIRP) of an antenna, determined by the relationship, $EIRP = P \cdot G$, where G is the antenna gain and P is the net power accepted by the antenna. The link budget is balanced with low-noise amplifiers in the receive (uplink) chain. Examples of coverage solutions are transmit diversity, high-gain antennas, higher order receive diversity, remote radio units, amplifier integrated antennas, and relays/repeaters. On downlink, increased EIRP, as provided by transmit diversity, high-gain antennas, amplifier integrated antennas, and relays/repeaters, gives increased signal strength at the user device. On uplink, higher order receive diversity, tower-mounted amplifiers, and high-gain antennas provide an increase in effective radio base station receiver sensitivity by extracting additional signal energy using statistical diversity gains and effectively larger antenna apertures.

4.2 Advanced Antenna Technologies

The main principles and concepts of advanced antenna systems are presented in terms of block diagrams and basic properties. Performance figures given are based on first order models and assume mostly that all base stations in the network are equal. The following advanced antenna concepts for radio base stations, in addition to a three-sector reference system, are described:

- Three-sector omnidirectional antenna
- Higher order receive diversity
- Transmit diversity
- Antenna beamtilt
- Modular high-gain antenna
- Higher order sectorization
- Fixed multibeam array antenna
- Steered beam array antenna
- Amplifier integrated sector antenna
- Amplifier integrated multibeam array antenna

One fundamental assumption for all concepts, except transmit diversity, is that no additional feedback is needed from the user device to the radio base station due only to the employment of any of the advanced antenna system concepts. This means that any user device can be served via a conventional sectorized system or an advanced antenna system, without being informed about the type of system it is connected to.

4.3 Three-Sector Reference System

The reference system is defined as a three-sector antenna system equipped in each sector with one downlink branch, and two uplink branches for space or polarization diversity reception. The reason for selecting this as a reference system is that it is the most common cellular base station antenna configuration in macro cell applications. The three-sector antenna system is used as a reference case when implementations and performance are discussed for the different advanced antenna concepts.

Receive diversity is a means to improve base station receiver sensitivity (*signal-to-noise ratio*) and thereby coverage of wireless communication systems. The sensitivity improvement comes from receiving signals using multiple antennas (*receiver diversity branches*), the principle being that the signals received on different antennas will fade more or less independently in relation to their fading correlation. Since the occurrence of two independent signals experiencing deep fades simultaneously is statistically less likely than the occurrence of correspondingly deep fades for a single signal, the receiver system will provide diversity gain. Assuming maximum ratio combining, there is an inherent gain, or power gain, from using multiple antennas that stems from the increased total antenna area (when going from one to two antennas). This area gain exists regardless of the statistical behavior of the received signals on the diversity branches. Two-way receive diversity is implemented in most cellular standards. To further improve the sensitivity in uplink, low-noise amplifiers are mounted close to the antenna, so-called tower-mounted amplifiers.

A fundamental aspect of antenna-based solutions is the type of interference distribution over a cell that results from the antenna radiation pattern properties in a sectorized system. A cell (also a *sector* when discussing macro sites) is a region that is serviced by the same radio resource, for example, as identified by the strongest broadcast/pilot signals in GSM and WCDMA. Two major antenna design parameters that influence the cell shape and the structure of the cell plan are the azimuth half-power beamwidth (HPBW) and the azimuth pointing direction.

Two different cell plans are commonly employed—*Ericsson cell plan* and *Bell cell plan,* as shown in Figure 4.1. The idealized cell shapes of

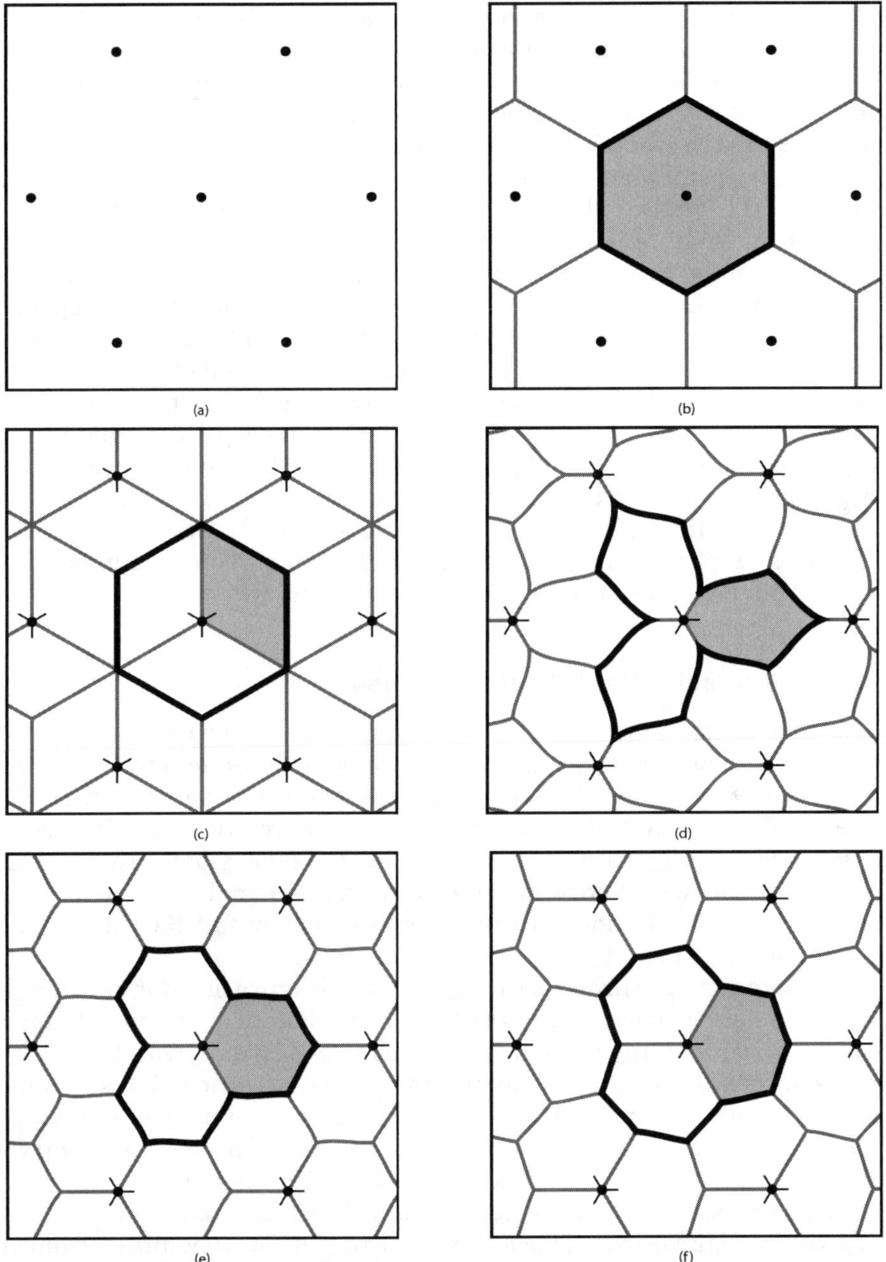

Figure 4.1 Three-sector cell plans: (*a*) site positions, (*b*) omnidirectional antennas, (*c*) Bell cell plan with 90° (but, in general, arbitrary) HPBW antennas, (*d*) Ericsson cell plan with 45° HPBW antennas, (*e*) Ericsson cell plan with 65° HPBW antennas, and (*f*) Ericsson cell plan with 90° HPBW antennas. The cell corresponding to one antenna is indicated by the shaded area. Dots indicate the locations of the base stations and line stubs show the antenna main beam pointing directions.

these cell plans are based on the assumption that identical base stations and antennas are placed on a regular hexagonal grid, with the three antennas oriented 120° apart in azimuth (in the horizontal plane), and that the propagation can be modeled by a uniform path loss model with no fading. In the case of the Bell cell plan, each antenna is directed toward a symmetry point among three base stations placed at the vertices of an equilateral triangle and the cell shape becomes independent of the half-power beamwidth. Each site in the Bell cell plan covers a hexagon. The cell shape of the Ericsson cell plan, on the other hand, is strongly dependent on the azimuth half-power beamwidth. Assuming a wave propagation attenuation proportional to the distance raised to 3.5, a half-power beamwidth of 65° will produce a hexagonally shaped cell rather than site. In a three-sector WCDMA system,[9] the beamwidth that maximizes the capacity in downlink is about 65° when each beam is pointing directly to the closest site (Ericsson cell plan), and about 75° when the beams from neighboring sites are directed toward the symmetry point in-between these sites (Bell cell plan) and similar results hold for a GSM system. The elevation properties of the radiation pattern will also have an impact on the cell shape, as discussed in the beam-tilt section.

4.4   Three-Sector Omnidirectional Antenna

The three-sector omnidirectional antenna configuration offers a low-cost solution for low-capacity requirements because it reduces the required number of power amplifiers by a factor of three, compared to the conventional radio base station arrangement. Therefore, this antenna is a candidate configuration for early and rapid deployment. In general, configurations with two or more sector antenna panels may be used to form combined radiation patterns with fractional or full 360° horizontal coverage.

A conventional three-sector antenna configuration for downlink transmission is shown in Figure 4.2*a*. Three identical sector antennas are used, sequentially rotated 120° about a vertical axis, with the sector antenna radiation patterns having identical polarization states over all azimuth angles in the main beam region. Each antenna is individually fed and has associated with it a separate signal path (*radio chain*), including a separate (downlink) power amplifier module.

The corresponding three-sector omnidirectional configuration is shown in Figure 4.2*b*. The antennas are fed (on downlink) from a common radio, with the power distributed from a dividing/combining network via dedicated feeder cables. In general, the lengths of the feeders are not identical, which results in differences in phase values. The same antenna arrangements are used on receive with the introduction of duplex filters or switches to separate transmit and receive signals.

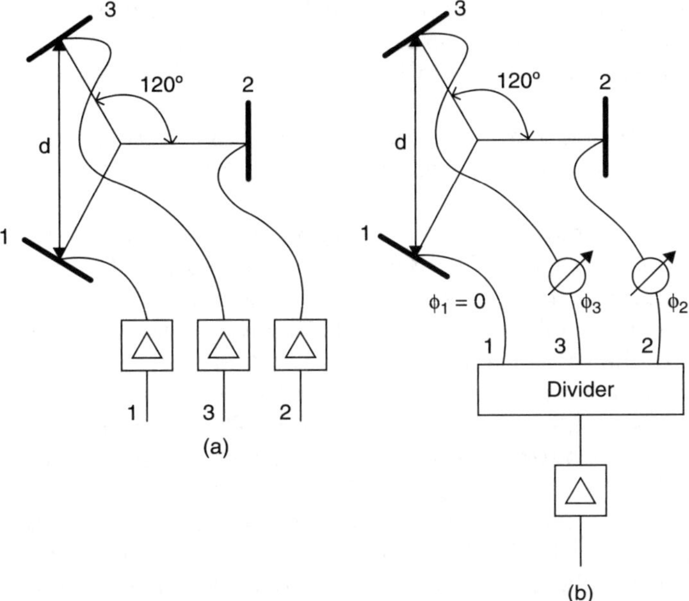

Figure 4.2 (*a*) Conventional three-sector antenna configuration. Each sector antenna is individually fed by a separate power amplifier. (*b*) Three-sector omnidirectional antenna configuration. All sector antennas are fed by a common power amplifier. Only downlink (transmit) signal paths are shown.

The sector antennas in the shown omni configuration are placed at the nodes of a horizontal equilateral triangle. Due to the separation distance, *d,* between the antennas, and since the radiation patterns of the sector antennas have identical polarizations, the free-space array radiation pattern resulting from a summation of the individual sector radiation patterns exhibits periodic amplitude oscillations or ripple. The period of this ripple is inversely proportional to the antenna separation distance.

The signal received by the user device is the sum of many multipath signal components with varying amplitude and phase. These components are due to the scattering occurring in the environment surrounding the user device. The result of the scattering is that the power from the base station antenna is averaged over an angular interval. As a consequence, the received signal strength at the user becomes less sensitive to amplitude variations of the base station antenna pattern with increasing radius of the scattering region. This also mitigates the effects of the unknown phase differences among the antennas, since the angle-dependence of the free-space pattern ripple is reduced.

The scattering behavior can be modeled by using power-averaged radiation patterns. The averaging is performed by calculating the mean value of the radiated power (uniformly weighted) over all angles within a sliding window of width ϕ_{int} (see Figure 4.3). The size of the averaging window is related to the distance R between the base station and the user device, and an equivalent local scattering radius r, by $\phi_{int} = 2r/R$ for $r \ll R$.

The null depths (ripple) in the three-sector omni radiation pattern does not lower the averaged, effective pattern below that of the individual sector radiation pattern crossover level, if the antenna separation d fulfills the rule-of-thumb $d > 1/\phi_{int} \approx R/(2r)$ (wavelengths). However, the actual level of the three-sector omni radiation pattern is decreased 5 dB relative to the sector pattern level because the power (from a single power amplifier) is divided over three antennas. This general drop in the three-sector omni pattern level can be reduced in the crossover region by up to 3 dB. To increase the three-sector omni pattern level by about 2 dB in the crossover region, the antenna separation distance d (in wavelengths) should fulfill the inequality $d > R/r$, which results in a three-sector omni radiation pattern level only 3 dB below the crossover level of the isolated sector patterns, rather than 5 dB below. This is particularly useful in scenarios where a three-sector site is not part of a regular cell plan because it provides a more direction-independent pattern and, hence, omni-like coverage.

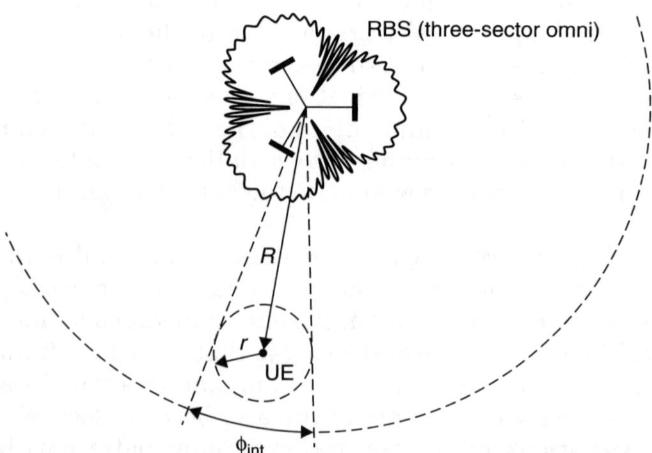

Figure 4.3 Scattering environment surrounding a user device (UE) for pattern averaging

4.5 Higher Order Receive Diversity

Coverage is an important issue, especially in sparsely populated areas. One method to increase uplink coverage is to employ higher order receive diversity in base stations in order to improve receiver sensitivity. In systems where coverage is uplink limited, a solution based on higher order receive diversity can provide a balance in the uplink-downlink performance. The four-branch receive diversity uses four complete RF receiver chains. The four antenna signals may be obtained by a combination of polarization and space diversity combined at baseband. Figure 4.4a shows an antenna configuration with two polarization diversity antennas, with a separation of 10 to 20 wavelengths. However, space diversity is not crucial for the performance in environments with rich multipath, and in these cases, those two polarization antennas can be combined into one unit under a single radome, as shown in Figure 4.4b.[10] The advantages of the combined antenna unit are its compact size and easier installation.

For higher order receive diversity systems, i.e., receive diversity systems with more than two antennas, the additional statistical diversity gain decreases with increasing number of diversity branches for many macro site propagation environments, whereas the gain related to the additional antenna area, power gain, will still be available. The resulting improvement of the receiver sensitivity performance may not be motivated as it comes at the cost of additional receiver chains.

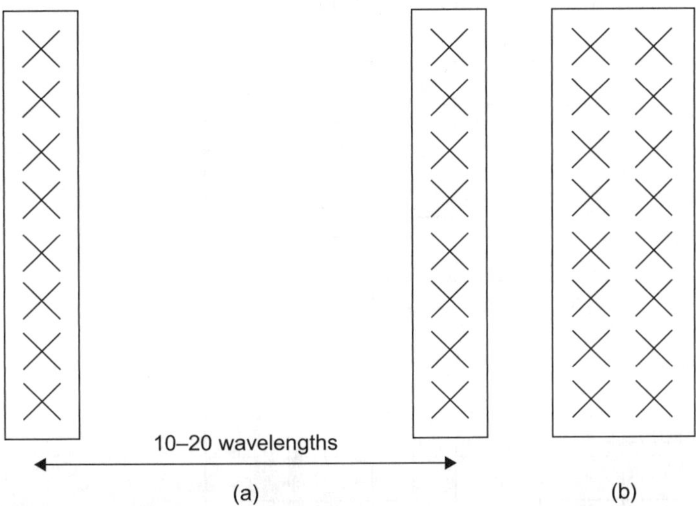

Figure 4.4 Four-branch receive diversity antenna configurations with dual-polarized antenna columns: (a) separate antenna units and (b) single antenna unit

4.5.1   Field Trial

A field trial of commercial two-branch and four-branch diversity implementations was conducted in a live GSM network.[11] Two statistical measures were recorded. The uplink signal strength, RxLev in dBm, was reported as the maximum value taken over all two or four receive branches, i.e., the signals were not combined before RxLev was reported. This gives a measure of the selection diversity gain. The active measurement range of RxLev was bounded from below at −110 dBm, i.e., all measured values less than −110 dBm were reported as RxLev = 0, which corresponds to −110 dBm. The uplink bit error rate, RxQual on a scale from 0 to 7, was measured after combining. A low value of RxQual corresponds to good signal quality, whereas a high value of RxQual means poor signal quality.

The selection diversity gain was about 1 dB with four-branch diversity compared to two-branch diversity, as measured by the improvement in uplink RxLev for the best branch. This corresponds well with simulated and measured results that show an improvement of 3–4 dB in receiver sensitivity, from which an estimated power gain of 2.5–3 dB should be subtracted. Figure 4.5 shows a plot of the Cumulative Distribution Function (CDF) of the uplink RxQual for measurement reports for all mobile traffic in a cell with RxLev = 0, i.e., for cases when the best branch has a received signal strength of −110 dBm or less. It is clearly seen that four-branch diversity provides better quality than two branches. For example, about 45% of the two-branch measurement reports show RxQual < 4 compared to 66% of the four-branch measurement reports, an improvement of almost 50%.

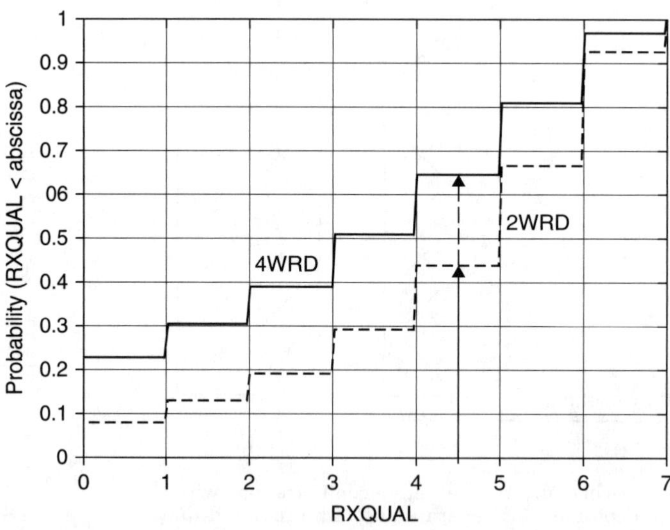

Figure 4.5   CDF of RxQual when RxLev equals 0 for a two-branch (2WRD) and a four-branch receive diversity (4WRD) field trial for all commercial traffic in a cell

4.6 Transmit Diversity

Transmit diversity is a concept used for improved downlink capacity in which multiple antennas are used. Radio base stations are already equipped with multiple antennas, which are used for receive diversity. The idea is to use these antennas to mitigate the effects of fast fading also in downlink. A number of transmit diversity schemes exist, such as space-time coding, delay diversity, antenna hopping, and phase hopping.[12]

Alamouti space-time coding is an open-loop transmit diversity scheme that uses two antennas to simultaneously transmit two signals to the user device.[13] This scheme, however, requires user devices with special space-time decoding algorithms. On the other hand, delay diversity, antenna hopping, and phase hopping are transparent to user devices and can be employed for existing devices. The delay diversity scheme transmits delayed replicas of the signal by utilizing two antennas at the base station. The user device needs to receive an uncorrelated signal from each transmit antenna to take advantage of the technique. The radio base station antennas need to be spaced sufficiently far apart or transmit with orthogonal polarizations. Delay diversity artificially creates increased time dispersion in the propagation channel, which is resolved by the equalizer in the user device. In the antenna hopping scheme, the transmitted signal is switched among the available base station antennas. The signal then passes through a propagation channel that varies even if the user device is stationary. With antenna hopping, no additional hardware, such as transceivers or combiners, is required in the base station, since only one antenna is transmitting at a time. The phase-hopping scheme also aims at creating a propagation channel that is nonstationary. This scheme transmits identical signals that are phase-offset across two antennas at the base station.

Most transmit diversity schemes use only two antenna branches. Each of these antennas uses only half of the power of a system with a single antenna in order to maintain constant transmit power. In the case when full transmit power is used on each antenna simultaneously, the transmit diversity schemes can be used to increase coverage. All schemes can be extended to more than two antennas.

4.7 Antenna Beamtilt

Beamtilt or adjustment of the main beam direction in elevation, typically between $0°$ to $10°$ downward, is used for many purposes. One main purpose is to obtain good cell-border coverage. The beam direction is adjusted to follow the surface of the ground to radiate as much power as possible toward the cell border. Another application is to increase capacity by reducing inter-cell interference, which is achieved by beam downtilt and the corresponding increased isolation between the cells. The main beam is directed such that the upper slope of the main beam radiation pattern is pointed toward the cell border. This means that interference signals toward neighboring cells are reduced at the cost of lower antenna gain for devices located at or near the cell border.[14] The reduction in gain at the cell border due to downtilt is often not a problem for densely planned systems, i.e., system with small inter-site distances where path loss is not a limiting factor, since capacity (and, hence, interference) is the main concern. In metropolitan environments with high antenna installations, beamtilting far beyond 10° below the horizon is sometimes applied. Beamtilt has been frequently used in large-scale Personal Digital Cellular (PDC) systems since their introduction, and it is frequently used today in many mobile communication cellular networks.[15]

Two main methods for tilting an antenna beam exist, and they are used either separately or in combination. The simplest method is to mechanically tilt the antenna while the more sophisticated method utilizes electrical beamtilt.[16] Mechanical beamtilt affects the antenna radiation differently along the horizontal plane. For example, if the main beam in a three-sector system is tilted downward, then the back-lobe is tilted upward while the wide angle sidelobes are almost unaffected. Electrical beamtilt, on the other hand, has essentially the same tilt effect on the radiation pattern for all azimuth angles for cylinder-shaped antennas, such as conventional sector antennas, when these are installed vertically. Plots of iso-path gain curves in Figure 4.6 show

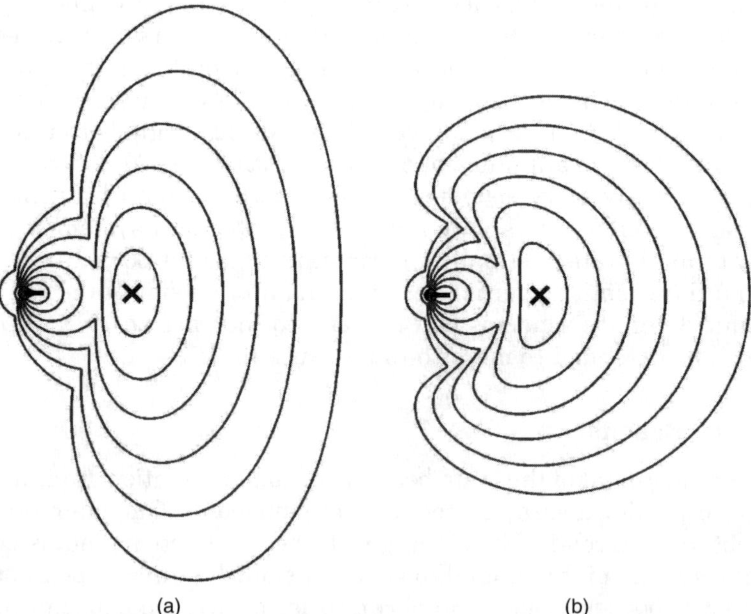

Figure 4.6 Iso-path gain curves on ground for an antenna downtilted 10°: (*a*) mechanical downtilt and (*b*) electrical downtilt. Main beam peak points at the × (5 dB steps between contours).

the difference in the resulting shape of the covered area for the two tilt methods (ignoring surrounding cells).

A combination of mechanical and electrical beamtilt is sometimes an attractive solution. By applying a fixed mechanical beamtilt, the requirements on the electrical tilt range offered by the antenna may be reduced and the hardware implementation becomes less complex. One application of combined mechanical and electrical tilt is for reducing backlobe interference for antennas that have a distinct backlobe that behaves similarly to the forward main beam when the element excitations are varied. This can be achieved by mechanically tilting the main beam, either upward or downward, and then offsetting the mechanical tilt by electrically tilting the beam in the opposite direction. The combined result is that the main beam gets the desired beamtilt angle (the difference between the mechanical and the electrical tilt angles) in the forward direction whereas the backlobe is tilted away from the horizontal plane, as shown in Figure 4.7.

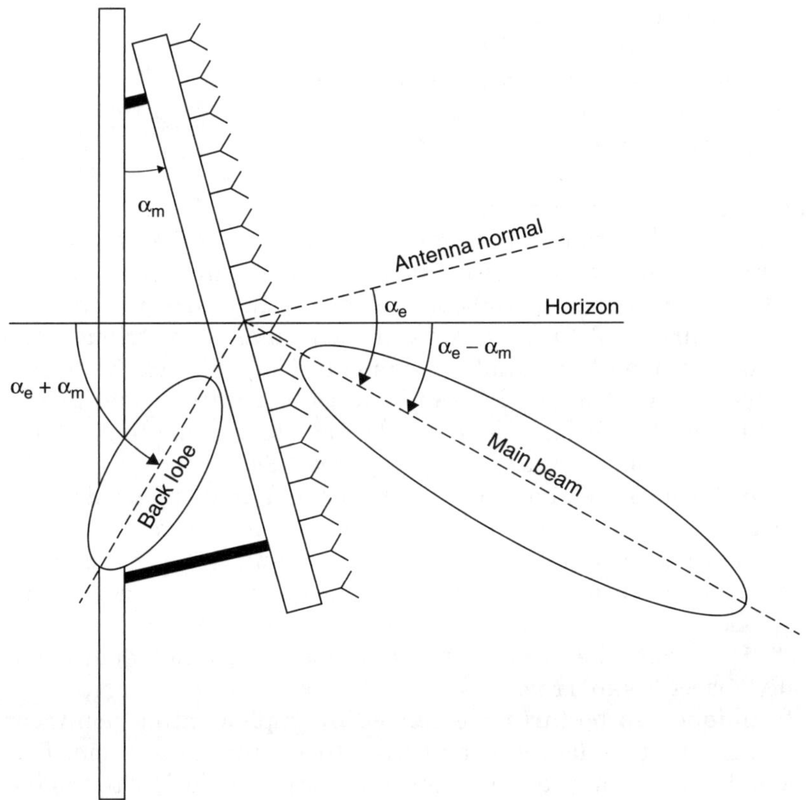

Figure 4.7   Main beam and backlobe directions for an antenna with simultaneous mechanical uptilt (α_m) and electrical downtilt (α_e)

The electrical beamtilt is accomplished by having a tilt unit that divides the available power (on transmit) and changes the phase or time delay to each antenna element or each subarray consisting of a number of antenna elements stacked vertically (see Figure 4.8). The design of an antenna with electrical downtilt is a trade-off between antenna performance and antenna system complexity and cost. Ideally, each antenna element would be individually controlled with respect to amplitude and phase of the excitation. With antenna elements spaced a fraction of a wavelength apart, this configuration would not incur grating lobe problems, but the solution would require a high-complexity tilt unit as well as a large number of cables in the antenna assembly. This is not considered a cost-effective solution and instead subarrays are controlled by the tilt unit.

The subarrays have fixed distribution networks that are designed to give a desired fixed tilt angle and beam shape. The subarrays may also be optimized to counteract the grating lobe effects, i.e., generation of undesirable copies of the main beam, that appear when controlling the excitation of the groups of antenna elements (the subarrays) separated by more than a wavelength. This is illustrated in Figure 4.9. The array antenna consists of 16 radiating elements that are grouped into 4 identical subarrays. A fixed network in each subarray distributes the power equally among the radiating elements. The subarray radiation patterns are displayed as gray lines in Figure 4.9. A grating lobe appears as the array antenna beam is scanned from 0° to 10° due to the large spacing between subarrays as seen in Figures 4.9*a*, 4.9*b*, and 4.9*c*. The grating lobe level can be reduced by introducing a fixed phase delay in the subarray distribution network. This is shown in Figures 4.9*d*, 4.9*e*, and 4.9*f*, where each subarray has a fixed scan angle of 5.5°. The total array radiation pattern now has a grating lobe level not exceeding the first sidelobe in the total array radiation pattern when scanned from 0° to 10°. The technique can be further extended with nonidentical or unequally spaced[17] subarrays.

The technique can be further extended by implementing nonlinear phase-shifters or time delays when tilting the beam in elevation. This allows an adaptive change of the beam shape or beamwidth of the radiation pattern with tilt angle during operation of a cellular network. The advantage of using antennas with a nonlinear type of phase-shifter is

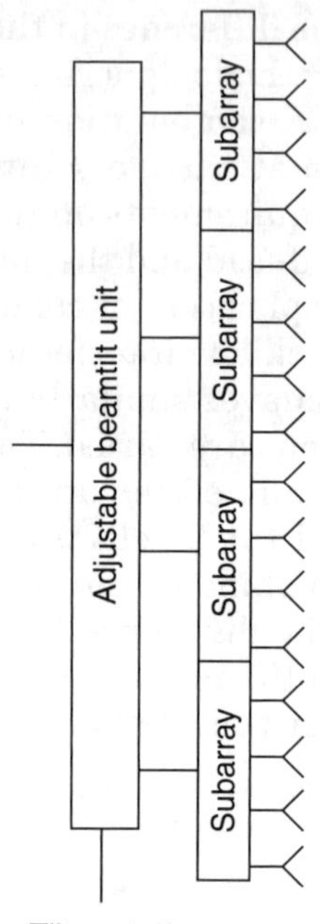

Figure 4.8 Array antenna consisting of four subarrays, an adjustable tilt unit, and phase-matched cables

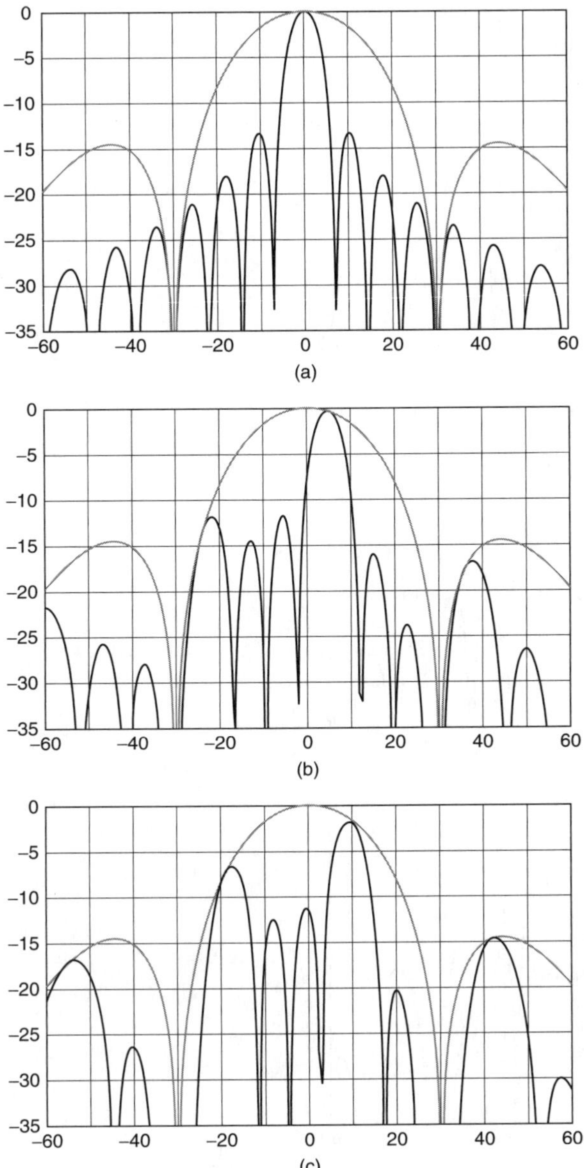

Figure 4.9 Elevation radiation pattern in dB as a function of the angle in degrees. The antenna consists of four subarrays with four elements each. Black lines indicate the total scanned radiation pattern and gray lines the fixed subpanel radiation pattern. (a) 0° scan angle, subarray at 0° fixed tilt angle; (b) 5° scan angle, subarray at 0° fixed tilt angle; (c) 10° scan angle, subarray at 0° fixed tilt angle; (d) 0° scan angle, subarray at 5.5° fixed tilt angle; (e) 5° scan angle, subarray at 5.5° fixed tilt angle; and (f) 10° scan angle, subarray at 5.5° fixed tilt angle.

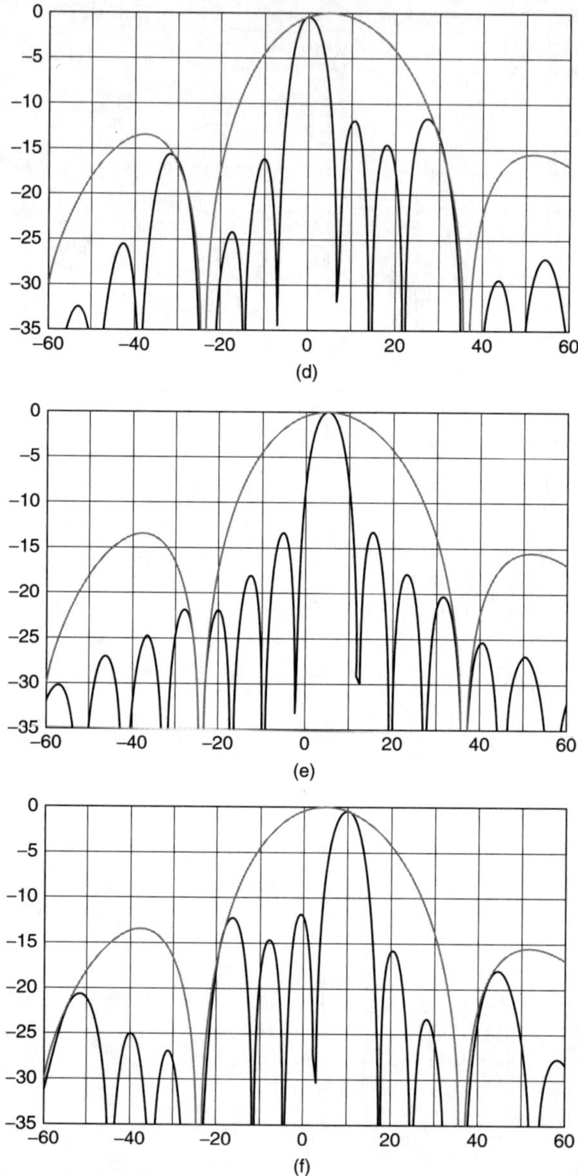

Figure 4.9 Elevation radiation pattern in dB as a function of the angle in degrees. The antenna consists of four subarrays with four elements each. Black lines indicate the total scanned radiation pattern and gray lines the fixed subpanel radiation pattern. (*a*) 0° scan angle, subarray at 0° fixed tilt angle; (*b*) 5° scan angle, subarray at 0° fixed tilt angle; (*c*) 10° scan angle, subarray at 0° fixed tilt angle; (*d*) 0° scan angle, subarray at 5.5° fixed tilt angle; (*e*) 5° scan angle, subarray at 5.5° fixed tilt angle; and (*f*) 10° scan angle, subarray at 5.5° fixed tilt angle. (*Continued*)

that the beam shape is tailored to the beamtilt angle. For example, a non-linear phase-shifter can provide high gain at the cell edge for small tilt angles to maximize area coverage and null-fill near the base station for larger downtilt angles, as shown in Figure 4.10, in order to avoid service areas inside the cell with poor path gain, which could be detrimental, especially for high data-rate users. Such an antenna can, therefore, be deployed for both coverage and capacity scenarios, even if the resulting system operating point is uncertain when the network is rolled out.

The optimal beamtilt angle depends on site location, antenna installation height, cell size (inter-site distance, ISD), and traffic load.[18] Basically, the smaller the cell size, the larger the beam downtilt angle should be. Beam downtilt has the most impact for high gain, narrow vertical beamwidth antennas. Highest impact is achieved in areas with small cells and/or high antenna installations. In larger cells, antenna downtilt is still useful in order to reduce local interference problems, to decrease the cell size, or to improve coverage (signal-to-noise ratio or SNR) inside the cell away from the cell border. However, this is at the cost of reduced coverage at the cell border.

For efficient use of the radio access network, antenna beamtilt is controlled adaptively from an operation center, the so-called remote (electrical) tilt. The purpose of a control system is to perform the antenna beamtilt setting, as requested from the operation center. The beamtilt is set per antenna or beam independently for optimal system performance.[19] Optimal tilt angle settings may differ with frequency band for multiband antennas with independent tilt control possibilities on a band-by-band basis. The tilt angle of each antenna can also be set locally at the base station by connecting a portable computer to the control system or manually on site without using a control system. Systems using antennas with remote tilt control may be adapted to the existing traffic situation or changes in the propagation environment on a short- or long-term basis.

Although the discussion has been focused on single beam antennas, beamtilt, which makes use of the elevation dimension, can be used in

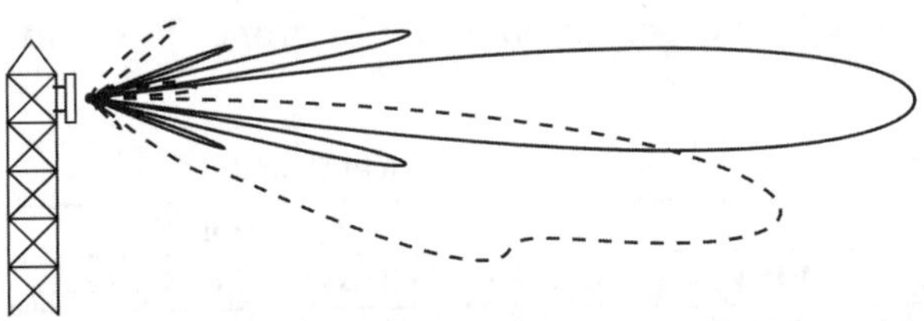

Figure 4.10 Elevation radiation patterns of an antenna with a tilt unit containing nonlinear phase shifter for two different tilt settings: untilted high-gain pencil beam (solid line) for maximum cell edge coverage and downtilted shaped beam (dashed line) for uniform cell coverage and, hence, capacity, with interference suppression toward neighbor cells accomplished by low sidelobe levels above the main beam.

sectorized systems as well as in multibeam antenna systems that are mainly focused on exploiting the azimuth dimension.

4.7.1 Case Study

Antenna beamtilt is an efficient way of also improving system capacity in re-use one systems, such as WCDMA, although there are fundamental differences in the interference situations between GSM and WCDMA. In WCDMA downlink, for example, the desired user and a user being interfered can be located fairly close to each other, such as on either side of the cell border.

The setting of the beamtilt angle has been evaluated in a re-use one three-sector network.[14] All users are assumed to be uniformly distributed in the network and all traffic assumed to be speech. Furthermore, there are identical antennas in each cell and an identical electrical beamtilt angle is applied to all base stations. Six different radiation patterns are used in the study (see Table 4.1) and two types of network layouts, one is the Ericsson cell plan and the other one is the Bell cell plan. The 65° azimuth half-power beamwidth (HPBW) pattern is evaluated in the Ericsson cell plan (Figure 4.1e), whereas the 90° azimuth patterns are evaluated in both the Bell cell plan (Figure 4.1c) and the Ericsson cell plan (Figure 4.1f). A 120° HPBW pattern is evaluated in the Bell cell plan only. The relative sidelobe level (SLL) is set to −15 dB in both the elevation and the azimuth plane. The antenna installation height is 30 m in all cases, and the site-to-site distance is 2000 m for high-gain (20 dBi) antennas and 800 m when low-gain (17 dBi) antennas are analyzed.

The evaluated performance is the WCDMA pole capacity defined as the maximum capacity possible while still fulfilling the desired quality for each user. This capacity is found by loading the system until no solutions to the power equation can be found, i.e., the SINR quality requirement is no longer fulfilled. At pole capacity, the required power asymptotically approaches infinity.

Pole capacity, the maximal theoretical load, is shown in Figure 4.11 for high- and low-gain antennas. One observation is that a proper beamtilt

TABLE 4.1   Antenna Radiation Pattern Data Used in the Electrical Beamtilt Study

Antenna	Gain (dBi)	Azimuth Cut		Elevation Cut	
		HPBW	SLL (dB)	HPBW	SLL (dB)
65 high gain	20	65°	−15	2.6°	−15
65 low gain	17	65°	−15	5.5°	−15
90 high gain	20	90°	−15	1.9°	−15
90 low gain	17	90°	−15	4.0°	−15
120 high gain	20	120°	−15	1.4°	−15
120 low gain	17	120°	−15	3.0°	−15

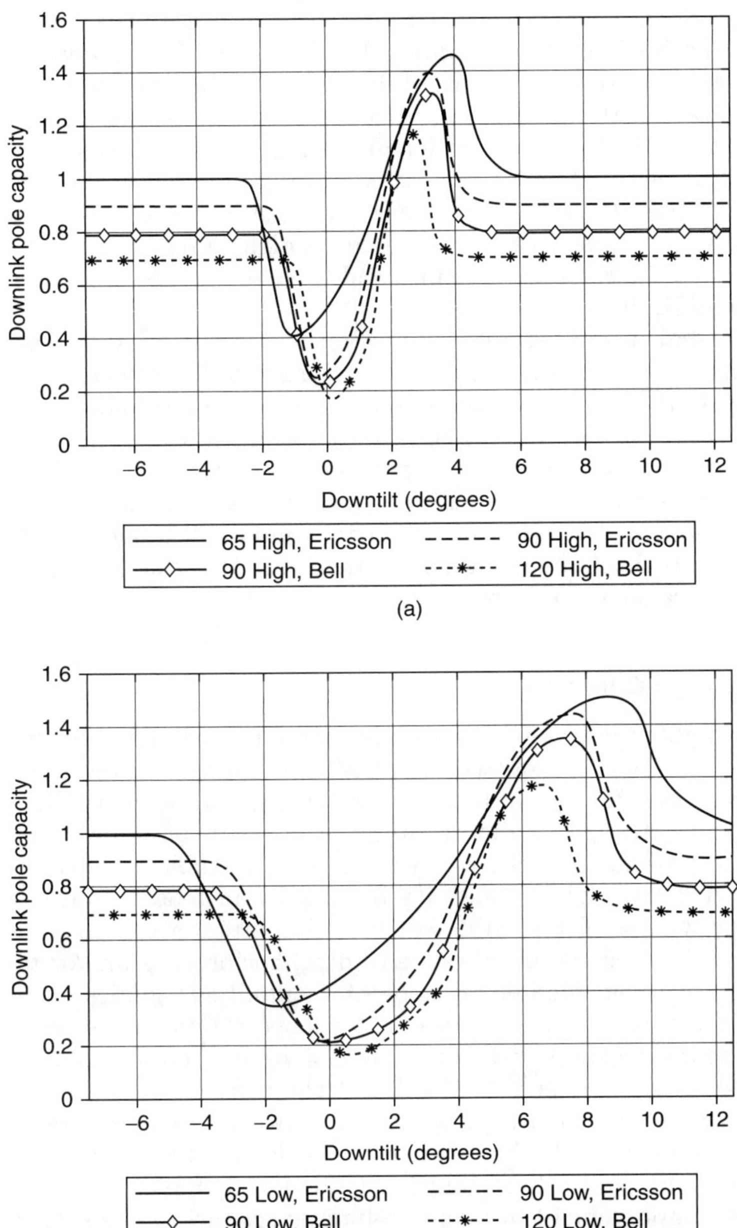

Figure 4.11 Pole capacity in downlink as a function of electrical beam-tilt angle: (*a*) high-gain antennas and (*b*) low-gain antennas

angle significantly improves pole capacity compared to no beamtilt, with a tilt angle of about one half-power beamwidth below the midpoint between two sites being close to optimal. A second observation is that the wider the elevation beamwidth, the less variation in capacity over beamtilt angles. A third observation is that the elevation and azimuth beamwidths have to be selected carefully to reduce variations in illumination of the cell (i.e., path gain), since these variations are of utmost importance for the pole capacity. All antennas within each category, high gain and low gain, respectively, have the same gain values, and the elevation beamwidths are chosen to fulfill that.

Electrical beamtilt and azimuth and elevation beamwidths are key factors for improving downlink performance in mobile networks. To achieve the potential gains, which depend on antenna parameters as well as cell characteristics, the settings of the beamtilt angles need to be optimized in each real network separately. This study focused on pole capacity performance on downlink. Considering that the uplink and lower traffic load will typically result in smaller tilt angles being optimal, a compromise between uplink and downlink performance, as well as between capacity and coverage, is necessary.

4.8   Modular High-Gain Antenna

The modular high-gain antenna concept is used for maximizing coverage (*cell range*) in environments with low traffic intensity. This antenna is characterized by offering an improved link budget in terms of high antenna gain in both uplink and downlink. A vertical combination of multiple sector antennas results in higher gain and enables the possibility of extended coverage without altering the azimuth beamwidth.[20] Mechanical as well as electrical downtilt of the antenna main beam are incorporated to reduce interference in neighboring cells and/or to maximize coverage area. Modularity allows for simplified logistics, with relaxed requirements on infrastructure and transportation means, with site-based antenna assembly using high-precision self-aligning mounting frames to ensure excellent electrical performance.

By combining commercially available hardware, such as standard sector antennas, combiners, and coaxial cables, into a high-gain antenna, a modular and robust solution is achieved that is suitable for link budget improvement in coverage-limited scenarios. The sector antennas used as building blocks, referred to as *subpanels,* are fed by a feed network, which includes power dividers, delay lines (phase shifters), and coaxial cables. The subpanels may have different radiation characteristics, and they are installed to get a desired radiation pattern behavior. Amplitude taper, potentially nonuniform, is provided by the divider/combiner, whereas phase taper is handled by delay lines or phase shifters with phase-matched cables running to each subpanel. A schematic illustration of high-gain antennas based on the modular principle is shown in Figure 4.12.

Subpanel tilt is applied either on a mechanical basis, using a physical tilting of each subpanel structure, or on an electrical basis, for subpanels with a means for progressive time delays or phase shifts of the excitation over the length of the subpanel array. Combinations of mechanical and electrical tilt are also used. For increased flexibility, remote control of tilt is possible, but for most extended coverage installations, this is not an essential feature. Tilt applied to individual subpanels is combined with delay compensations in the feed network to avoid grating lobe appearing due to the large subpanel separations.

An advantage with individual subpanel mechanical tilt is that it gives a more compact installation than if the entire structure is mechanically tilted. This reduces the wind load, resulting in less torque, and provides an overall antenna configuration that is essentially conformal with the mounting structure. Similarly, electrical tilt in the feed network reduces the amount of hardware compared to a solution with electrical

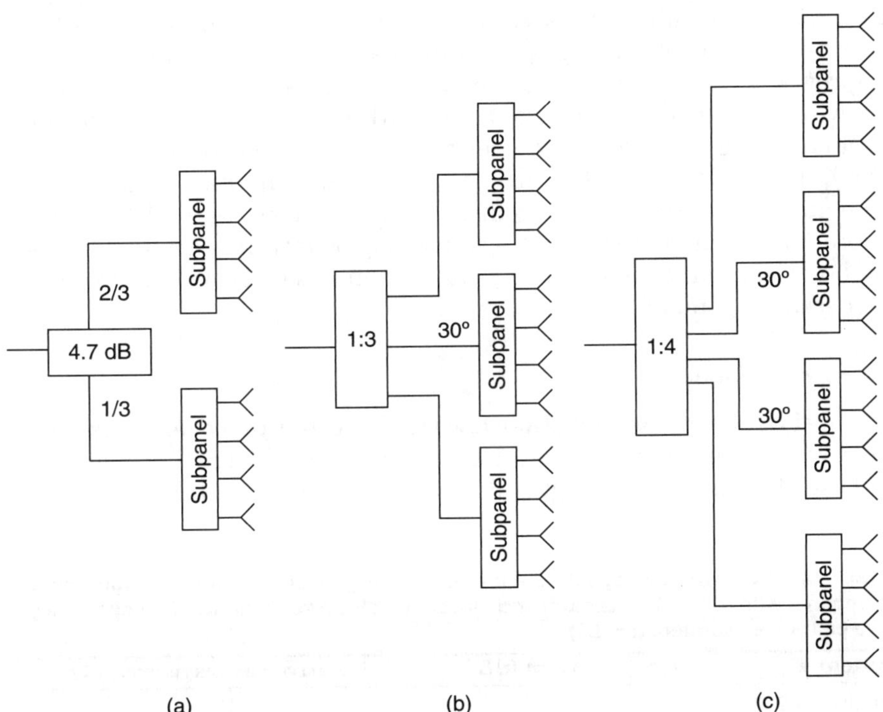

Figure 4.12 Modular high-gain antennas made up of standard sector antennas (subpanels), power dividers/combiners, and phase-matched cables

tilt implemented on an antenna element basis over the full length of a high-gain antenna. Of course, the combination of electrical tilt in the feed network with electrically tiltable subpanels provides full electrical tilt, with the added benefit of independent control of the tilt settings.

The polarization of subpanels using the same corporate feed network is typically the same. However, combining subpanels of different polarizations is also possible, for example, to avoid deep pattern nulls. Still, the main part of the power is fed to subpanels with identical polarization in order to achieve the desired improvement in antenna gain and, as a consequence, coverage. Similarly, subpanels that differ with respect to other properties, for example, size and gain, can be combined. Null-filling for identical copolarized subpanels can be achieved by nonuniform phase and amplitude weighting, depending on the number of subpanels. Filling the first pattern nulls to about −20 dB will only incur a small reduction, a few tenths of a dB, in peak antenna gain.

Based on the gain improvements for the configurations with multiple subpanels, an increase in coverage is achieved. To estimate the general behavior, a basic range-dependent model for the relative coverage area is written as $\Delta A = \Delta G^{2/\alpha}$, where ΔG is the relative antenna gain and α is the path-loss exponent. The relative coverage area with different high-gain antenna configurations for α equal to 3.5 is shown in Table 4.2.

Introducing modular high-gain antennas, Figure 4.13, into a network improves the link budget and increases coverage up to 70%. It is ideal for wide-area coverage in rural areas with low traffic density, over flat ground, along highways, and over sea. The antenna concept improves the link budget 2.5 dB to 4 dB equally in both uplink and downlink. The number of sites may be reduced with this concept, which reduces overall deployment cost. Existing sites may be retrofitted with modular high-gain antennas without any changes in base station equipment or feeder arrangement.

4.8.1 Case Study

In rural deployments and other low traffic scenarios, coverage is a key performance indicator and solutions that minimize the required number of base station sites are highly desirable to reduce capital expenditure

TABLE 4.2   Relative Coverage Area With Different High-Gain Antenna Configurations (Subpanel with 65° and 8° azimuth and elevation half-power beamwidth, respectively, and path-loss exponent $\alpha = 3.5$)

Antenna	Gain (dBi)	Relative coverage area (%)
Single subpanel	18	100
Dual-subpanel	20.5	140
Tri-subpanel	22	170

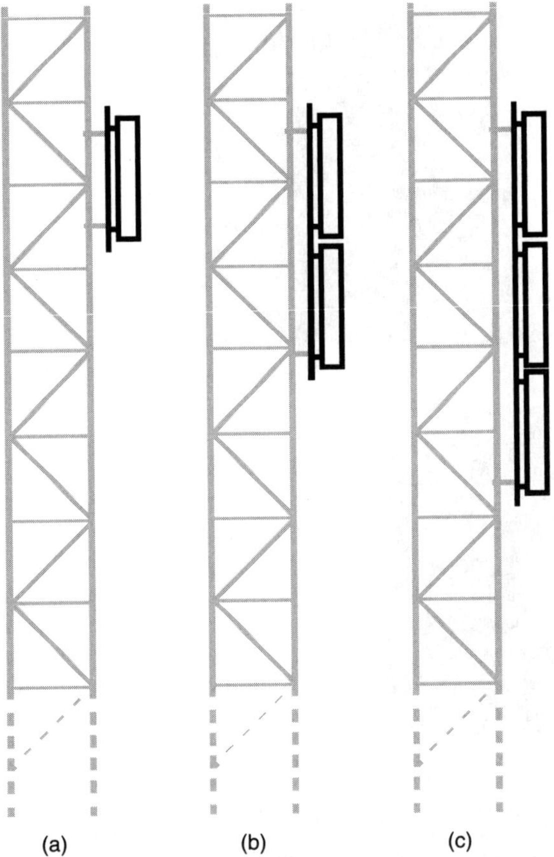

Figure 4.13 Modular high-gain antenna configurations:
(*a*) single subpanel, (*b*) dual subpanels, and (*c*) triple
subpanels

and operational cost. A study of the potential for coverage improvement in a realistic environment by increasing base station antenna gain was performed. Changes in antenna gain have a balanced impact on uplink/downlink link budget performance. With the azimuth half-power beamwidth set to a fixed value of 65°, the antenna gain increase must come from a decrease in elevation half-power beamwidth, i.e., from a corresponding increase in antenna vertical size.

Plots of predicted coverage for three different antenna installations over hilly terrain are shown in Figure 4.14 for a three-sector site. The coverage for a standard sector antenna panel with 8° elevation half-power beamwidth is compared with the coverage for two high-gain antennas. The high-gain antennas comprise, respectively, two and three vertically stacked subpanels (identical to the sector antenna) combined

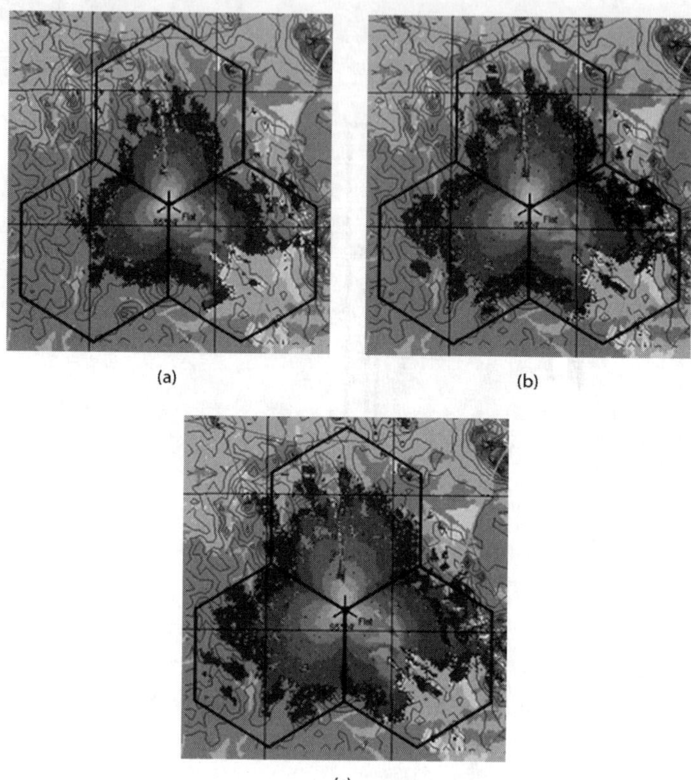

(a)

(b)

(c)

Figure 4.14 Coverage of a three-sector high-gain antenna installation over hilly terrain: (*a*) single subpanel with gain G_0 dBi, (*b*) dual-subpanel with gain G_0 + 3dB, and (*c*) tri-subpanel with gain G_0 + 5 dB

with feed networks. This generates gain increases of 3 dB and 5 dB relative to the standard sector antenna panel (ignoring feed network losses), resulting in relative coverage increases of 40% and 60% in this scenario, as given in Table 4.3. The modular high-gain antenna concept is shown to provide link budget improvement over a large part of a serviced cell, which shows the potential for larger site separations and fewer installations.

TABLE 4.3   Calculated Relative Coverage Area With Different High-Gain Antenna Configurations (subpanel with 65° and 8° azimuth and elevation half-power beamwidths, respectively, over hilly terrain)

Antenna gain (dBi)	Coverage area(10^3 km^2)	Relative coverage
G_0	0.7	1
G_0 + 3	1.0	1.4
G_0 + 5	1.15	1.6

4.8.2 Field Trial

The performance of a modular high-gain antenna, as shown in Figure 4.15, was verified in a field trial in a live GSM 1900 network.[20] A cell with small terrain variations served by a base station with a 28m high tower was selected for the trial. The performance of two different antennas was compared: a reference sector antenna with 18 dBi gain (with 2° electrical beamtilt) and a dual-subpanel modular high-gain sector antenna with 22.8 dBi gain.

Figure 4.16 shows a sample of measured received signal strength on downlink along a radial road emanating from the trial site for a 22km long measurement segment. A very small part of the cell, the first kilometer of the measurement segment near the base station, experiences better signal strength from the reference antenna. Even so, in this part of the cell, the high-gain antenna provides a very good signal with approximately −60 dBm or better signal strength, in part as a result of radiation pattern null-filling. Test results show an

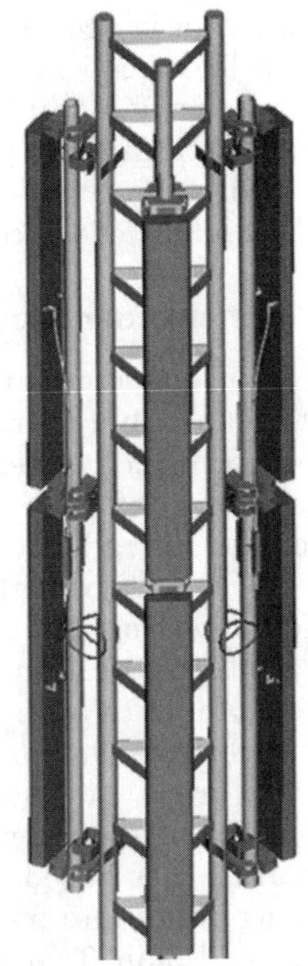

Figure 4.15 Three-sector modular high-gain antenna with two subpanels installed in a tower

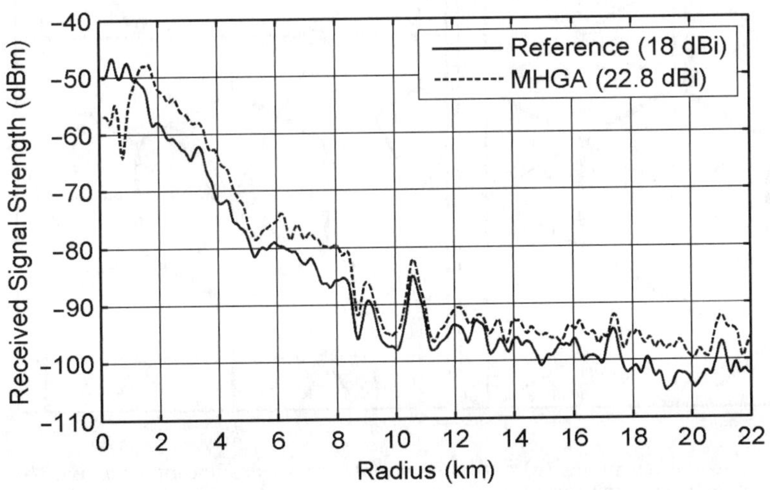

Figure 4.16 Measured downlink received signal strength for reference sector antenna and modular high-gain sector antenna (MHGA). The modular high-gain antenna coverage at small radii is maintained by null-filling.

average relative coverage improvement of about 5 dB for almost the entire cell, corresponding to the full gain difference between the antennas, thus verifying that the modular high-gain antenna retains its gain characteristic in a real propagation environment. Assuming a radial cell size of 15km, the trial shows a coverage improvement of about 5 dB in more than 95% of the cell area.

4.9 Higher Order Sectorization

Higher order sectorization[21] is conceptually straightforward to implement because there is no principal change in the base station structure compared to the reference system. Moreover, the radio network algorithms are unaffected, which means the interface between the base station and the radio network controller (RNC) is unchanged. The transmission capacity over the back-haul interface, however, needs to meet the demands from increased traffic over the air interface offered by the increased sectorization.

The changes in a system with higher order sectorization, compared to the reference system, are basically that there are $N > 3$ independent coverage areas, instead of three, with unique cell identities that are serviced by the same radio base station (or site). The most common configuration is based on dividing the azimuth angular interval surrounding a site into fractions of 360°, typically 360°/N per sector in a regular cell plan. Two principal six-sector configurations in a regular hexagonal cell plan are shown in Figure 4.17. Since the sectors become

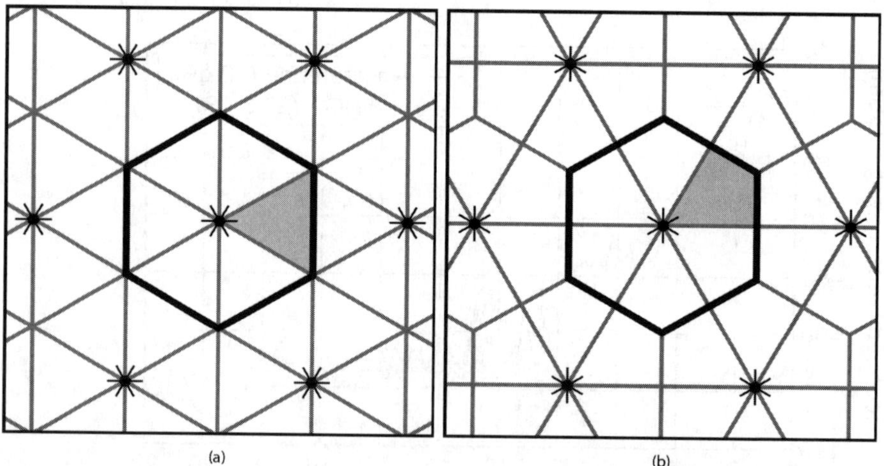

(a)   (b)

Figure 4.17 Six-sector cell plans: (*a*) antennas oriented toward adjacent sites and (*b*) antennas oriented toward the incenter of a equilateral triangle with adjacent sites at its vertices. The cell corresponding to one antenna is indicated by the shaded area. Dots indicate the locations of the base stations and line stubs show the antenna main beam pointing directions.

narrower than in the reference three-sector case when the number of cells is increased, the azimuth beamwidth of the antenna pattern needs to be reoptimized. The beamwidth shall be narrow to give good spatial filtering effect in azimuth but wide enough to offer sufficient gain over the entire sector. The optimization leads to the use of antennas with a decreased half-power beamwidth. Whereas in a three-sector WCDMA system, the beamwidth that maximizes the capacity in downlink is about 65° (Ericsson cell plan) and about 75° (Bell cell plan), in a six-sector system the optimum beamwidth is about 35°.[9,22] The results for the uplink are similar to the downlink case, and the same optimal beamwidths apply.

The main application of higher order sectorization is for increasing system capacity in interference-limited radio network scenarios, i.e., scenarios where the signal-to-interference ratio (SIR) is limiting performance. The increase in capacity comes from the improved spatial filtering offered by antennas with narrower azimuth beamwidth that, in turn, allows more cells per unit area. From a capacity performance point of view, and based on a first-order model, a higher order sectorized system is able to serve $N/3$ times as many users as the reference system in both uplink and downlink due to the improved spatial filtering in azimuth. In practice, the performance gain is somewhat lower than this due to, for example, angular spread in the propagation environment. Angular spread also leads to an increased fraction of users being in handover, thus requiring resources in terms of hardware and output power.

Spatial filtering has a negligible effect in a noise-limited scenario, unless it is accompanied by a change in the effective SNR. If the vertical dimension of the antennas used in the reference, three-sector system is the same as that of the antennas in the higher order sectorized system, the latter will have higher gain. Also, the total power resource may increase with higher order sectorization. This occurs if identical radio chains, including power amplifiers, are used in each sector for both types of sectorization. Increases in antenna gain or available power translate into increased signal-to-noise ratio and, hence, potential coverage improvements.

Increased antenna gain is useful not only for improved coverage but also, to some extent, for improved capacity. In WCDMA, increased antenna gain reduces the output power required for fulfilling network quality requirements on downlink, and thus, as power is a limited resource in the base station, the reduction can be used for increasing the traffic.

Resources other than the number of radio chains are affected by increased traffic load. These resources are, for example, baseband processing, bus capacity, and back-haul communication capacity between the base station and the higher level control equipment, such as radio network controller units.

4.9.1 Case Study

A case study of a real WCDMA network deployment was undertaken to determine the capacity and coverage performance of an urban cellular network when a limited set of sites are upgraded from three-sector to six-sector sites.[21] This is compared to a homogenous and regular hexagonal deployment, where a full network migration is performed. The conclusion is that the performance gain shown in homogenous deployments is maintained in a real network.

The antennas used are characterized by a half-power beamwidth scaled proportionally to the number of sectors per site, i.e., three- and six-sector sites are configured with 65° and 33° half-power azimuth beamwidth antennas, respectively. The vertical size of the antennas is kept constant for all antennas, i.e., the directivity increases by 3 dB for each halving of the azimuth half-power beamwidth. The resulting capacities are equivalent to a capacity increase of 86% when doubling the cells from three to six. Thus, the expected capacity increase per site is about 1.8 times for doubling the number of cells per site.

Irregular deployment and nonuniform traffic distributions that characterize a real network are suspected to have an impact on network performance. A higher order sectorization case study addressing a real network containing over a hundred sites in a major city serves as an example of a possible and viable way for system improvement for capacity increase. Three different radio network configurations are evaluated. The first studied scenario (*baseline*) consists of the original network deployment where all sites, even the higher order sectorization evaluation sites, are equipped with three-sector sites. Each cell has a maximum downlink power of 20 W. The other two scenarios, H20 and H10, are network configurations based upon five higher order sectorization evaluation sites that are migrated to six-sector sites. In the H20 scenario, the higher order sectorization sites are equipped with six cells with 20 W downlink power each, whereas in the H10 sites the downlink power is 10 W per cell. The total power per site is then 60 W in the latter case, i.e., equal to the total power of the baseline three-sector site.

Figure 4.18 shows the speech capacity in the different network scenarios and for the three different cell groups. Since areas covered by different cell groups are not the same in the different scenarios, it is interesting to evaluate the capacity per site. The capacity increase relative to the reference scenario for the different cell groups are summarized as follows. All active cells: +28% (H20) and +23% (H10); higher order sectorization (H.O.S.) cells: +77% (H20) and +65% (H10); and other cells: +4% (H20) and +5% (H10). The expected relative increase of 1.8 times for a doubling of the number of sectors is thus confirmed in the realistic network environment study. Higher order sectorization

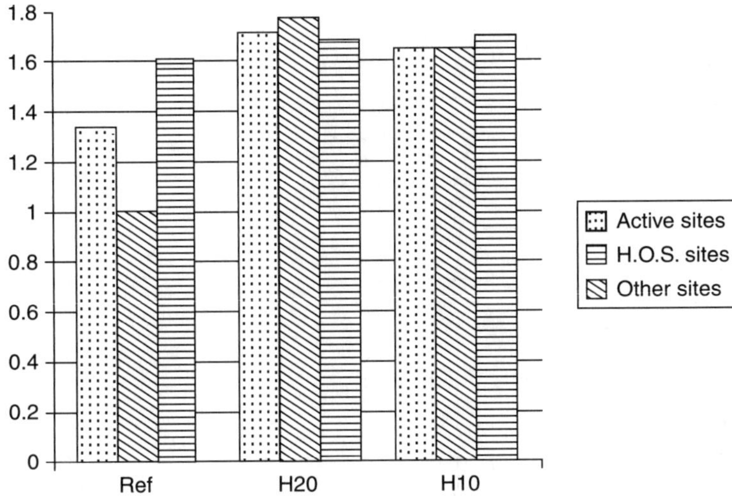

Figure 4.18 Scenario comparison of speech capacity per site in a real network when a limited number of sites with higher order sectorization are introduced.

provides a feasible option for increasing the capacity of third-generation networks without finding new sites, a possibly expensive and time-consuming process, provided that the practical installation issues are solved cost effectively.

4.10 Fixed Multibeam Array Antenna

As an alternative to increasing the number of sites or cells per site as the means to enhance capacity in cellular networks, multibeam array antennas may be introduced. Fixed multibeam systems are characterized by the use of a set of fixed azimuth beams in each cell for transmission and reception.[7,23–27] Each beam in the set has predefined properties such as pointing direction and beam shape and covers only part of the cell. The beams are generated either by radio frequency (RF) beamformer or at baseband or by a combination thereof. In the case when the fixed beams are formed at baseband, signal coherency is required from baseband to the antenna aperture. If the fixed beams are formed at RF, then coherency is required from the beamformer to the antenna aperture. In such an implementation, the need for calibration of the radio chains is eliminated because the antenna unit, including the beamforming network, is manufactured with required coherency. A block diagram showing a fixed multibeam antenna system for one cell is presented in Figure 4.19.

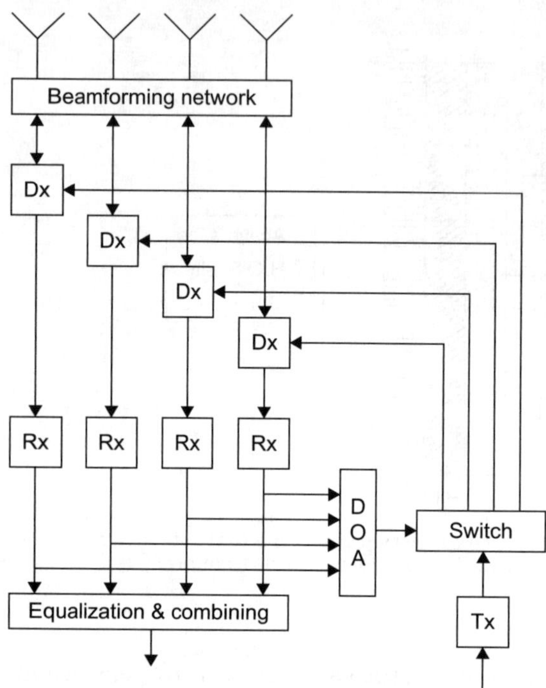

Figure 4.19 Block diagram of a fixed multibeam array antenna. The example shows a system where the beams are formed at RF in both uplink and downlink in a beamforming network.

The cell serviced by a multibeam antenna is defined by a broadcast channel that is transmitted using a beam with wider beamwidth and coverage area than the individual array antenna beams. This wider beam may be generated by a separate sector antenna (column) or by transmitting the broadcast channel over a set of the array antenna beams such that the combined pattern represents the desired cell coverage. A simple and effective solution uses an additional column of radiating elements next to the array antenna columns. The deviation between the sector antenna radiation pattern and the array antenna multibeam envelope has to be limited since the cell is defined by the sector pattern coverage; i.e., beam tracking is essential. For proper system behavior, ensuring that the mutual coupling effects between the two antennas do not distort the tracking between the array antenna and sector antenna coverage is important. A deviation could lead to a mismatch between traffic and control channel coverage, in the worst case resulting in dropped calls.

A common implementation of a fixed multibeam array antenna system is to use a Butler matrix beamforming network to generate orthogonal beams at RF within the antenna unit. In such an implementation, the number of beams generally equals the number of antenna elements or columns. A Butler matrix can be thought of as a hardware realization of the Fast Fourier Transform (FFT); hence, the beam orthogonality is a consequence of the property that the inner product of the output port excitations of the Butler matrix is a Kronecker delta-function δ_{mn}, where m and n are indexes of the Butler matrix input port. When the Butler matrix output ports are connected to equidistant antenna elements or, more commonly, columns of antenna elements in an array antenna, the radiation pattern (array factor) beam peak corresponding to a given input port coincides with pattern nulls for all other input ports. The orthogonal beams of a four-column array antenna with a horizontal spacing of half a wavelength is shown in Figure 4.20. The theoretical sidelobe level is –13 dB and the cross-over level between two adjacent beams is –4 dB. However, the sidelobe level can be reduced with an amplitude taper implemented with a modified Butler matrix at the expense of a deeper crossover level.[28] Antenna parameters such as radiating element half-power azimuth beamwidth, number of fixed azimuth beams, column spacing, and column excitation weights may be

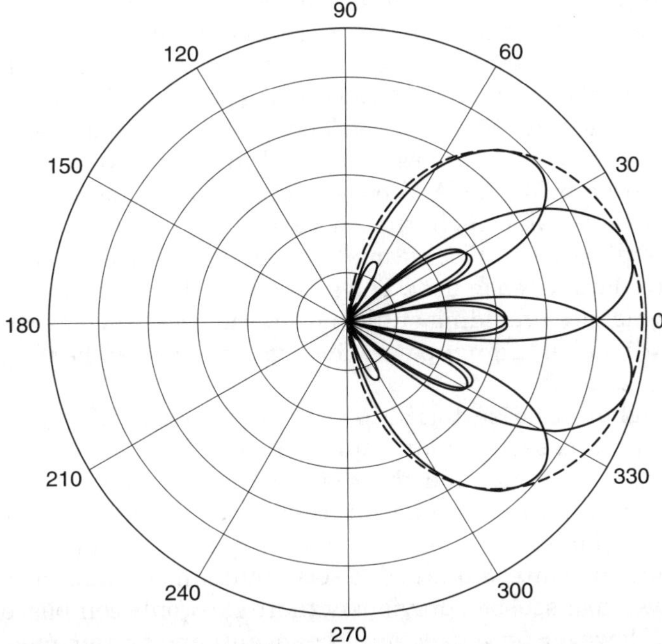

Figure 4.20 Four orthogonal beams (solid) in azimuth used in one sector of a fixed multibeam three-sector antenna system. A sector beam pattern (dashed line), normalized to coincide with the array antenna beam peaks, is also shown (5 dB grid steps radially).

optimized in order to minimize the total base station power required to guarantee an acceptable quality of service.[29]

A more general implementation of a fixed multibeam antenna system may have more beams than antenna elements or columns, thus providing non-orthogonal beams, particularly if the beams are formed at baseband. The number of beams may be arbitrarily large, thus making a fixed multibeam system and a steered-beam system less distinguishable.[30]

The principal function in uplink for the fixed multibeam array antenna system, just as for the reference sector system, is to combine all energy from the desired signal arriving at the antenna. This process may comprise the combination of signals from different cells, as in a softer handover situation in WCDMA, as well as the combination of signals from the diversity antennas. The same types of receive diversity used in sector-antenna systems, for example, spatial or polarization diversity, may also be used in fixed multibeam array antenna systems.

In downlink, the desired action is to transmit the signal to the user device as efficiently as possible by selecting one of the available beams. This means that information concerning the angular location of the user device, such as direction-of-arrival (DOA) information, is needed, at least on a beam resolution level. Other information such as traffic load per beam is used in the beam selection process to improve system performance regarding amount of traffic served. The typical situation is that the downlink transmission is performed via only one of the beams within each cell included in the active set. The limitation to one beam only is imposed to minimize interference spread in the network. In a soft handover situation in WCDMA, the transmission is performed via one beam in each cell, for all cells in the active set.

As is shown in Figure 4.19, DOA estimation can be performed using the uplink received data for each user. The estimated DOA information and other similar metrics providing sufficient information, such as the beam with highest received signal power for a given user, can be used for downlink beam selection.

In contrast to higher order sectorization, the introduction of a fixed multibeam array antenna system may require changes on the interface between the radio base station and the radio network controller. For WCDMA, one example of information that needs to be available in the radio network controller, to allow efficient use of system resources, is the spatial distribution of interference. In a sector antenna system, such as the reference system discussed previously, there exist only cell-based quality measures. However, for a fixed multibeam antenna system measures are needed on a beam-by-beam basis since, due to the spatial resolution, some beams might carry a high traffic load whereas others carry less. In a beam carrying heavy traffic, it may be necessary to block new users whereas in other beams adding users will not be a problem. Another issue for the radio network algorithms in WCDMA networks is the handling of multiple scrambling codes. As downlink transmission with different codes is not orthogonal, it is necessary to allocate scrambling codes based on the location of the user within the cell. Thus, the radio network controller needs spatial information for each user device such as an identification of the beam showing the lowest path loss.

Fixed multibeam array antennas have also been shown to enhance the quality of services and the network capacity of Enhanced GPRS (EGPRS) to provide packet-switched mobile services. For the 1/3 reuse scheme, a capacity gain of over 200% is realized in a homogeneous downlink system employing a relatively low complexity array antenna system compared to a three-sector configuration.[31] The investigated array antenna system comprises eight interleaved beams in azimuth generated by two four-by-four Butler matrices. The matrices are connected to four antenna columns with orthogonally polarized radiating elements.

An alternative to planar multibeam array antennas is to use a cylindrical array antenna. Combined with a Butler matrix feeding network, the antenna can produce a fixed set of radiation patterns with narrow beams in directions around the cylinder axis. With two Butler matrices back-to-back, simultaneous fixed multibeam patterns and cell-defining omnidirectional patterns can be provided.[32]

4.10.1 Field Trials

The performance of fixed multibeam antenna systems have been evaluated in a number of field trial activities in GSM and TDMA (IS-136) systems.[33–39] The results show considerably increased capacity when using fixed multibeam array antenna systems.

Fixed multibeam array antennas were tested as an option to enhance capacity in a GSM network.[33] A two-dimensional antenna array, a planar array antenna, Figure 4.21, was developed for the field-trials, which were performed in a live network operating in the 900 MHz frequency band. Evaluations of the field trial results indicate that adaptive antenna systems are able to give considerably increased performance. Using dual-polarized antennas makes polarization diversity schemes possible and only a single array antenna is needed in each direction from a base station, i.e., for each cell, which minimizes installation and aesthetical issues on site.

Within the scope of the trial, the array antenna was used both in transmit and receive mode. Therefore, the antenna was required to work over the full 900-MHz frequency band, with a total bandwidth of about 10%, which was achieved with a design using aperture-coupled microstrip patches. A principle view of the array antenna is shown in Figure 4.22.

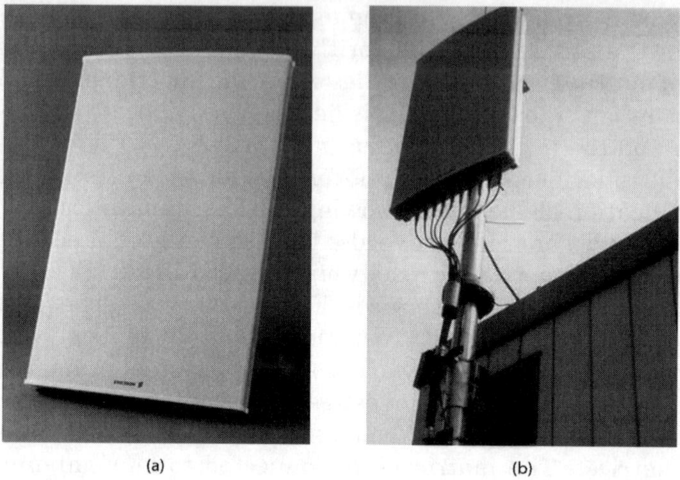

Figure 4.21 Fixed multibeam array antenna at 900 MHz: (*a*) antenna unit, (*b*) roof-top mounting (Courtesy of Ericsson)

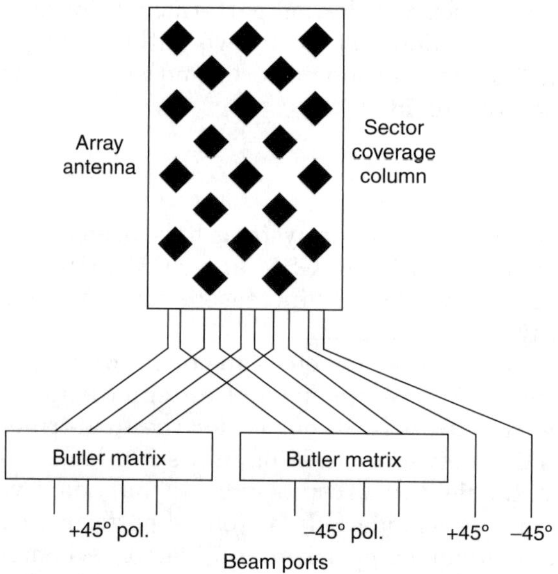

Figure 4.22 Block-diagram of dual-polarized fixed multibeam array antenna configuration

It is a dual-polarized fixed multibeam array antenna, having four azimuth beams in each of two orthogonal polarizations.[40] The polarization orientations are slant linear ±45° generated by microstrip patches located in four columns. For each polarization, the radiating elements in each column are combined using a fixed vertical distribution network.

A sparse element grid is implemented in order to minimize feed network losses and coupling effects among radiating elements. On the other hand, grating lobes are avoided to maintain beam pattern control at all beam positions. The columns are spaced half a wavelength apart with radiating elements positioned in a triangular grid. By using the triangular grid, grating lobes come close to the visible space only for the outermost beam positions, where achieved gain is not as critical as for the center beams.

The horizontal beamforming networks used in the antenna are Butler matrices with equal number of antenna ports and beam ports, with one Butler matrix per polarization connected to the four array columns of radiating elements, generating four beams with +45° polarization and four beams with −45° polarization. Beam generation from the Butler matrix results in low loss patterns but with a cross-over depth of approximately −4 dB between adjacent beams, as stated previously. By interleaving the beams of the two polarizations, every other beam has opposite polarization that results in significantly reduced cross-over depths between adjacent beams, as shown in Figure 4.23.

Base stations transmit control channels simultaneously over the entire ±60° azimuth sector region in a three-sector site scenario. In order to satisfy this requirement, a separate sector antenna function is introduced as part of the fixed multibeam array antenna. By putting the sector antenna next to the array antenna, the two antennas are still in principle functionally separated, even if they are mechanically one unit with a common radome. Combined with dual polarization, only a single antenna unit, comprising both array antenna and sector, is needed in

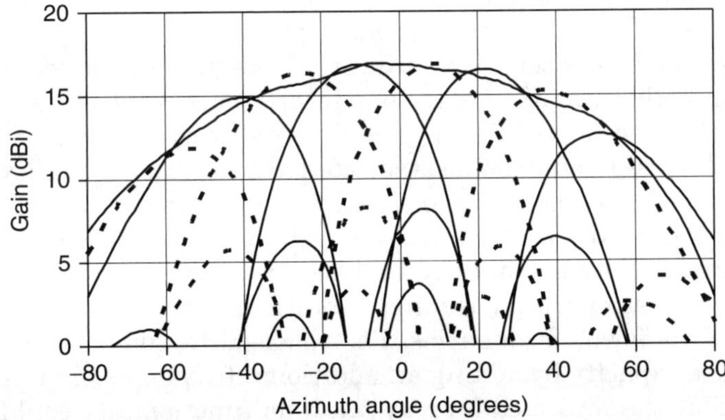

Figure 4.23 Measured interleaved azimuth array beam patterns together with the sector antenna beam. Fixed directive beams with +45° and −45° linear polarization are shown with solid and dashed lines, respectively.

every sector, which has the advantage of smaller visual impact and a simpler installation procedure. The resulting measured radiation pattern for the sector beam next to the GSM 900 array multibeams is included in Figure 4.23. Due to the horizontal beamforming, the array antenna provides approximately 5 dB higher antenna gain than the sector antenna, which has a gain of 12.7 dBi. In the plot, the relative level of the sector beam pattern is adjusted for easier visual comparison between the sector pattern and the array antenna pattern envelope.

Three GSM 900-MHz base stations comprising fixed multibeam array antennas were installed and evaluated in the live GSM network.[34] One transceiver was used for transmitting the control channel in the whole sector; i.e., it was connected to a sector antenna in the downlink and to the array antenna in the uplink. Another transceiver was connected to the sector antenna in both uplink and downlink. This transceiver worked like an ordinary sector transceiver and was used as a reference during the performance measurements. The remaining three transceivers were connected to the multibeam array antenna in both uplink and downlink. The array antenna system is characterized by having very effective uplink diversity, obtained by the many degrees of freedom including polarization, angle, and pattern.

The performance test was designed to introduce interference into the system gradually. This was done in order to have full control over the quality of the network, as ensuring that customers experience high quality throughout a test is important. Gradually increasing interference was generated by applying different frequency plans. The number of frequencies used was decreased, meaning that a higher fraction of the frequencies experienced a re-use corresponding to re-use one.[41]

The quality improvement of multibeam array antennas increases with increasing interference compared to the sector-only antennas, as displayed in Figure 4.24. The least dense frequency plan with nine deployed frequencies, which leads to essentially no internal interference, provides similar quality for both configurations. When interference is introduced, the two configurations start to differ. The quality level corresponding to 12 intra-cell handovers/(Erlang*hour) is selected to be a suitable quality level for comparison, corresponding to a satisfactory network quality. As shown in the figure, the sector antenna system needs 120% more frequencies to provide the same quality as the array antenna system at this quality level. In other words, the fixed multibeam array antenna system requires 55% fewer frequencies. The released frequencies can be used to increase capacity by adding an additional frequency plan. It was also found that the multibeam array antenna functionality could be used to obtain valuable information about the distribution of traffic in the network.

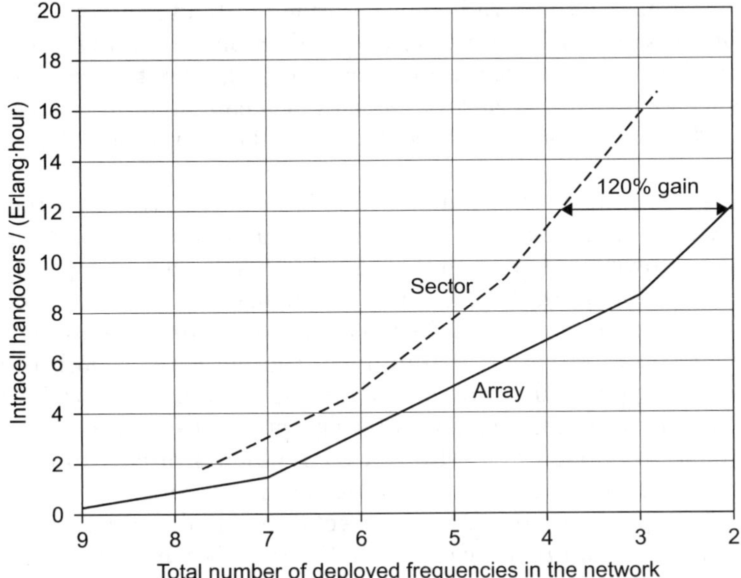

Figure 4.24 Quality level in terms of intra-cell handovers for different frequency plans when multibeam array antennas and sector antennas are evaluated in a commercial network.

The performance of fixed multibeam antenna systems has also been evaluated in a number of field trial activities in TDMA (IS-136) systems.[35,36] The results in Hagerman et al.[35] show considerably increased capacity using fixed multibeam antenna systems in live networks. An SIR improvement of 4 dB was reported with planar array antennas with four fixed beams per sector.

4.10.2 Migration Strategy

Cellular networks usually consist of a mixture of large macro cells and smaller micro cells. This implies that multibeam antennas can be introduced for increasing the network capacity in a number of different ways depending on the network configuration. The preferred migration strategy is to gradually install multibeam antennas in networks as the need for capacity increases.[42,43] One advantage is lower initial deployment costs. The basic idea is to use the multibeam antennas in hot-spots and interference-limited cells. These antennas are employed at base stations that serve a lot of traffic, which severely disturb, through interference, a large number of neighboring cells. An example for GSM is large macro cells with many transceivers using sector antennas mounted in high masts.

Multibeam antennas at these radio base stations mitigate the interference situation considerably. The interference reduction is used to tighten the frequency plan in the macro layer either by adding more transceivers at selected sites or by using fewer frequencies or a combination of these alternatives. The frequencies so released can be allocated in added cells or for new services.

An interference-limited GSM macro cell network consisting of three-sector sites with mixed sizes and a heterogeneous traffic distribution covering a downtown area was analyzed.[43] In this example, capacity is gained by introducing multibeam array antennas in a step-by-step migration of radio base stations in the network. Increasing the number of transceivers and improving the quality in the network in both the target cells equipped with multibeam antennas and surrounding cells using sector antennas are possible. The simulations show that a large increase in capacity is achieved by replacing only a limited number of existing installations with multibeam antennas, as shown in Figure 4.25. Replacing only 12% of the cells in a real network with multibeam antennas improved capacity 40%. In this way, additional new site locations can be avoided even though a significant capacity increase is achieved. Two reference cases, cell split from three- to six-sector sites and the introduction of multibeam antennas, show linear increases in capacity in idealized homogeneous networks. This shows the particular advantage of selective use of multibeam antennas in heterogeneous scenarios.

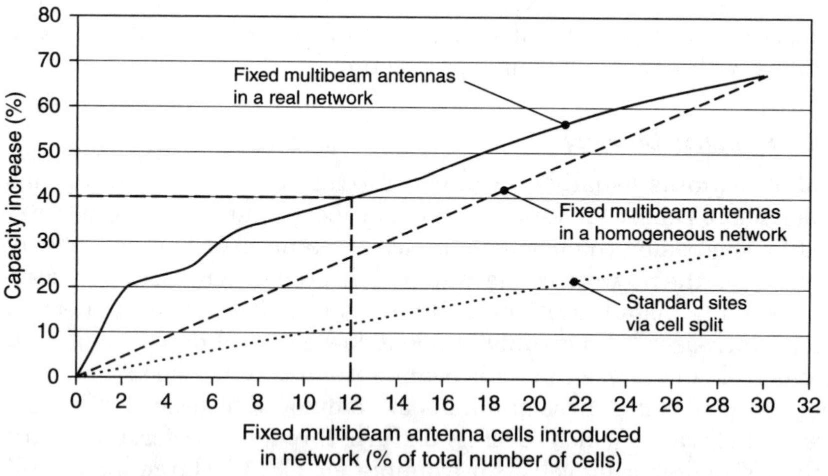

Figure 4.25 Capacity gain in a GSM network as a function of relative number of cells migrated to use four-column array antenna compared to six-sector migration

4.11   Steered Beam Array Antenna

A steered beam antenna is characterized by the use in downlink of a beam shape, on a per-user basis, that is adapted to the existing traffic situation in the network. This adaptation is in terms of adjusting only pointing direction for simpler systems but also beam shape and sidelobe levels for more advanced antenna systems.[7,44–46] As each user is served by a unique beam, the beamforming is performed at baseband because this is the only place in the transmission chain where user signals are separately controllable. Baseband beamforming means that coherency, i.e., sufficiently small errors in amplitude, phase, and time between signal paths, is needed from baseband to the antenna aperture. Calibration is likely necessary to fulfill the coherency requirements for most implementations.

A steered beam system in uplink is similar to a fixed multibeam reception system. However, the configuration depends on whether or not a Butler matrix is used in the antenna unit. One major purpose of using a Butler matrix in a steered beam implementation is to relax coherency requirements.

For systems with uplink and downlink separated in frequency more than a fraction of the correlation bandwidth of the propagation channel, for example, frequency-division duplex (FDD) systems, reciprocity will typically not hold; i.e., the uplink signal state over the antenna (amplitude and phase distribution) cannot be directly applied to the downlink signals. Either a frequency compensation corresponding to the duplex distance has to be applied, if at all possible, or, which is more likely, higher level information about the user device may be extracted from the uplink signals, such as knowledge about the spatial location of the user. The beam used for downlink transmission may then be based on the direction information for the user device of interest. Other information that is useful is the spatial distribution of traffic load, which may contain information about other user devices within the cell, for example, their location, data rates, and power allocation, and possibly also information about user devices in adjacent cells served by the same base station.

Figure 4.26 shows a functional chart of one implementation of a steered beam antenna system. This system employs beamsteering on downlink and has fixed beamforming on uplink that performs angular prefiltering of the signals to facilitate direction of arrival (DOA) estimation.

This antenna system requires one individual transmitter for each antenna element or column as well as phase coherency of the branches on both the receiving and transmitting sides. The advantage is that downlink beamforming is not limited to a fixed set of beams or beam shapes.

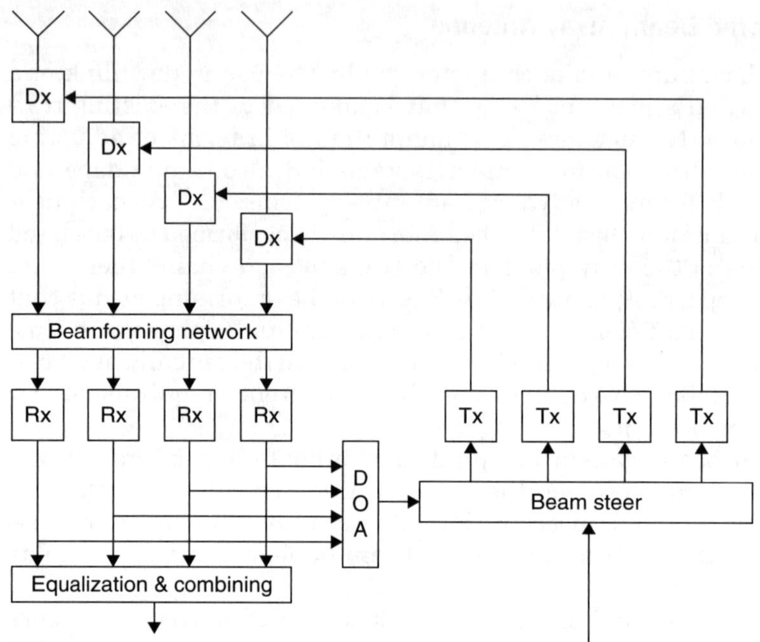

Figure 4.26 Implementation of a steered beam array antenna system

In addition, if the directions to the desired user as well as to interferers are known, more advanced features such as interference nulling in downlink can be introduced. The angular resolution of the directional information depends on system complexity but is on the order of 20% or less of the azimuth half-power beamwidth of the steered beam.

4.12 Amplifier Integrated Sector Antenna

In normal macro base station installations, the radio base station equipment is attached to passive antennas in a mast or tower. The transmit power is generated by high-power amplifiers in the radio base station cabinet located far from the antenna installation. Thick feeder cables are used for interconnecting the units in order to minimize cable losses. With the introduction of active parts in the antenna unit, the effects of downlink feeder cable loss can be reduced by applying signal power amplification after, rather than before, the feeder cables. This also leads to reduced-size base station solutions, combining high levels of effective radiated power and low power consumption. Schematic block diagrams in transmit mode of a conventional passive antenna installation and an active antenna installation with integrated power amplifiers

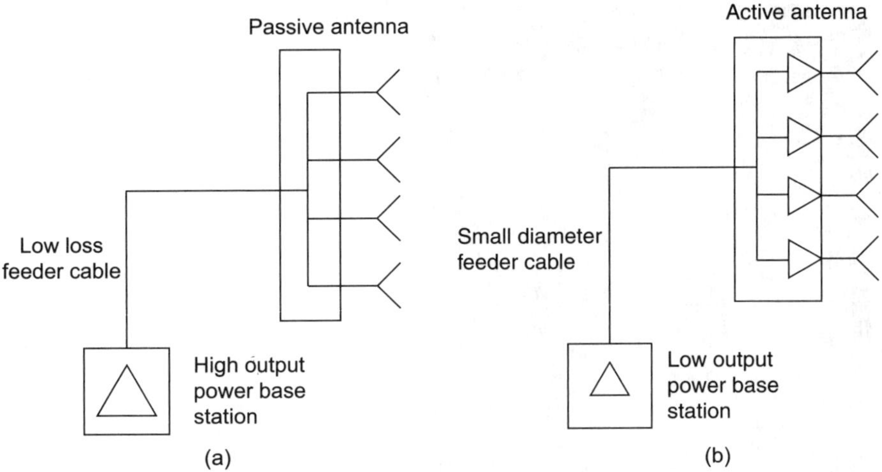

Figure 4.27 Base station installations in transmit mode: (*a*) conventional with passive antenna and (*b*) antenna with integrated power amplifiers ("active")

distributed along the antenna aperture are shown in Figure 4.27. On receive, integrated low-noise amplifiers are used, which are not shown in the figure.

Base station installations using integrated amplifiers in the antennas have a number of advantages that will lead to a cost-effective deployment solution. The total efficiency becomes high due to low losses between the distributed power amplifiers and the radiating elements, and all amplifiers in parallel are active, which results in an architecture with little or graceful performance degradation in case of an amplifier failure. In addition, a smaller ground unit size is needed as lower output power from ground equipment is required, and thinner feeder cables with higher acceptable loss can be installed, which results in lower cost and easier installation.

4.12.1 Case Study

The coverage of a micro base station connected to an amplifier integrated sector antenna can be extended to get wide area macro cell coverage without any requiring increased equipment space. An amplifier integrated sector antenna with both power amplifiers and low-noise amplifiers integrated in the antenna unit is shown in Figure 4.28. The antenna unit is less than 1-m high and has an azimuth half-power beamwidth of 65°. The ±45° dual-polarized radiating elements are low-profile microstrip patches that cover both transmit and receive frequency bands. In the transmit part of the antenna unit,

Figure 4.28   Roof-top mounted sector site consisting of antenna unit with integrated amplifiers, micro base station, and battery backup unit (Courtesy of Ericsson)

power amplifiers are distributed along the antenna structure to generate power where it is most efficient, namely close to the radiating elements. The power amplifiers are individually connected to subarrays of dual-polarized radiating elements. At the back of the transmit array, cooling flanges are arranged from the bottom to the top of the antenna unit to keep the temperature of the amplifiers low, even at high ambient temperatures. The receive part of the antenna uses polarization diversity and low-noise amplifiers connected directly to each of the sum ports of the receive antenna elements. The antenna unit is connected to a micro base station and a battery backup unit. All units are compact and easy to install. Because an equipment room is not needed, site acquisition becomes easy. The flexibility of the concept makes it possible to build a rooftop three-sector site where the three base stations are placed together in one spot or distributed over the building. Three-sector sites can also be built in rural areas without the need for heavy construction work, using existing structures and towers.

4.13   Amplifier Integrated Multibeam Array Antenna

Multibeam array antennas, which employ distributed power amplifiers close to the radiating elements of the antenna array, improve the radiated power efficiency. An example shown in Figure 4.29 consists of a dual-polarized antenna array with four columns for the multibeam radiation patterns and a fifth column for sector coverage. Five amplifiers are mounted for each polarization. Four of these are connected to the multibeam array whereas the remaining amplifier is used for sector beam amplification. The antenna unit can be configured both as a column power amplifier array, as in Figure 4.29a, and as a hybrid power amplifier array, as in Figure 4.29b.

In the column power amplifier array, a separate power amplifier is connected directly to each column of radiating elements, whereas the hybrid power amplifier array has a separate amplifier configuration located

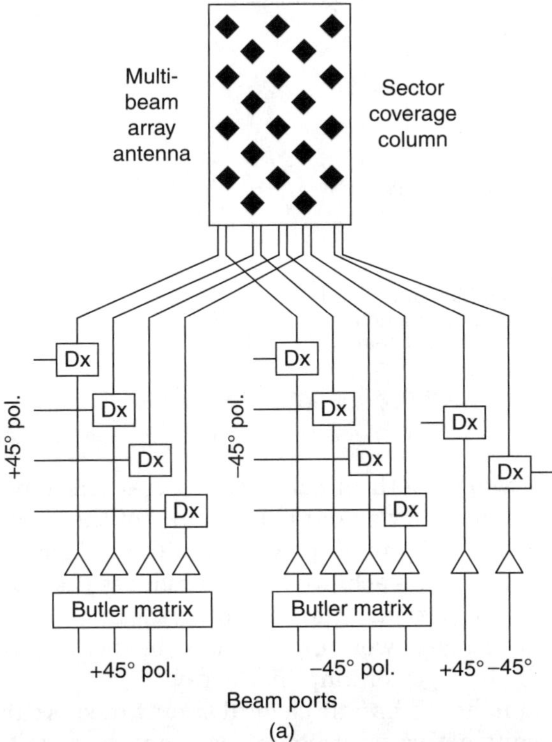

Figure 4.29 Block diagram of an amplifier integrated multibeam array antenna: (a) column power amplifier array configuration and (b) hybrid power amplifier array configuration

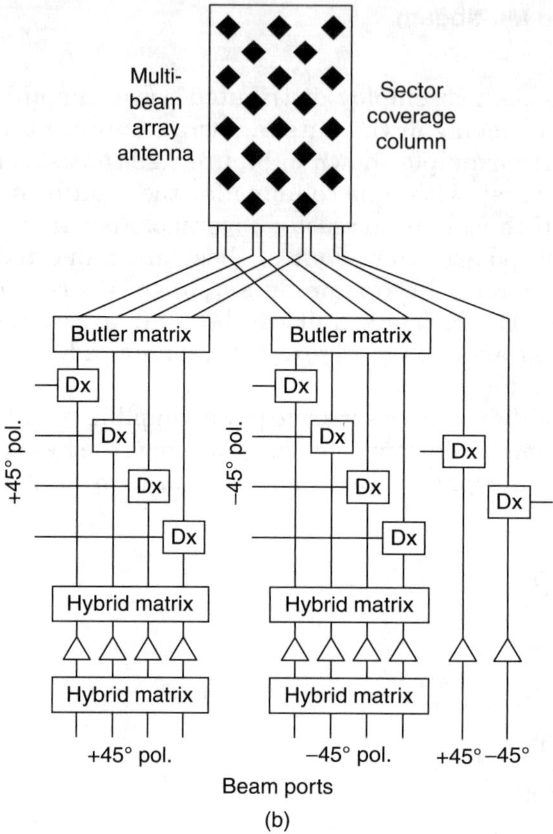

Figure 4.29 Block diagram of an amplifier integrated multibeam array antenna: (*a*) column power amplifier array configuration and (*b*) hybrid power amplifier array configuration (*Continued*)

before the beamforming network. Both of these configurations have the important characteristic that every signal fed into any of the array antenna beam ports is divided equally to all power amplifiers. Thereby, an even load over all power amplifiers is achieved, which reduces peak-to-average power ratio and intermodulation requirements on the amplifiers. In principle, also the sector coverage power amplifier may be included in the pool of amplifiers in the hybrid power amplifier array.

In a GSM system, the system has to handle several simultaneous carriers since the purpose of using active multibeam array antennas is to increase capacity. Multicarrier amplifiers are, therefore, necessary.

One of the potential advantages with active multibeam array antennas is that intermodulation products from the different amplifiers may be designed to be uncorrelated. In that case, the transmitted intermodulation signal is not directed into a narrow high-gain beam. The suppression of radiated power for the intermodulation signal with respect to main beam power for the carrier signal will then be enhanced, and the requirements on the individual power amplifiers are relaxed.

4.14 Conclusion

Advanced antennas may be deployed network-wide or at select base station sites to improve capacity and extend coverage of current and future cellular networks. The discussed advanced antenna technologies are categorized in Table 4.4 in terms of uplink or downlink, capacity or coverage, or a combination thereof according to their primary advantage and usage in macro base station installations.

The use of advanced antennas is a way to achieve higher network capacity by reducing interference spread in downlink and by utilizing spatial separation in uplink. Various capacity-oriented antenna concepts have been evaluated and tested in live cellular networks, proving that advanced antennas increase the capacity of existing mobile communication networks. No new additional sites are introduced by a grow-on-site strategy with site-by-site migration of advanced antennas.

Coverage solutions serve to increase the range of a network. Today, the use of diversity and modular high-gain antennas already represent cost-effective coverage solutions. Amplifier integrated antennas offer an attractive solution, having simple installation with small volume, high overall power efficiency, and high levels of equivalent radiated power. These parameters are of great interest, especially in urban areas where a denser network can be foreseen with increasing capacity and higher data rate needs.

TABLE 4.4 Summary of Advanced Antenna Technologies

Advanced Antenna Technology	Downlink	Uplink	Capacity	Coverage
Three-sector omnidirectional antenna	•			•
Higher order receive diversity		•		•
Transmit diversity	•		•	•
Antenna beamtilt	•	•	•	•
Modular high-gain antenna	•	•		•
Higher order sectorization	•	•	•	
Fixed multibeam array antenna	•	•	•	
Steered beam array antenna	•		•	
Amplifier integrated sector antenna	•			•
Amplifier integrated multibeam array antenna	•		•	

Most of the presented advanced antenna solutions are compatible with the second- and third-generation user devices in use throughout the world. This means that advanced antennas provide direct benefits both during roll out of new networks and during migration of existing networks and sites because the improvements are not dependent on a change of equipment in the installed user base of more than four billion mobile communication devices.

References

1. J. C. Liberti, Jr. and T. S. Rappaport, *Smart Antennas for Wireless Communications: IS-95 and Third Generation CDMA Applications*, Upper Saddle River, NJ: Prentice Hall, 1999.
2. S. R. Saunders, *Antennas and Propagation for Wireless Communication Systems*, Chichester, England: John Wiley & Sons, 1999.
3. L. C. Godara (ed.), *Handbook of Antennas in Wireless Communications*, Boca Raton, FL: CRC Press, 2002.
4. F. Gross, *Smart Antennas for Wireless Communications*, New York: McGraw-Hill, 2005.
5. K. Fujimoto (ed.), *Mobile Antenna Systems Handbook*, 3rd edition, Norwood, MA: Artech House, 2008.
6. J. H. Winters, "Smart antennas for wireless systems," *IEEE Personal Communications,* vol. 5, no. 1 (February 1998): 23–27.
7. S. Andersson et al., "Adaptive antennas for GSM and TDMA systems," *IEEE Personal Communications,* vol. 6, no. 3, (June 1999): 74–86.
8. J. Barta, M. Ericsson, B. Göransson, and B. Hagerman, "Interference distributions in mixed service WCDMA systems—Opportunities for advanced antenna systems," in *Proceedings IEEE 53rd Vehicular Technology Conference (VTC-Spring 2001),* Rhodes, Greece, 263–267, May 2001.
9. F. Athley, "On base station antenna beamwidth for sectorized WCDMA systems," in *Proceedings IEEE 64th Vehicular Technology Conference, (VTC-Fall 2006),* Montréal, Canada, 2267–2271, September 2006.
10. H. Holma and A. Tölli, "Simulated and measured performance of 4-branch uplink reception in WCDMA," in *Proceedings IEEE 53rd Vehicular Technology Conference (VTC-Spring 2001),* Rhodes, Greece, 2640–2644, May 2001.
11. M. Olsson, B. Hagerman, M. Riback, M. Hesse, and B. Niksic, "Field trial results of 4-way receive diversity in a live GSM network," in *Proceedings IEEE 63rd Vehicular Technology Conference (VTC-Spring 2006),* Melbourne, Australia, 2747–2751, May 2006.
12. T. Tynderfeldt and M. Olsson, "Transmit diversity options for GSM/EDGE," in *Proceedings IEEE 59th Vehicular Technology Conference (VTC-Spring 2004),* Milan, Italy, 490–494, May 2004.
13. S. M. Alamouti, "A simple transmit diversity technique for wireless communications," *IEEE Journal on Selected Areas in Communications,* vol. 16, no. 8 (October 1998): 1451–1458.
14. L. Manholm, M. Johansson, and S. Petersson, "Influence of electrical beamtilt and antenna beamwidths on downlink capacity in WCDMA: Simulations and realization," in *Proceedings of International Symposium on Antennas and Propagation (ISAP),* Sendai, Japan, 641–644, August 2004.
15. Y. Yamada, Y. Ebine, and M. Kijima, "Low sidelobe characteristics of a dual-frequency base station antenna in the case of electrical beam tilt use," in *IEEE International Antennas and Propagation Symposium Digest,* Orlando, FL, 2718–2721, July 1999.
16. I. Forkel, A. Kemper, R. Pabst, and R. Hermans, "The effect of electrical and mechanical antenna down-tilting in UMTS networks," in *Proceedings IEE 3rd International Conference on 3G Mobile Communication Technologies (IEE Conf. Publ. 489),* London, UK, 86–90, May 2002.

17. Y. Yamada, S. Takubo, and Y. Ebine, "An unequally spaced array antenna for mobile base stations," in *IEEE Antennas and Propagation International Symposium Digest*, Boston, MA, 432–435, July 2001.
18. J. Niemelä, T. Isotalo, and J. Lempiäinen, "Optimum antenna downtilt angles for macrocellular WCDMA network," *EURASIP Journal on Wireless Communications and Networking*, vol. 5, no. 5 (2005): 816–827.
19. L. Zordan, N. Rutazihana, and N. Engelhart, "Capacity enhancement of cellular mobile network using a dynamic down-tilting antenna system," in *Proceedings IEEE 49th Vehicular Technology Conference (VTC 1999)*, Houston, TX, 1915–1918, May 1999.
20. M. Johansson, S. Petersson, and S. Johansson, "Modular high-gain antennas," in *IEEE Antennas and Propagation International Symposium Digest*, San Diego, CA, paper 530.8, July 2008.
21. B. Hagerman et al., "WCDMA 6-sector deployment - Case study of a real installed UMTS-FDD network," in *Proceedings IEEE 63rd Vehicular Technology Conference (VTC-Spring 2006)*, Melbourne, Australia, 703–707, May 2006.
22. B. C. V. Johansson and S. Stefansson, "Optimizing antenna parameters for sectorized W-CDMA networks," in *Proceedings IEEE 52nd Vehicular Technology Conference, (VTC-Fall 2000)*, Boston, MA, 1524–1531, September 2000.
23. B. Hagerman, K. J. Molnar, and B. D. Molnar, "Evaluation of novel multi-beam antenna configurations for TDMA (IS-136) systems," in *Proceedings IEEE 49th Vehicular Technology Conference (VTC 1999)*, Houston, TX, 653–657, May 1999.
24. Y. Li, M. J. Feuerstein, and D. O. Reudink, "Performance evaluation of a cellular base station multibeam antenna," *IEEE Trans. on Vehicular Technology*, vol. VT-46, no. 1, (February 1997): 1–9.
25. B. Johannisson, "Adaptive base station antennas for mobile communication systems," in *Proceedings of the IEEE Conference on Antennas and Propagation for Wireless Communications (APWC'98)*, Waltham, MA, 49–52, November 1998.
26. M. Ericsson, A. Osseiran, J. Barta, B. Göransson, and B. Hagerman, "Capacity study for fixed multi beam antenna systems in a mixed service WCDMA system," in *Proceedings 12th IEEE International Symposium on Personal, Indoor and Mobile Radio Communications (PIMRC)*, San Diego, CA, USA, A-31–A-35, October 2001.
27. A. Osseiran and A. Logothetis, "Smart antennas in a WCDMA radio network system: Modeling and evaluations," *IEEE Trans. on Antennas and Propagation*, vol. AP-54, no. 11 (November 2006): 3302–3316.
28. S. Gruszczyński, K. Wincza, and K. Sachse, "Reduced sidelobe four-beam N-element antenna arrays fed by 4 x N Butler matrices," *IEEE Antennas and Wireless Propagation Letters*, vol. 5 (2006): 430–434.
29. A. Osseiran and A. Logothetis, "A method for designing fixed multibeam antenna arrays in WCDMA systems," *IEEE Antennas and Wireless Propagation Letters*, vol. 5, (2006): 41–44.
30. B. Göransson, B. Hagerman, S. Petersson, and J. Sorelius, "Advanced antenna system for WCDMA: Link and system level results," in *Proceedings 11th IEEE International Symposium on Personal, Indoor and Mobile Radio Communications (PIMRC)*, London, UK, 62–66, September 2000.
31. B. K. Lau, M. Berg, S. Andersson, B. Hagerman, and M. Olsson, "Performance of an adaptive antenna system in EGPRS networks," in *Proceedings IEEE 53rd Vehicular Technology Conference (VTC-Spring 2001)*, Rhodes, Greece, 2354–2358, May 2001.
32. S. Raffaelli, M. Johansson, and B. Johannisson, "Cylindrical array antenna demonstrator for WCDMA applications," in *Proceedings International Conference on Electromagnetics in Advanced Applications (ICEEA)*, Torino, Italy, September 2003.
33. S. Andersson et al., "Ericsson/Mannesmann GSM field-trials with adaptive antennas," in *Proceedings IEEE 47th Vehicular Technology Conference (VTC 1997)*, Phoenix, AZ, 1587–1591, May 1997.
34. H. Dam et al., "Performance evaluation of adaptive antenna base station in a commercial GSM network," in *Proceedings IEEE 49th IEEE Vehicular Technology Conference (VTC 1999)*, Houston, TX, 47–51, May 1999.
35. B. Hagerman et al., "Ericsson-AT&T Wireless Services joint adaptive antenna multi-site field trial for TDMA (IS-136) systems," in *Proceedings Sixth Annual Smart Antenna Workshop*, Stanford, CA, July 1999.

36. C. C. Martin, N. R. Sollenberger, and J. H. Winters, "Field test results of downlink smart antennas and power control for IS-136," in *Proceedings IEEE 49th Vehicular Technology Conference (VTC 1999)*, Houston, TX, 453–457, May 1999.
37. A. Kuchar, M. Taferner, M. Tangemann, and C. Hoek, "Field trials with GSM/DCS1800 smart antenna base station," in *Proceedings IEEE 49th IEEE Vehicular Technology Conference (VTC 1999)*, Houston, TX, 42–46, May 1999.
38. P. E. Mogensen et al., "Preliminary measurement results from an adaptive antenna array testbed for GSM/UMTS," in *Proceedings IEEE 47th Vehicular Technology Conference (VTC 1997)*, Phoenix, AZ, pp. 1592–1596, May 1997.
39. J. Strandell et al., "Experimental evaluation of an adaptive antenna for a TDMA mobile telephony system," in *Proceedings 8th IEEE International Symposium on Personal, Indoor and Mobile Radio Communications (PIMRC)*, Helsinki, Finland, 79–84, September 1997.
40. B. Johannisson and A. Derneryd, "Array antenna design for base station applications," in *Proceedings of Antenna Applications Symposium*, Monticello, IL, 98–106, September 1999.
41. S. Engström et al., "Multiple reuse patterns for frequency planning in GSM networks," in *Proceedings IEEE 48th IEEE Vehicular Technology Conference (VTC 1998)*, Ottawa, Canada, 2004–2008, May 1998.
42. F. Kronestedt and S. Andersson, "Migration of adaptive antennas into existing networks," in *Proceedings IEEE 48th IEEE Vehicular Technology Conference (VTC 1998)*, Ottawa, Canada, 1670–1674, May 1998.
43. H. Aroudaki and K. Bandelow, "Effects of introducing adaptive antennas into existing GSM networks," in *Proceedings IEEE 49th IEEE Vehicular Technology Conference (VTC 1999)*, Houston, TX, 670–674, May 1999.
44. L. C. Godara, "Application of antenna arrays to mobile communications, Part I: Performance improvement, feasibility, and system considerations," *Proceedings. IEEE,* vol. 85, no. 7, July 1997: 1031–1060.
45. L. C. Godara, "Application of antenna arrays to mobile communications, Part II: Beam-forming and direction-of-arrival considerations," *Proceedings IEEE,* vol. 85, no. 8, August 1997: 1195–1245.
46. A. Osseiran, M. Ericsson, J. Barta, B. Göransson, and B. Hagerman, "Downlink capacity comparison between different smart antenna concepts in a mixed service WCDMA system," in *Proceedings IEEE 54th IEEE Vehicular Technology Conference (VTC-Fall 2001)*, Atlantic City, NJ, 1528–1532, October 2001.

Antenna Issues and Technologies for Enhancing System Capacity

Yasuko Kimura, NTT DoCoMo
Zhi Ning Chen, Institute for Infocomm Research

5.1 Introduction

Over the past three decades, we have all witnessed the rapid development of mobile communications from devices to systems to infrastructure. Japanese companies have played an important role in the development of mobile communication technologies and services. Therefore, this chapter first briefly reviews the historic evolution of technological innovation and mobile communication systems in Japan. After that, general information regarding wireless access systems is introduced with basic but important concepts. Following the discussion of design considerations for base station antennas in mobile communications systems, five typical base station antennas will be discussed in depth.

5.1.1 Mobile Communications in Japan

As one of the key contributors, Japan has experienced all the stages of development of modern mobile communications. Figure 5.1 shows the evolution of technological innovations and generations of mobile communication systems in Japan.[1–4]

In 1979, the Japanese company Nippon Telegraph and Telephone (NTT) Public Corporation started the first commercial land-mobile communication service in the Tokyo district with the first-generation (1G) automobile telephone. The system operated with analog transmission technology. In 1984, NTT achieved countrywide service in Japan. One year later, with the reduction in the size and weight of mobile phones,

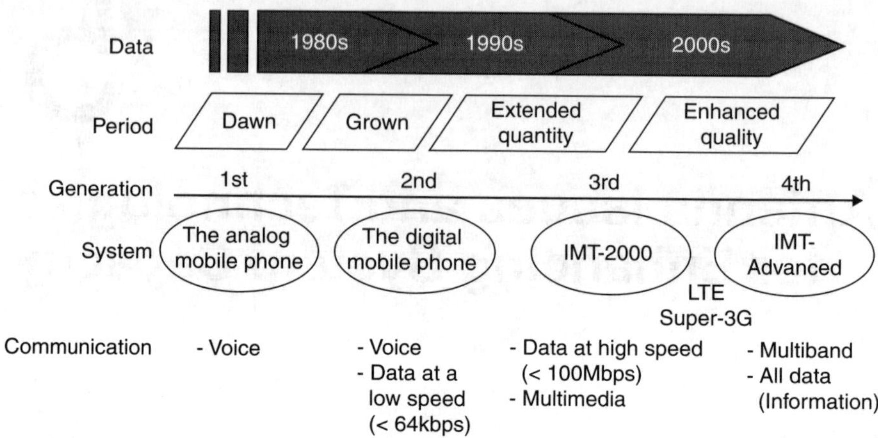

Figure 5.1   The evolution of Japan's mobile access systems

people could carry them; however, the phones still weighed around 3 kg. In 1987, people could actually use portable telephones, partially because their size and weight were reduced dramatically. For example, handsets then weighed approximately 700 g. NTT had a monopoly on the market until 1985 when Japanese regulations allowed new operators to enter the mobile communication market. At that time, the service systems provided by operators differed from each other. In 1G systems, users couldn't employ roaming service because each operator had its own distinct wireless interface and signal system. If a person used Operator A's phone system, he or she couldn't talk to his or her friend who was using Operator B's system. Therefore, a strong demand arose for unifying the systems to allow operators to offer users roaming service. The Ministry of Internal Affairs and Communications of Japan proposed standards and received positive feedback.

As a result, in 1993, operators accepted a new standard, which was called *personal digital cellular (PDC) technology* and which was used as the second-generation (2G) technology, replacing the 1G analog network. The 2G digital cellular network in Japan operated in 800-MHz and 1500-MHz bands. Seamless roaming service was now available. However, the 2G systems in Japan couldn't be used by those employing systems developed by foreign operators. Therefore, the next goal was to develop the technology to provide users with a mobile communication service that could be used anywhere in the world with one mobile terminal.

In Japan, the mobile communication system has been developed to align with the global International Mobile Telecommunications-2000 (IMT-2000) standard for third-generation (3G) mobile telecommunication services and equipment. The 3G services were designed to offer broadband cellular access at the targeted speeds of 2 Mbps indoors. The IMT-2000 operates in the 2-GHz band. NTT DoCoMo, which took over the mobile communication business from NTT in 1992, started the first commercial Wideband Code-Division Multiple-Access (W-CDMA) 3G network in the world in 2001, which made it possible for users to employ one mobile terminal using a worldwide standard system. The 3G system is much more powerful in providing multimedia, personalized, and globalization services. The 3G system operates in the 800-MHz, 1.7-GHz, and 2-GHz bands for increasing system capacity.

Currently, the research and development of mobile communications in Japan are focused on advancing Long Term Evolution (LTE, Evolved UTRA, or UTRAN in the 3G band) and the fourth-generation (4G) system.[5,6] The LTE, which is also called *Super-3G,* allows the 3G system to work with the 4G system, which is called the *IMT-Advanced system* in Japan. In May 2003, NTT DoCoMo conducted a field trial of a 4G mobile communication system in Yokosuka, Japan, with a preliminary/temporary license.

5.1.2　Wireless Access System

5.1.2.1　Concept　The main traffic carried by mobile communication systems until the 2G system was developed was voice communications. However, the 3G system requires high-speed transmission service because this system has changed to data communications. To realize the high data-rate transmission in 3G networks, a variety of wireless access systems have been investigated. Figure 5.2 shows the concepts of 2G and 3G frequency reuse systems in Japan. In this concept, each base transceiver station (BTS) has been designed to cover a single cell. Figure 5.2*a* shows an example of a PDC system, specifically, the 2G digital cellular telephone communication technology used in Japan. In the PDC system, one frequency channel is assigned to the individual target cell, which is different from those used in the surrounding cells, for avoiding co-channel interference.

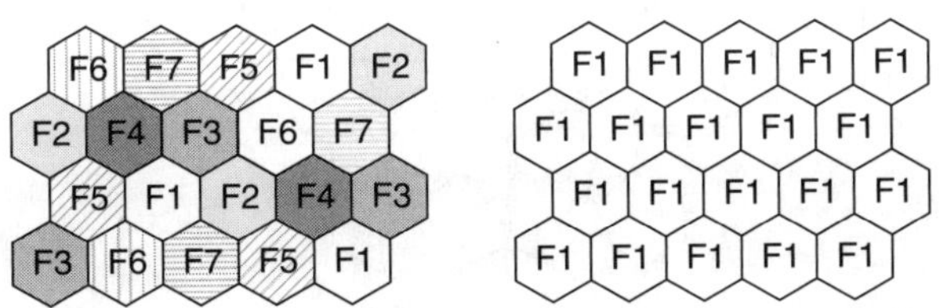

(a) PDC repetitive frequency　(b) IMT-2000 repetitive frequency

Figure 5.2　Wireless access system

Figure 5.2*b* is another example of the W-CDMA system, which is the 3G network technology for multimedia cellular phone communication. This technology has been used in Freedom of Mobile Multimedia Access (FOMA) for the 3G services being offered by NTT DoCoMo since 2001. In a W-CDMA system, the data at the same frequency channel can be transmitted simultaneously. In this scheme, all users in the individual cell employ the same frequency channel. Therefore, frequency planning is unnecessary. System capacity decreases, however, due to interference because the system shares the same transmission medium for frequency and time domains for all users, including those in other cells. To further increase the system capacity in IMT-2000, reducing the area of interference, namely reducing the overlapping area between adjacent cells, is necessary. This results in the reduction of the Diversity Hand Over (DHO) and of interference with other cells. As a result, base station antennas are designed to have narrow beamwidths to fit the shapes of cells with reduced overlapping areas.

5.1.2.2 Sectoring for CDMA System

All mobile communication networks operate with very limited frequency resources. Therefore, frequency resources must be effectively used to ensure that all subscribers are able to receive a consistent quality of service (QoS). Increasing the number of sectors in each cell is an effective way to increase frequency usage efficiency. The multiple sectors per cell scheme in mobile communication networks has been proposed to increase frequency usage efficiency instead of the single sector per cell scheme. Using multiple sectors per cell is also conducive to increasing W-CDMA system capacity and data traffic per subscriber. Usually, three-sector or six-sector per cell structures are used in mobile communication networks, as shown in Figure 5.3. As mentioned previously, the number of sectors per cell is determined by capacity and traffic.

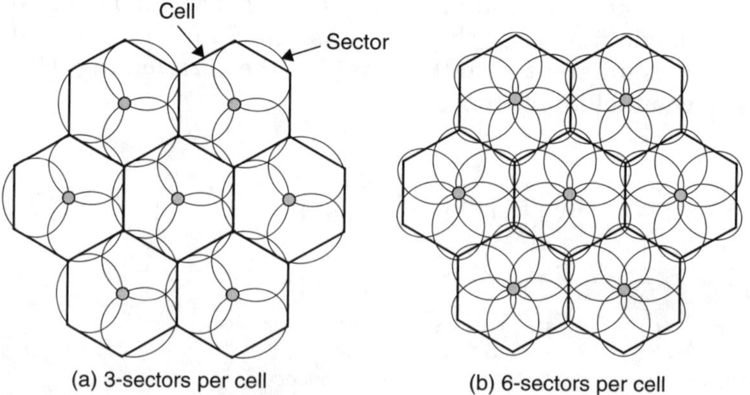

Figure 5.3 Sector structure

Figure 5.4 shows the relationship between capacity and half-power beamwidth (HPBW) in a W-CDMA system for a three-sector or six-sector per cell scheme.[7] Each sector division angle is 120° for a three-sector per cell scheme or 60° for six-sector per cell scheme. However, Figure 5.4*a* shows that peak system capacity in the uplink can be achieved for an HPBW of 60°–90° for the three-sector per cell scheme and 35°–45° for the six-sector per cell scheme, respectively. Similarly, from Figure 5.4*b*, the peak system capacity in the downlink occurs when the HPBW varies

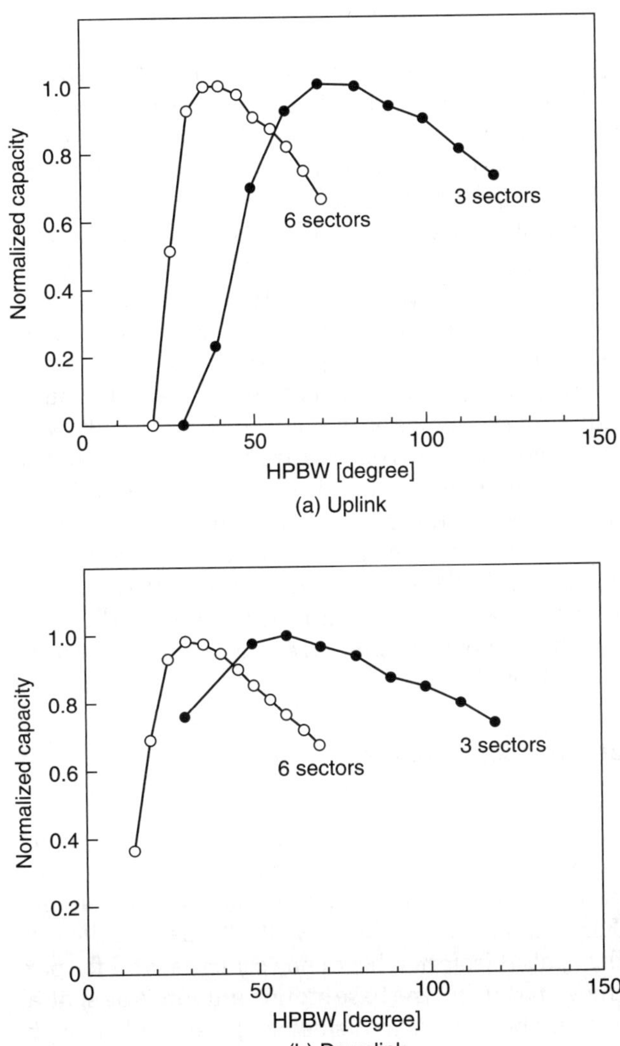

Figure 5.4 Relationship between capacity and HPBW in W-CDMA

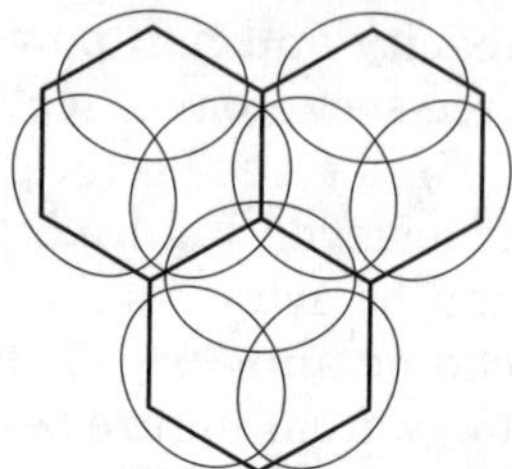 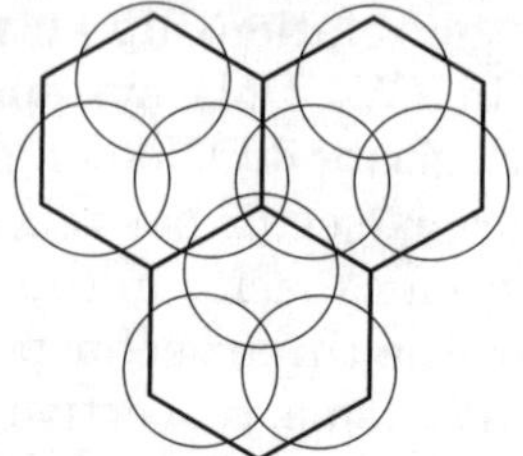

(a) Sector division angle = HPBW   (b) Sector division angle > HPBW

Figure 5.5   HPBW in horizontal planes

from 50°–80° for the three-sector per cell scheme and from 30°–45° for the six-sector per cell scheme, respectively. These observations suggest that using beamwidths that are narrower than those of the sectors can increase system capacity. For example, in a three-sector per cell scheme with an HPBW of 120°, which equals a sector division angle, system capacity in the uplink is just 70% of the system capacity with an HPBW of around 75°. Because HPBW is designed to be equal to the sector division angle, as shown in Figure 5.5a, the advantage is that a uniform power-level distribution will be achieved across the sector. However, unfortunately, such an area design usually leads to many large overlapping portions between the adjacent sectors, which causes severe interference, reducing system capacity. For an HPBW narrower than the sector division angle, as shown in Figure 5.5b, the size of the overlapping areas is significantly reduced. Subsequently, system capacity can be increased by suppressing the interference among the adjacent sectors. The drawback of such an area design is, however, that some portions will suffer low-power levels. Therefore, the optimal HPBW of the base station antennas will increase system capacity significantly, but a tradeoff between the size of the overlapping areas and the uniformity of power-level distribution is required.

5.2.   Design Considerations for Antennas from a Systems Point of View

All mobile cellular communication networks require increased system capacity, improved coverage, better QoS, and reduced transmitted power. The base station antennas in these networks can be designed to enhance network performance. For example, when diversity effect is used in the networks for better QoS, the spatial isolation between sectors is vital for performance and can be achieved by using the base station antennas as spatial filters. It is well-known that the HPBW and sidelobe levels will strongly influence the spatial filtering performance of base station antennas. Thus, the radiation performance of a directive base station antenna in a W-CDMA network is an important design parameter. For instance, the HPBW of the antennas in both downlink and uplink significantly affects the power and interference distribution in the network, as mentioned previously. As a result, the optimization of the HPBW and sidelobe levels are the most important considerations in base station antennas design in W-CDMA networks, although gain is another crucial parameter as well.[8,9]

Furthermore, radiation patterns in the horizontal and vertical planes have different requirements. In the horizontal planes, usually the HPBW plays the important role in determining coverage and controlling the possible interference among adjacent cells or sectors. Sidelobes are hardly the problem because only single elements or an array with few, for instance, two or four, elements are used in the horizontal plane. In the vertical plane, the number of vertical array elements is determined by the required gain of the base station antenna. Moreover, controlling sidelobe levels becomes an antenna design consideration because the sidelobe levels will affect the inference performance.

In addition to the electrical specifications, which are determined by system requirements, design considerations for base station antennas primarily include

- Mechanical strength

- Compact size for limited installation space

- Low construction cost

- Light weight

- Less environment influence

- High water resistance/weather proofing

- Low material and manufacturing cost

In addition, base station antennas are always installed outdoors in high places like on the roofs of buildings or on antenna towers where wind pressure affects the mechanical strength of antenna installation.[10] The wind pressure can be calculated as in Eq. 5.1:

$$P = Q \times C \times A \tag{5.1}$$

where P is the wind pressure, Q the dynamic wind pressure, C the pressure coefficient, and A the exposed surface area of antenna to the wind direction. For example, C is 1.4 for flat plates and 0.9 for cylinders at the base station antenna. The cylinder is 0.64(= 0.9/1.4) of the flat plate value.[11,12] Cylindrical radomes are, therefore, more commonly used.

In base station antenna engineering, slim antennas reduce wind pressure. In Japan in particular, some places experience extremely powerful winds produced by typhoons. The record of the most powerful wind pressure is up to 85.3 m/s from a typhoon. Base station antennas must be windproof and waterproof against typhoons.

In the early days of the 3G network, system capacity was sacrificed to using slim antennas that had the same coverage area as the 2G network.[10] However, the HPBW, which is narrower than those in the 2G network, was used due to the increased number of subscribers.[7,8] Due to these reasons and the effects of HPBW on system capacity, as shown in Figure 5.4 and Figure 5.5, the ratio of the HPBW to the sector division angle of around 0.75 was used to increase system capacity and reduce the areas without low-power levels. Thus, the HPBW of 90° and 45° are for the three- and six-sector per cell structures, respectively.[13]

Moreover, multiband antennas have been used to reduce the space required for antenna installation. The limited space available for antenna installation, especially in metropolitan areas, has been a big problem in Japan. A compact design for antennas is strongly desired.

5.3   Case Studies

Based on the design considerations just mentioned, five typical antenna designs have been selected for study. Techniques for controlling the HPBW for optimal system capacity and miniaturizing the width of the antenna are introduced in the design case studies.

5.3.1   Slim Antenna

5.3.1.1   Reduction in Width of Reflector   Usually, a base station antenna is designed with a reflector behind the radiators to suppress possible back radiation for avoiding possible interference among antennas or sectors. Figure 5.6 shows three 120° beam antennas with reflectors behind the reflectors and radomes that protect the antennas.[14] The reflectors are installed in parallel to the arms of dipoles with a spacing of $\lambda/4$ (λ is the wavelength at operating frequency) to keep high gain. The diameter of the antenna of the flat reflector is 0.53λ, as shown in Figure 5.6a. The width of a corner reflector antenna can be shrunk but with the same HPBW as the antenna with a flat reflector. For example, the diameter of the corner reflector antenna with the same HPBW of 120° as the antenna with a flat reflector is reduced to 0.3λ from 0.53λ, as shown in Figure 5.6b. A semi-cylindrical reflector can be used to reduce the diameter of the antenna reflector further. Such a diameter can be reduced to only 0.267λ, as depicted in Figure 5.6c, so the slimmest design can be achieved.

Figure 5.7 compares the radiation patterns for antennas with a flat reflector and a semi-cylindrical reflector. The radiation patterns are roughly identical and the latter pattern has an approximately 3-dB improvement in back radiation. Therefore, the diameter of the antenna can be reduced by about 50% using a semi-cylindrical reflector, as shown

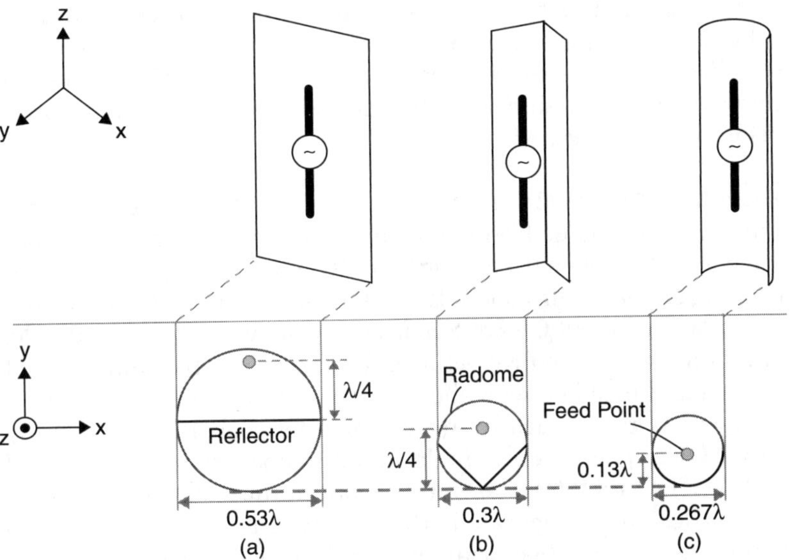

Figure 5.6   Structures of 120° beam antennas

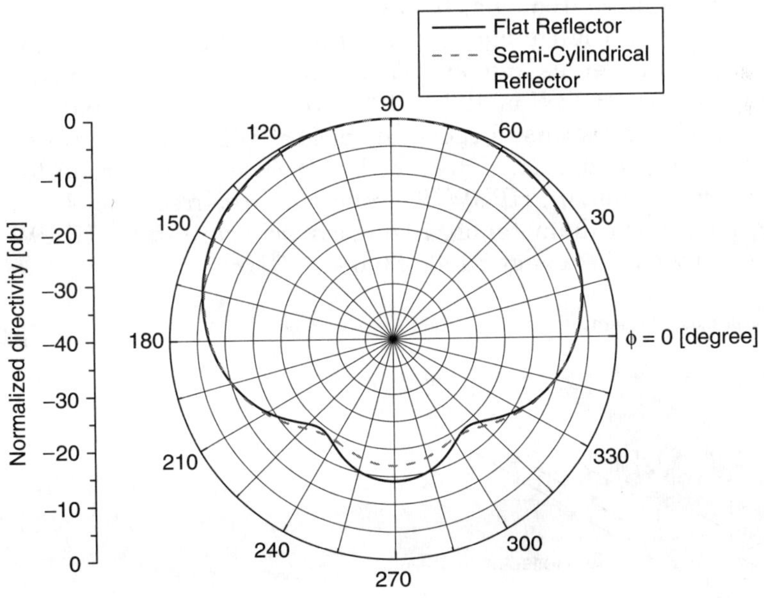

Figure 5.7   Radiation patterns

in Figure 5.6*c*, although the radiation performance of the antenna remains the same as the antenna with a flat reflector, as depicted in Figure 5.6*a*. From this, we can conclude that the diameter of the antenna can be narrowed by optimizing the reflector's shape while also maintaining radiation performance.

5.3.1.2 Triple-Band Slim Antenna

As a design example, Figure 5.8 shows the configuration of a triple frequency-band antenna. The antenna consists of one printed dipole with a length of ~0.5λ at 1.5 GHz with a parasitic conducting element with a length of ~0.5λ at 2 GHz and two shorted stubs with lengths of about 0.2λ at 800 MHz at the two ends of the dipole. The antenna has one curved reflector and a radome with a diameter of 110 mm.[10] The distance between shorted stubs is 150 mm. Therefore, if this antenna is stacked vertically, the distance between antennas is only about 1λ at 2 GHz. The antenna is designed to operate in the bands of 800 MHz, 1.5 GHz, and 2 GHz for mobile communication services.

The dipole is designed to operate in the 1.5 GHz band. The stubs, whose ends are grounded to the reflector, are used to improve impedance matching in the 800-MHz band. The parasitic element resonates at 2 GHz. The current flow on the dipole in each band is shown in Figure 5.9. Figure 5.10 shows the measured return loss characteristics with the required return loss of −14 dB in the required three frequency bands. From the diagram, we can confirm that this antenna achieves good impedance matching across all three required frequency bands.

Figure 5.11 shows the change in the HPBW of the radiation patterns for the antenna versus frequency. With the optimized reflector, the antenna achieves an HPBW of 120° ± 10° across the range of 0.6–2.1 GHz, a very wideband response. In general, the HPBW becomes narrower when the frequency is higher. However, the structure of this reflector features wideband-constant HPBW. Therefore, a three-frequency-band antenna with a radome diameter of 110 mm can achieve a stable HPBW of 120° across a wide frequency band of 0.6–2.1 GHz.

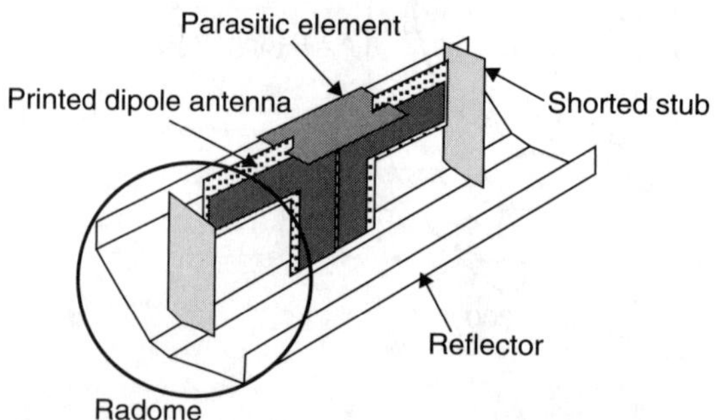

Figure 5.8   Triple frequency-band antenna

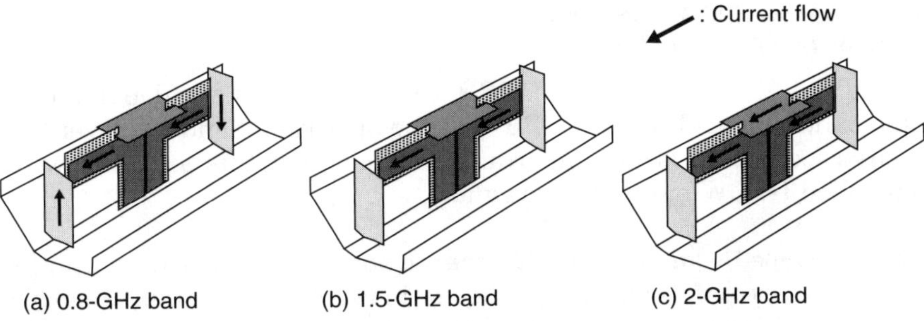

Figure 5.9   Current flow of a triple frequency-band antenna

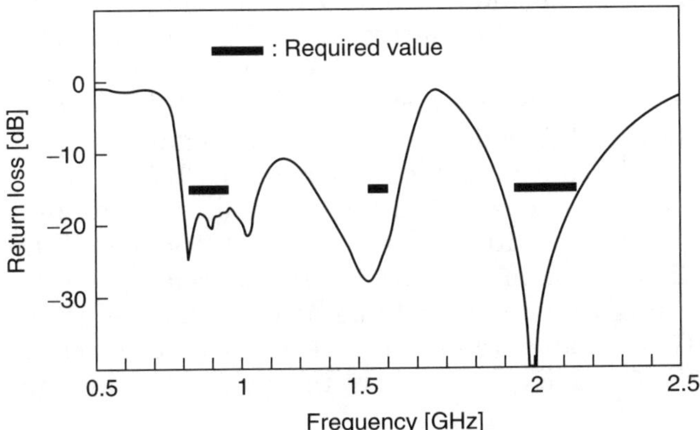

Figure 5.10   Return loss response against frequency

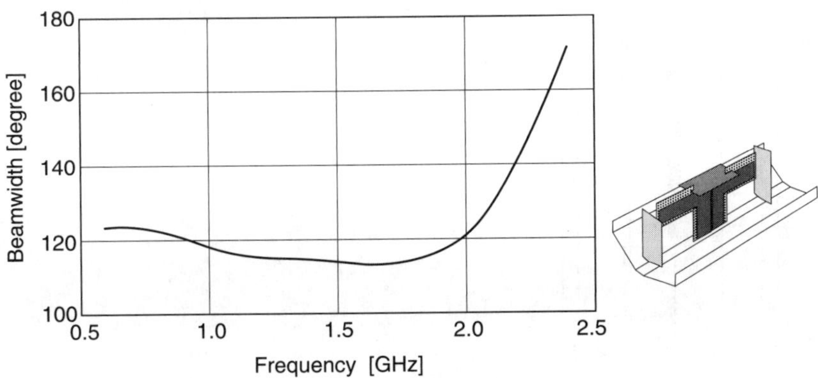

Figure 5.11   HPBW of triple frequency-band antenna

5.3.2 Narrow HPBW Antenna with Parasitic Metal Conductors

In addition to stabilizing the HPBW response across a wideband/multiple bands, controlling the HPBW in the operating bands to cover the desired sector area is also important. To accomplish this, antennas with controlled HPBW have been designed.

5.3.2.1 Single-Beam Antenna In order to narrow the HPBW, we usually have to increase the size of the reflector in certain dimensions. To avoid a significant increase in the size of the antenna, two slender metal conductors are used. Consider a 0.5λ dipole with a semi-cylinder reflector as shown in Figure 5.12. Metal conductors having the same length as the reflector are symmetrically positioned in front of the reflector with a distance S between the dipole and the slender metal conductors. The slender metal conductors are used to narrow the HPBW in horizontal planes. In Figure 5.13, the change in the HPBW versus the spacing S between the dipole and the metal conductors is shown. HPBW can be varied from 110° (without metal conductors) to 60° by changing the spacing S. Slender metal conductors can achieve narrower HPBW.

Figure 5.14 compares the measured and simulated radiation patterns for the antennas with and without parasitic metal conductors. The comparison reveals that the simulated results are in good agreement with the measured results. An antenna with parasitic metal conductors has a narrower HPBW than an antenna without the metal conductors. The achieved HPBW is 90°, which is required by the system. The antenna

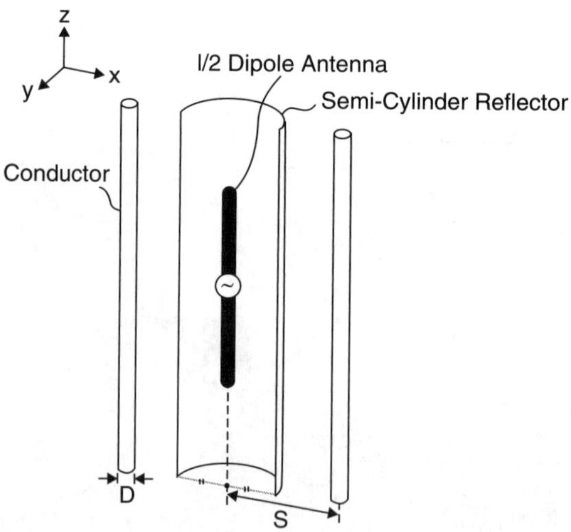

Figure 5.12 120° beam antenna with parasitic conductors

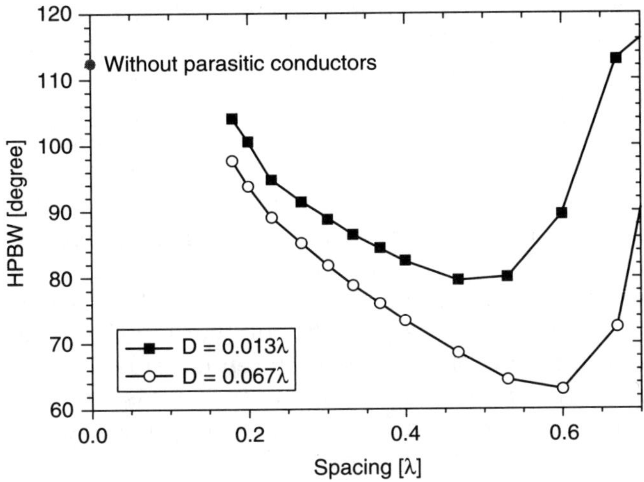

Figure 5.13   The change in HPBW with parasitic conductors

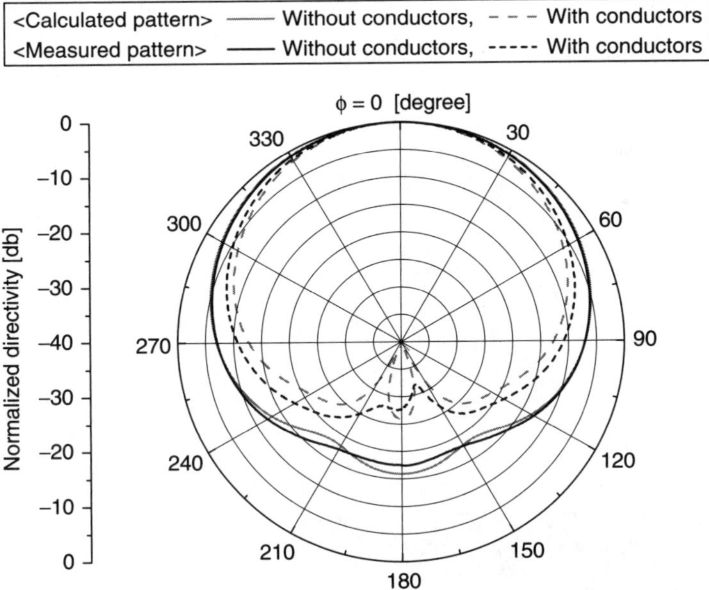

Figure 5.14   The change in radiation patterns with parasitic conductors

aperture of a 90° HPBW is only about 0.3λ, whereas the aperture of the antenna shown in Figure 5.6c is 0.267λ, and the radius of the metal conductors is 0.13λ. In addition, the antenna with the metal conductors achieves a 10 dB lower back radiation than that of the antenna without the metal conductors.

5.3.2.2 Triple-Band Antenna   As just mentioned, a 90° HPBW increases the capacity for a three-sector CDMA system, as shown in Figure 5.5*b*. However, in the 800-MHz and 1.5-GHz bands, the PDC systems, as shown in Figure 5.2*a*, have no interference issues. A 120° HPBW is, therefore, good for PDC systems in the 800-MHz and 1.5-GHz bands. The optimal HPBWs for the CDMA systems operating at 2 GHz and PCD systems operating at 800 MHz and 1.5 GHz are 90° and 120°, respectively.

When we installed the antenna elements, mentioned in Section 5.3.2.1, in the antenna radome to form an antenna array, the HPBW of the antenna arrays in each frequency band was 120° for a three-sector zone scheme. Due to the recent increase in the number of 3G system subscribers, however, the HPBW of triple-band antennas must be narrowed to 90° in the 2-GHz band for 3G systems to increase system capacity. To achieve an HPBW of 90° in 2-GHz bands, we propose using metal conductors near the antenna elements.[15]

Figure 5.15 shows an illustration and a photo of a triple frequency-band antenna (Figure 5.8) with metal conductors. Figure 5.16 shows

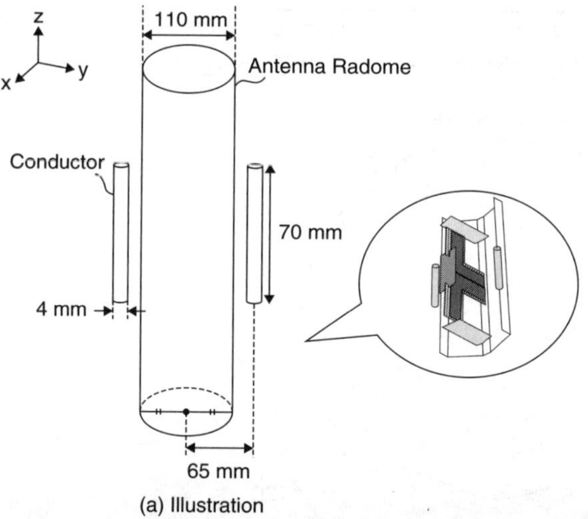

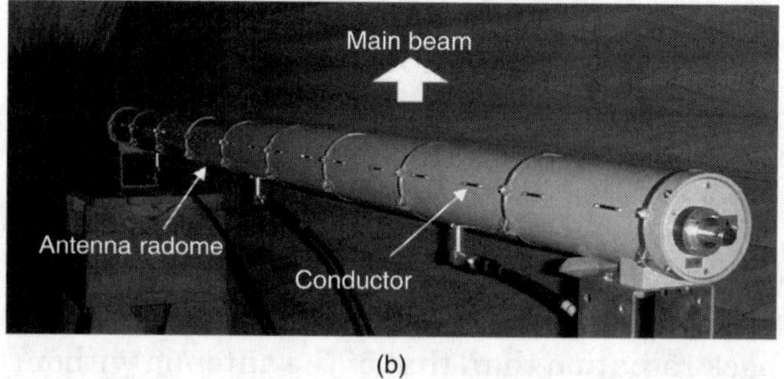

Figure 5.15   Triple frequency-band antenna with parasitic conductors

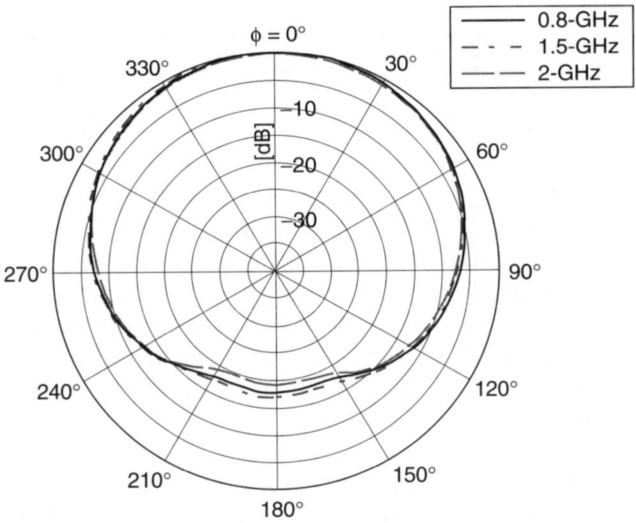

(a) Measured patterns without parasitic conductors

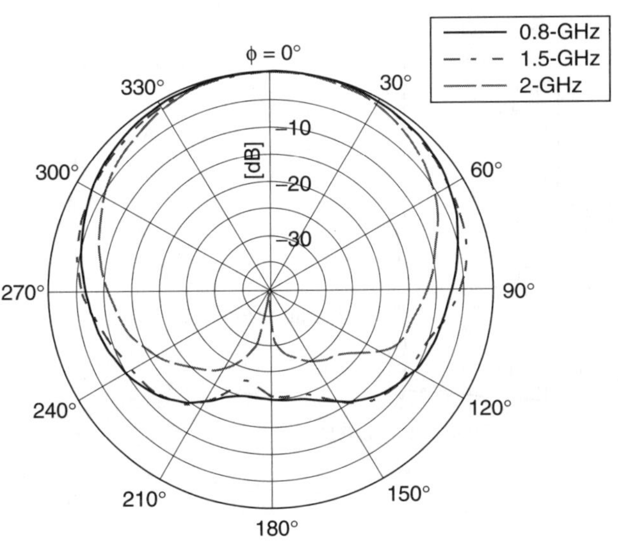

(b) Measured patterns with parasitic conductors

Figure 5.16   Measured patterns of triple frequency-band antenna

the measured results of horizontal radiation patterns with and without parasitic conductors. This figure indicates that the HPBW of the triple frequency-band antenna changes only in the 2-GHz band. Also, Figure 5.16 shows that the optimized structure and arrangement of the parasitic conductors achieved the HPBW of 90° in the 2-GHz band

TABLE 5.1   Measured HPBW

HPBW	800-MHz band	1.5-GHz band	2-GHz band
Without conductors	113.9°	112.5°	113.4°
With conductors	123.9°	129.8°	83.6°

without influencing the HPBW performance in the 800-MHz and 1.5-GHz bands. Table 5.1 shows the change in HPBW in 800-MHz, 1.5-GHz, and 2-GHz bands. Therefore, if the parameters of the metal conductors are selected correctly, only the HPBW in the highest frequency band can be changed as desired. The overall surface width of the antenna is 118 mm; the antenna aperture is 110 mm; and the metal conductors are 4 mm. The overall size of this antenna is slightly larger than the antenna without the conductors.

5.3.2.3 Dual-Beam Antenna Figure 5.17 shows a dual-beam antenna for spatial diversity.[16,17] The two antennas were installed on a bent

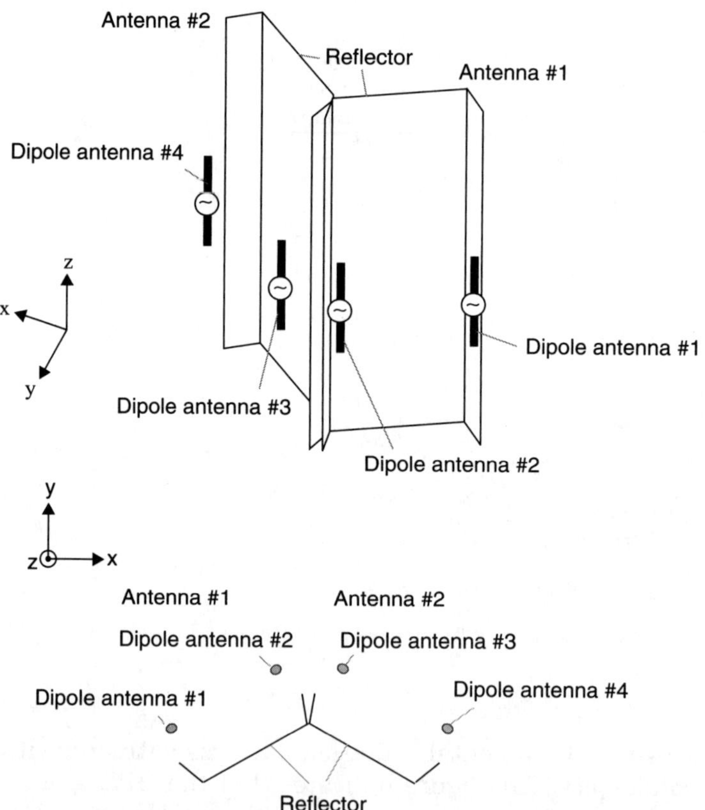

Figure 5.17   Dual-beam antenna

planar reflector with an angle of 60°. The three rims with a height of about 0.14λ are used to increase the isolation between the antennas as well as the reflection of radiation. The HPBW of this antenna is 60° for a six-sector radio zone. The difference between the beam directions of Antennas #1 and #2 is 60° for service in adjacent zones. To reduce the space for antenna installation, one radome is used to cover two antennas. The HPBW of the antennas are also controlled using metal conductors, as shown in Figure 5.18a.[18] The radiation patterns are shown in Figure 5.19. This figure shows a 60° HPBW dual-beam antenna in which the HPBW can be narrowed to 45° using metal conductors. Figure 5.18b shows a photograph of the antenna prototype shown in Figure 5.18a.

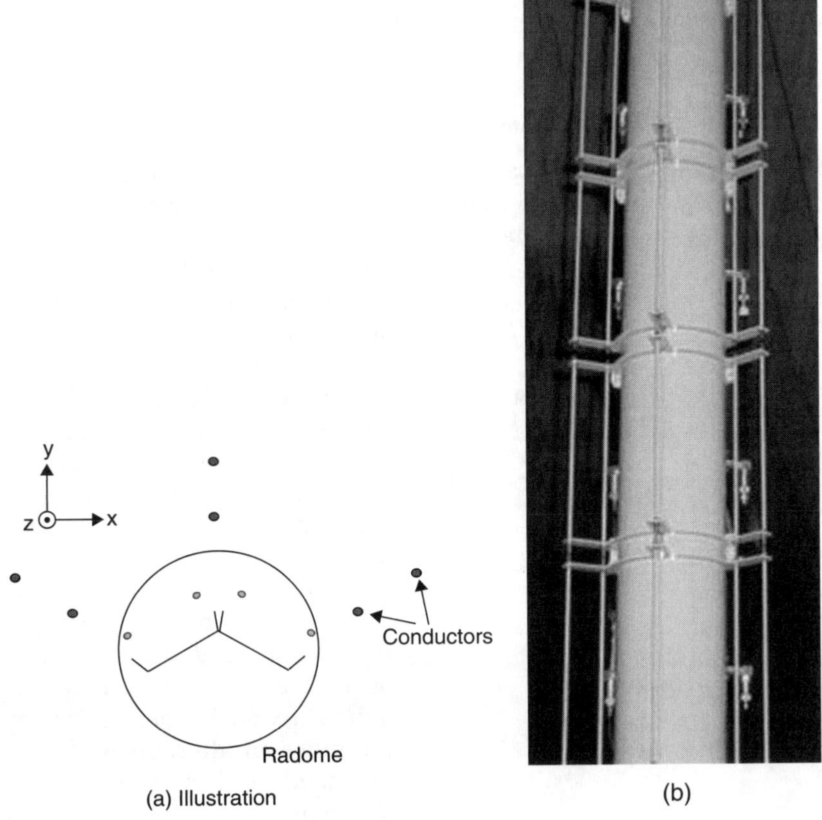

Figure 5.18 Dual-beam antenna with parasitic conductors

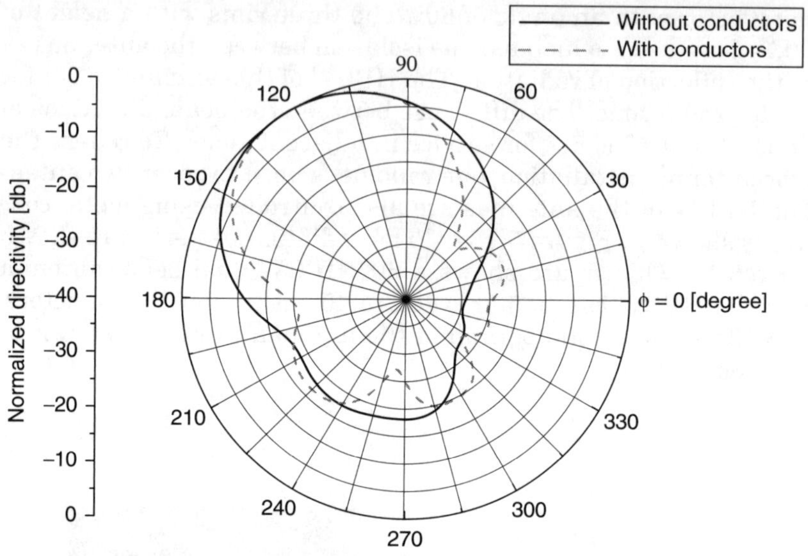

Figure 5.19 Radiation patterns of dual-beam antenna

5.3.3 SpotCell (Micro-Cell) Antenna

In addition to sectors, base stations are often installed in hotspot areas. The purpose is to plug small black-hole areas that are caused mainly by surrounding environments. Usually, omnidirectional antennas are used in micro-cells or spot cells. Figure 5.20 shows an omnidirectional antenna with a parasitic conducting cylinder operating in the 900-MHz band.[19–21] The antenna has been etched onto a 30-mm wide slab of dielectric substrate with a dielectric constant of 3.6. A rectangular slot with a width of 15 mm has been cut onto the ground plane and fed at its bottom side by a microstrip feed line that has been etched onto the dielectric substrate. Such a design is good for mass production with low material and manufacturing loss. The slot is asymmetrically cut by the edge of the dielectric

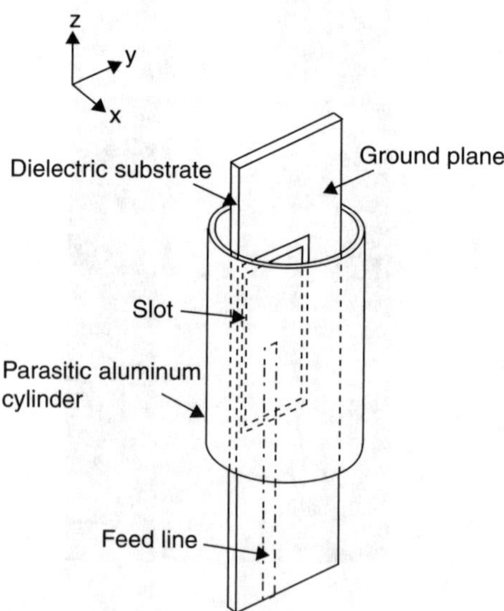

Figure 5.20 Omnidirectional antenna

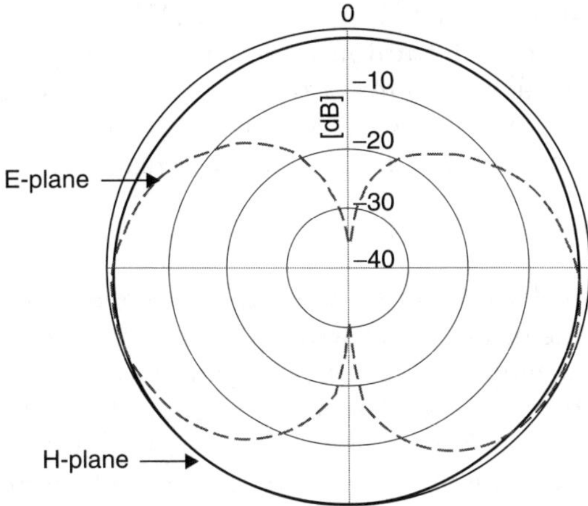

Figure 5.21   Measured patterns of an omnidirectional antenna

substrate to keep the space for the feed line that connects other elements in an array. The parasitic aluminum cylinder with a diameter of about 30 mm is used for omnidirectional radiation.

The measured radiation patterns in both vertical and horizontal planes at 900 MHz are shown in Figure 5.21. Omnidirectional radiation in horizontal planes has been observed. The minimum diameter of the antenna (including radome) is determined by the external diameter of the parasitic cylinder and the width of the dielectric substrate. Here, the width of the dielectric substrate or the inner diameter of the cylinder is about 30 mm. Thus, the minimum diameter of the radome is slightly greater than 30 mm.

Figure 5.22 shows the measured impedance bandwidths for a return loss of < −14 dB with the changed slot antenna element length, which resonates at selected frequencies. Bandwidth varies from 10–14% across

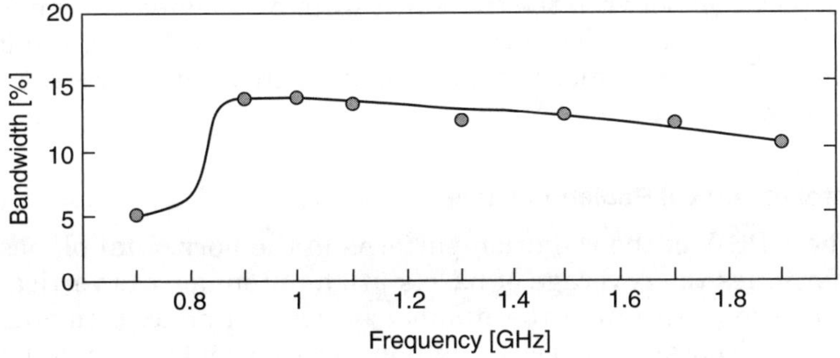

Figure 5.22   The change of HPBW

900–1900 MHz. The narrowest bandwidth occurs at 800 MHz because of the strong coupling between the slot antenna and parasitic cylinder. However, this antenna can achieve omnidirectional radiation as a dielectric substrate of 30-mm wide (0.09λ at 900 MHz) is used. The bandwidth reaches about 15% although the width is electrically large.

5.3.4 Booster Antenna

In mobile communication networks, booster systems are used to prevent possible oscillation caused by the interference waves between transmitting and receiving antennas. To enable the mobility of communications, a frequency offset booster has been proposed in order to realize the reradiation system in a blocked area, such as in the shadow of a mountain or building. Such a system has a back-to-back antenna arrangement, where one antenna or array points toward the base station whereas the other one points to mobile terminals. It is necessary to identify an absolute value for the mutual coupling between the antennas such that it is greater than amplifier gain.[22]

Figure 5.23 shows the structures of four- and sixteen-element choke-loaded patch antenna arrays, where the chocks are around dielectric substrates. The antennas are used to increase the front-to-back (FB) radiation ratio to reduce the mutual coupling in the back-to-back antenna arrangement. Figure 5.23a shows the configuration of a four-element chock-loaded patch antenna array, and Figure 5.23b shows a sixteen-element chock-loaded patch antenna array. Both antenna arrays are etched onto 1.2-mm thick dielectric substrates.

The radiation patterns in both E- and H-planes for both four- and sixteen-element chock-loaded patch antenna arrays are shown in Figure 5.24. Table 5.2 tabulates the measured HPBW and FB ratio. The FB ratios of the four- and sixteen-element choke-loaded patch array antennas are 37 dB/40 dB in E-planes and 32.5 dB/40 dB in H-planes, respectively. The HPBW of two four- and sixteen-element antennas are 39.8°/25.7° in E-planes and 36.4°/21.5° in H-planes, respectively. When the separation between the transmitting and receiving antennas increases to 4λ, the measured mutual coupling between the front and back arrays in the four-element choke-loaded patch array is less than −84 dB, as shown in Figure 5.25.

5.3.5 Control of Vertical Radiation Pattern

Whereas the HPBW of the radiation patterns in the horizontal planes mainly determines the coverage of base station antennas, the vertical radiation patterns determine the number of vertical array elements that are needed to obtain the desired gain in coverage and to design the required HPBW. Employing a shaped-beam pattern for a base station

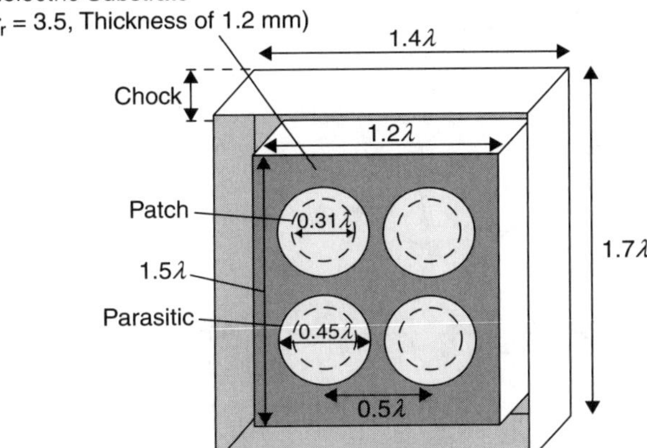

(a) Structure of four-element choke loaded patch array antenna

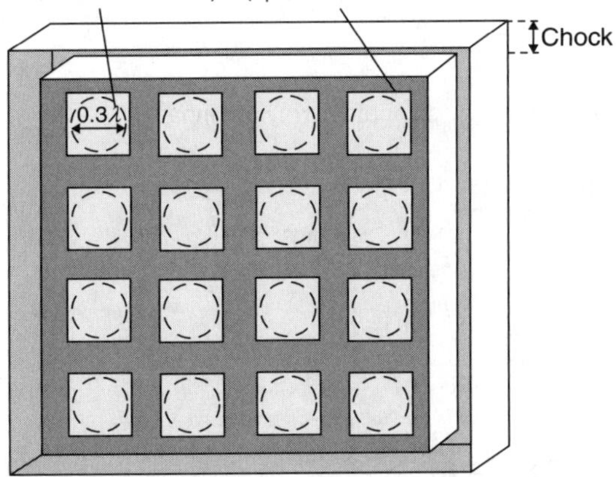

(b) Structure of sixteen-element choke loaded patch array antenna

Figure 5.23 Booster antennas

TABLE 5.2   HPBW and FB of Measured Results

	Four-element booster antenna		Sixteen-element booster antenna	
	HPBW	FB	HPBW	FB
E-plane	39.8°	37 dB	25.7°	40 dB
H-plane	36.4°	32.5 dB	21.5°	40 dB

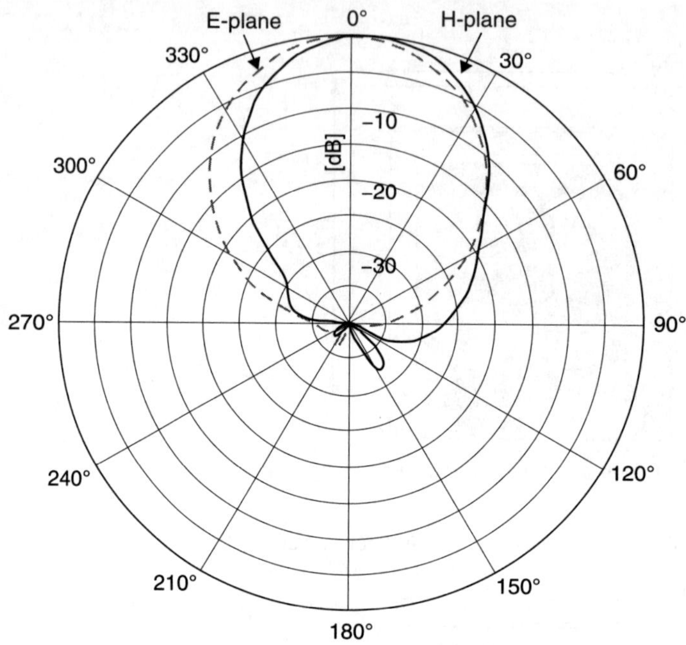

(a) A four-element choke loaded patch array antenna

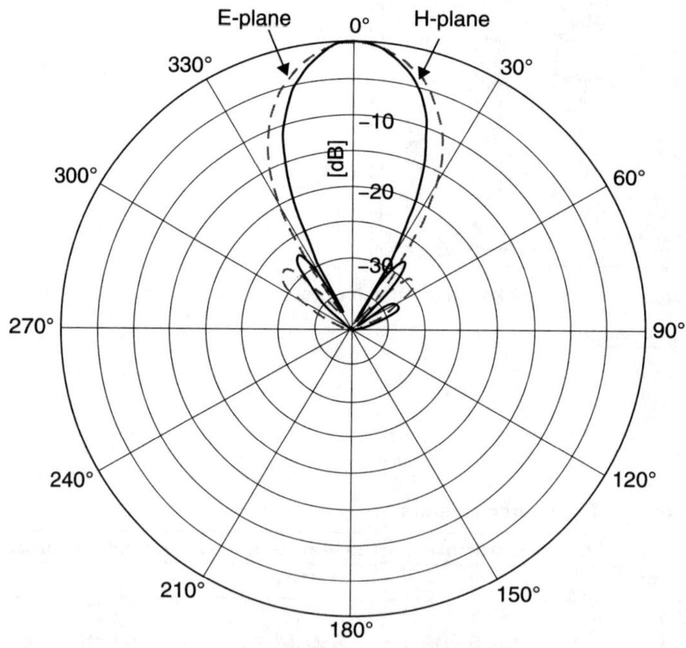

(b) A sixteen-element choke loaded patch array antenna

Figure 5.24   Measured pattern of booster antennas

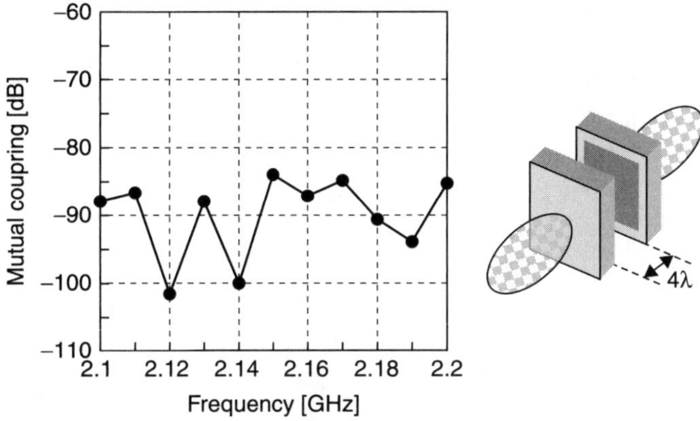

Figure 5.25 Mutual coupling characteristics of booster antennas

antenna—for instance, a cosecant beam pattern—is preferable because the shaped-beam pattern in a service area is capable of providing a sufficiently constant received power level for mobile users and uniform receiving power levels in spite of the cell edge while imparting only a slight interference power level to other adjacent cells.[23] The cosecant beam pattern can be achieved when the elements of the array are fed with linearly varied amplitudes and phases.

In the design of cosecant beam patterns, a downward tilting beam is designed to direct sidelobes to the horizon for preventing radiated power to other cells and receiving the signals with constant power levels in its own cell, as indicated in Figure 5.26. The beamtilting angle can be remotely controlled by using an electrical phase-tilt control box from a base station. Figure 5.27 shows a base station antenna array with a remote electric phase-tilt control box. In order to form a complete cosecant beam pattern with fine tilting angle θ_t, many elements and phase shifters

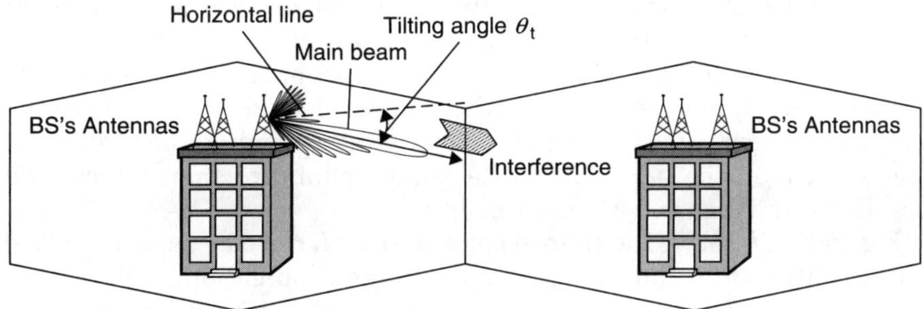

Figure 5.26 Radiation pattern of array antenna

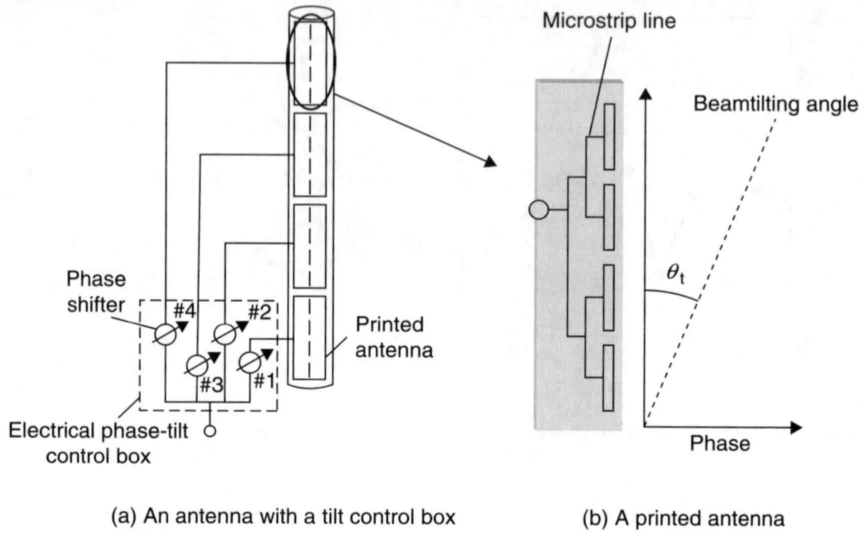

(a) An antenna with a tilt control box   (b) A printed antenna

Figure 5.27   Control of vertical radiation patterns

are required.[24] This will, however, increase size, weight, and cost of the antenna system because the phase shifter is usually very expensive, large, and heavy. To reduce the number of phase shifters, subarray blocks are used and connected to the phase shifters, as shown in Figure 5.27, where four subarrays form the antenna array. Each subarray fed by one phase shifter has four array elements fed in-phase.

Figure 5.28 shows the shaped vertical radiation patterns by using equiphase, namely the elements with the same phase difference and an operative method. With the operative method, the phases of the top and bottom elements in a vertically stacked array antenna are controlled in order to achieve beam-shaping. These phases form an S shape. The achieved tilting angle θ_t of the main beam is 5° without use of any electrical phase control, as shown in Figure 5.28a. The shaped beam pattern by equiphase indicates a bilaterally symmetric pattern with a central focus on the main beam. However, the shaped beam pattern using the operative method is similar to a cosecant beam. The sidelobe levels toward the ground side increase, whereas the sidelobe levels toward the sky side decrease. It is important to note that the suppression of sidelobe levels between the horizon and the main beam direction reduces the possible interference with other cells.

Figure 5.28b shows the shaped beam pattern formed by using electrical phase controls that can change the tilting angle more than 3°. Therefore, a total tilting angle is up to 8° compared to that shown in Figure 5.28a. The phase of each subarray is controlled by a phase shifter in order

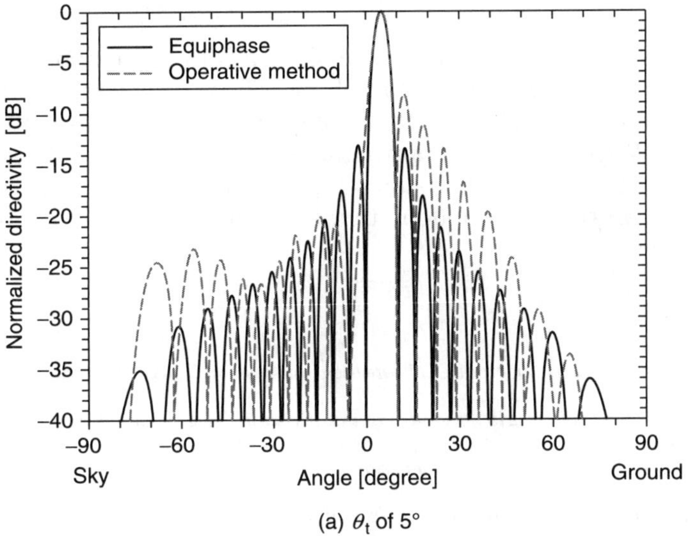

(a) θ_t of 5°

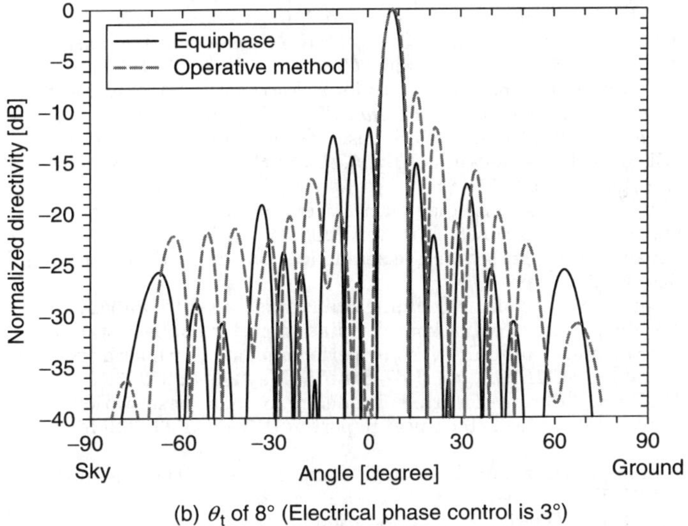

(b) θ_t of 8° (Electrical phase control is 3°)

Figure 5.28 Vertical radiation patterns

to change the main beam direction. The phase difference between the elements varies. Therefore, the radiation pattern has grating lobes, in particular, for large remote tilting angles. However, using the operative method for the shaped radiation pattern still keeps a cosecant beam, as shown in Figure 5.28*b*.

5.4 Conclusion

The important design issues of antennas and arrays in cellular phone base stations have been addressed in this chapter from a systems point of view for enhancing the performance and capacity of mobile communication systems. We presented the antenna designs, which have been used to enhance the capacity of systems and maximize the usage of transmitted power in cellular phone networks, from an antenna engineering perspective.

References

1. K. Tachikawa (ed.), *W-CDMA for mobile communication system,* Tokyo: Maruzen, 2001.
2. K. Tachikawa (ed.), *The latest digital mobile communication,* Tokyo: Kagaku Shinbun-sha, 2002.
3. K. Kinoshita (ed.), *IMT-2000 for third generation mobile communications,* Ohmsha, 2001.
4. T. Masamura (ed.), *Mobile Communications,* Tokyo: Maruzen, 2006.
5. Y. Yamao, N. Umeda, T. Otsu, and N. Nakajima, "Fourth generation mobile communications system," *IEICE,* vol. J-83-B, no. 10 (October 2000): 1364–1373.
6. T. Togi, H. Atarashi, and H. Yoshino, "ITU radiocommunication assembly (RA-07) report," *NTT DoCoMo Technical Journal,* vol.10, no. 1, (July 2008): 68–72.
7. M. Iwamura, Y. Ishikawa, K. Ohno, and S. Onoe, "Optimal beam-width of base station antennas for W-CDMA," *IEICE General Conference,* B-5-157 (1999).
8. F. Athley, "On base station antenna beamwidth for sectorized WCDMA systems," *2006 IEEE 64th Vehicular Technology Conference* (Fall), (September 2006): 1–5.
9. J. Niemela and J. Lempiainen, "Impact of the base station antenna beamwidth on capacity in WCDMA cellular networks," *The 57th IEEE Semiannual Vehicular Technology* (Spring), (April 2003): 80–84.
10. Y. Ebine, "Design of base station antennas for next generation cellular mobile radios (IMT-2000)," *Technical Report of IEICE,* AP2000-4 (April 2000): 23–30.
11. H. Kawai, "Wind resistant design law and recommendations for loads on buildings," *GBRC,* vol. 32 (October 2007): pp. 4–10.
12. MLIT, Enhancement Ordinance of the Building Standards Law # 87, Japan, 2000.
13. Y. Kimura and Y. Ebine, "Control of horizontal radiation pattern of base station antenna for cellular mobile communications by performing approach arrangement of slender metal conductors," *IEICE,* vol. J-87-B, no. 5 (May 2004): 673–684.
14. Y. Sugimoto and Y. Ebine, "Design of a triple frequency resonant base station antenna for cellular mobile radios," Technical Report of IEICE, AP99-262 (March 2000): 81–85.
15. Y. Kimura and Y. Ebine, "Triple-band base station antennas comprising slender metal conductors," *PIERS 2004 Nanjing,* (August 2004): 297.
16. Y. Yamaguchi and Y. Ebine, "Design of a base station antenna with 60 degrees beam-width in the horizontal plane for cellular mobile radios," *IEICE,* vol. J-86-B, no. 6.
17. Y. Kimura and Y. Ebine, "Performance of low side-lobe and narrow beam of a 60° beam antenna with closely arranged conductive poles," Technical Report of IEICE, AP2005-63 (July 2005): 149–154.
18. Y. Kimura and Y. Ebine, "Narrowed beam performance of a dual-beam base station antenna for six-sector by metal conductors arrangement," *IEICE,* vol. J-89-B, no. 9, (September 2006): 1824–1827.
19. N. Goto, M. Nakagawa, and K. Ito (eds.), *Antenna and Wireless Handbook,* Tokyo: Ohmusha, 2006.

20. Y. Danjou and M. Karikomi, "An omnidirectional antenna composed of a ring slot and a metallic cylinder," IEICE Society Conference, B-1-40 (1997).
21. Y. Yokoo and H. Arai, "Multi-frequency loop slot antenna with omnidirectional pattern," *IEICE,* vol. J-88-B, no. 9, (September 2005): 1718–1727.
22. N. Michishita, H. Arai, and Y. Kimura, "Mutual coupling characteristics of choke loaded patch array antenna," *IEICE Trans. Commun.,* vol. E88-B, no.1 (January 2005): 411–415.
23. Y. Kimura, Y. Ebine, and T. Imai, "Relationship between vertical beam shaping and downlink capacity for W-CDMA base station antenna," EuCAP2006, ESA SP-626 (October 2006).
24. IEICE, *Antenna Engineering Handbook,* Tokyo: Ohmsha, 1980: 207.

New Unidirectional Antennas for Various Wireless Base Stations

Hang Wong and Kwai-Man Luk

City University of Hong Kong

6.1 Introduction

The greatest triumph in mobile communication service boosts the development of the third-generation (3G), WiFi, WiMax, UWB, and forthcoming 4G systems, and creates high demand for wideband unidirectional antennas to accommodate several wireless communication systems with excellent electrical characteristics such as wide impedance bandwidth, low cross-polarization, low back radiation, a symmetric radiation pattern, and stable gain over the operating band for cost-effectiveness, space utilization, and environmental friendliness. Among many types of antenna elements, there are at least three conventional means for implementing wideband low-profile antennas with a directional pattern. These approaches are (1) directed dipoles, (2) wideband patch antennas, and (3) complementary antennas. Dipole antennas are popularly used in wireless communication systems because of their several advantages: reasonable bandwidth, good radiation characteristics, ease of construction, and the possibility of obtaining directional or bidirectional radiation patterns.[1-3] Many researchers have focused on developing wideband dipole antennas for present and future communication systems because of the continuously expanding range of wireless telecommunication services for voice and data transmission. Some methods have been suggested to achieve wide impedance bandwidth for dipole antennas such as flared dipole arms,[4] bowtie-shaped dipoles,[5-6] flat dipoles,[7] and the use of parasitic elements.[8] These designs can obtain a bandwidth of more than 30% to 100%, but a wideband balun must be included.

However, wideband dipole antennas do not have a very stable radiation pattern over the operating bandwidth. Their radiation patterns can change substantially depending on the frequency.

The second popular unidirectional antenna is the microstrip/patch antenna. Many publications are available regarding the designs of wideband patch antennas, for instance, a patch with an L-probe feed[9–13] or an aperture coupled feed,[14–16] double resonances by stacked patches[17–21] or a U-slot patch[22–27], and so on. These techniques give the patch antenna a very wide impedance bandwidth (from 20% to 40%) enhancement, which is sufficient for covering many wireless communication systems. However, this class of wideband antennas[9–27] has several weaknesses: high cross-polarization and large variations in gain and beamwidth over the operating band. Although some techniques, such as anti-phase cancellation,[28] a twin-L probes coupled feed,[29] an M-probe feed,[30–34] and so on, were suggested for suppressing cross-polarization, these antennas still are weak in terms of gain and beamwidth variations with frequency as well as different beamwidths in the E- and H-planes.

In order to achieve stable radiation over the entire operating band, the third approach of using complementary antennas combining an electric dipole and magnetic dipole should be explored. This idea for a complementary antenna with equal E- and H-plane patterns was revealed in 1954 by Clavin.[35] An electric dipole has a figure-8 radiation pattern in the E-plane and a figure-O pattern in the H-plane, whereas a magnetic dipole has a figure-O pattern in the E-plane and a figure-8 in the H-plane. If both electric and magnetic dipoles can be excited simultaneously with appropriate amplitude and phase, a unidirectional radiation pattern with equal E- and H- planes can be obtained. A practical design was proposed by Clavin again in 1974.[36] Another design, which consists of a passive dipole placed in front of a slot, was also reported by King.[37] Similarly, this idea, based on a slot-and-dipole combination,[38–40] was realized by other investigators; nonetheless, all of these designs[35–40] are either narrow in bandwidth or bulky in structure. They may not fulfill the requirements for current wireless communication systems. Recently, a new wideband unidirectional antenna element composed of a planar dipole and a vertically oriented shorted patch antenna has been presented.[41] This antenna is based on the complementary concept of exciting the electric dipole and the magnetic dipole simultaneously. And it has many advantages, including a simple structure, wide bandwidth, low cross-polarization, a symmetrical radiation pattern, and in particular, very low back radiation. Because of the low back radiation, the gain and beamwidth of the antenna are not noticeably changed with frequency, and the gain and efficiency of the antenna are higher than that of many other antenna elements available in the literature. This antenna finds numerous applications in modern wireless communications.

This chapter focuses on the development of wideband directional antennas with low cross-polarization, a stable radiation pattern, steady gain, and a simple architecture. First, microstrip/patch antennas fed by a twin-L probe design and by an M-probe design are introduced. Then, a differential-plate fed patch antenna is demonstrated. Finally, a wideband complementary antenna composed of a planar dipole and a shorted patch antenna is discussed.

6.2 Patch Antennas

6.2.1 Twin L-Shaped Probes Fed Patch Antenna

One popular technique for enhancing the bandwidth of patch antennas is to use parasitic elements, as in co-planar[42–44] or stacked geometry.[17] Another common method for enhancing bandwidth is to employ a coaxial probe to feed a slotted patch on an electrically thick substrate (0.08–0.1λ) of low dielectric constant.[22,45] Such designs yield an impedance bandwidth up to 30% (SWR $\leq$ 2). However, no noticeable improvement in gain is seen. On the other hand, although the gain can be improved to around 10 dBi (estimated)[18] by enlarging the separation between the driven patch and the stacked parasitic patch to approximately 0.3λ, the trade-off is narrow in bandwidth, around 2–3%. The gain can also be increased in co-planar geometry by placing the parasitic patches adjacent to the fed patch to form an array,[46] but its gain-bandwidth is not wide enough for most applications. In addition, the major disadvantage of using parasitic elements is an unfavorably large antenna size. To enhance both bandwidth and gain for a single antenna element, a twin L-shaped probe fed patch antenna is suggested.[29] This idea of using parallel feeds[47] is a simple twin-feed structure with two in-phased L-probes[9] to further enhance the gain to 10 dBi while maintaining wideband performance. More importantly, the achieved 1-dB-gain bandwidth is wide enough to cover the operating bandwidth.

A twin L-shaped probe fed patch antenna that operates at a center frequency of 5 GHz is shown in Figure 6.1. The copper patch has a thickness of 0.3 mm (0.005λ), a width W = 44 mm (0.733λ), and a length L = 22 mm (0.367λ). The aspect ratio $\frac{W}{L}$ of the patch is equal to 2.0. The patch is supported by two small cubic foam-spacers ($\varepsilon_r \cong 1$) of thickness H = 6 mm (0.1λ). The fundamental mode (TM$_{01}$) of the patch is simultaneously excited by the two in-phased L-shaped probes (with probe radius = 0.5 mm), which are separated by S = 28.6 mm (0.477λ) and connected to the microstrip feed network mounted on the other side of the ground plane. The simple T-shaped power divider is etched on a Duriod substrate with ε_r = 2.33 and a thickness of 1.5748 mm (0.062 inch). The square

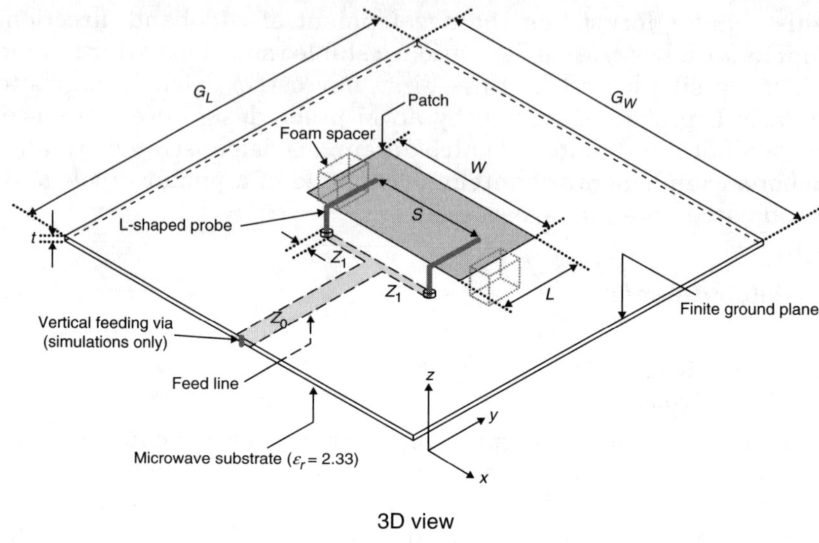

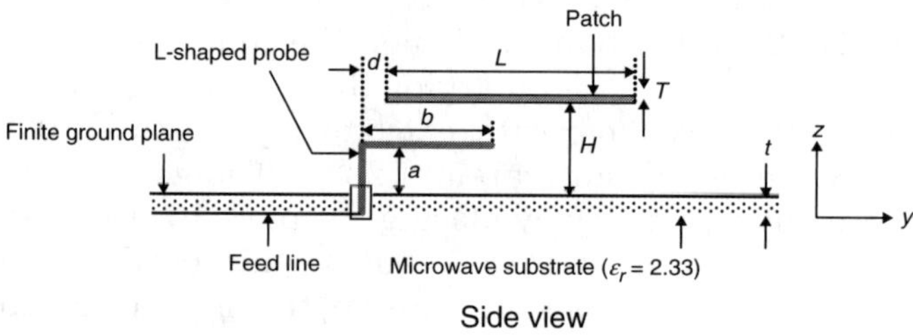

Parameters	L	H	W	T	a	b	v	d	S	t	G_L	G_W
Value/mm	22	6	44	0.3	4.5	12	2	0	28.6	1.5748	100	100
	(0.367λ)	(0.1λ)	(0.733λ)	(0.005λ)	(0.0075λ)	(0.2λ)	(0.033λ)	(0λ)	(0.477λ)	(0.026λ)	(1.667λ)	(1.667λ)

Figure 6.1   Geometry of twin L-probe coupled patch antenna: 3D view and side view[29] (© 2005 IEEE)

ground plane has a side length of 100 mm (1.667λ). The two L-shaped probes have the same size vertical arm, $a = 4.5$ mm (0.075λ), and horizontal arm, $b = 12$ mm (0.2λ). The characteristic impedance of the feed line is $Z_0 = 50$ Ω and $Z_1 = 100$ Ω, with a line width of 4.877 mm and 1.41 mm, respectively. A SMA launcher is connected to the end of the feed line.

In order to demonstrate the effectiveness of the twin-feed design, two prototypes are fabricated for comparison: the single L-probe fed and the twin L-probe fed patch antennas. With zero inserted distance ($d = 0$) again, the arm lengths of this single L-probe are slightly adjusted ($a = 3.5$ mm and

$b = 8$ mm) for achieving the best impedance matching. The performances of the two prototypes are measured by an HP8510-C Network Analyzer, a compact range with ORBIT/FR MiDAS Far-Field Antenna Measurement and Analysis System. As for the gain measurement, a NARDA-643 standard gain horn is used.

Figure 6.2 shows comparisons of the measured results of gain and SWR of both prototypes. As seen from the SWR curves, both antennas have a wide impedance bandwidth of 25% (SWR $\leq$ 1.5) from 4.42 to 5.7 GHz. From the gain curves, obviously the proposed antenna has a better gain of 10 dBi, which is stable across the operating bandwidth with a 1-dB-gain bandwidth of 26%. Compared with the maximum gain of the single-L-probe coupled patch of 8 dBi, there is an improvement of about 2 dB. The improvement is much more significant at the upper region, 5–5.7 GHz, of the operating band, about 3- to 6-dB difference in gain is observed. The increase in gain is mainly due to the effective suppression of cross-polar radiation.

Figure 6.3 shows the measured radiation pattern at 5.0 GHz for both antennas. The cross-polar level in the H-plane of the single L-probe coupled patch is relatively higher than the twin L-probe case. This higher cross-polar radiation is due to a stronger high-order mode radiation from the patch as well as unwanted radiation from the vertical arm. However, there is significant suppression in the cross-polar radiation when the twin L-probe fed technique is employed. The cross-polarization can be suppressed to less than −20 dB. For both antennas, the broadside patterns are stable across the operating band, and the average beamwidth in the H-plane is around 56°, which is slightly narrower than that in the E-plane of around 60°.

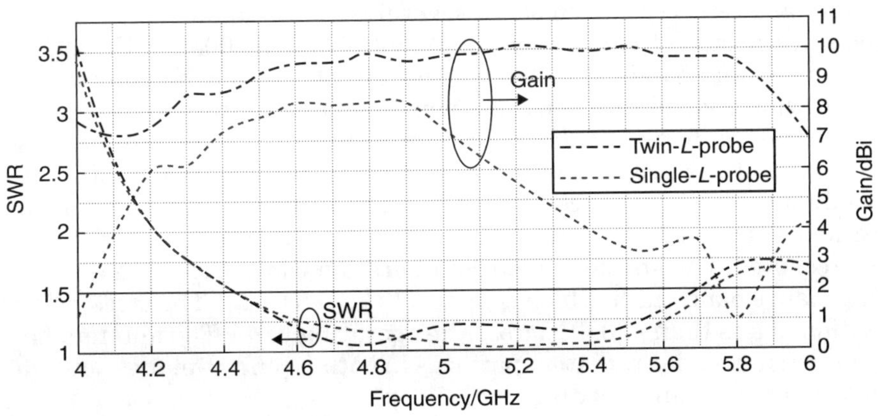

Figure 6.2 Measured results of SWR and gain of both single and twin L-probe cases[29] (© 2005 IEEE)

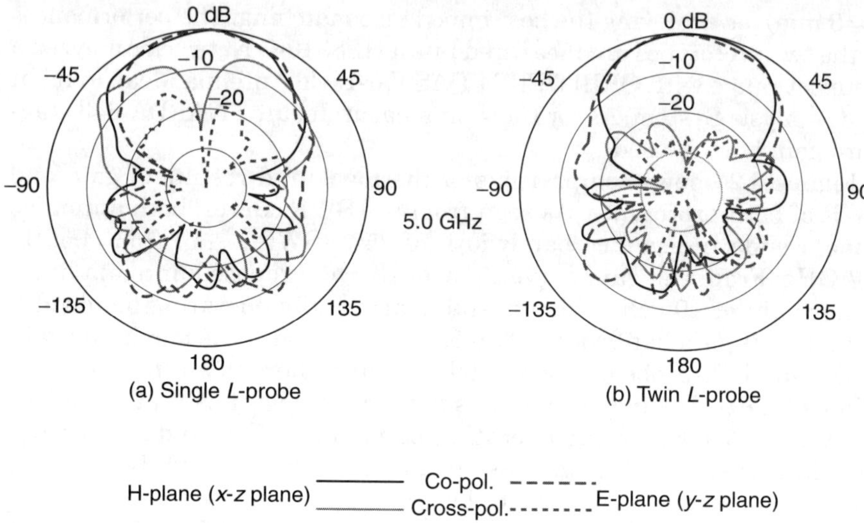

Figure 6.3   Measured radiation patterns at 5.0 GHz of (*a*) the single L-probe coupled patch antenna and (*b*) the twin L-probe coupled patch antenna[29] (© 2005 IEEE)

6.2.2   Meandering-Probe Fed Patch Antenna

A meandering-probe fed patch antenna was first proposed in 2004.[30,31] This wideband feeding mechanism not only can enhance the impedance bandwidth of a patch antenna, but also can suppress the cross-polarization level of the antenna. By aligning the center of the meandering probe with the center of the patch, the radiation pattern is symmetric in both E- and H-planes across the operating band. This antenna also has low back radiation and high gain characteristics. In some applications, antennas are necessary to serve several wireless communication systems at the same time, such as GSM1800, CDMA1900, and IMT-2000. The impedance bandwidth of meandering-probe fed patch antennas[32–34] is not wide enough to cover these wireless communication systems, so techniques to further enhance the impedance bandwidth are necessary. One of the proper solutions of using stacked radiating elements to enhance the impedance bandwidth of the patch antenna is demonstrated in this section. A meandering-probe fed stacked patch antenna is presented. The stacked patch configuration can greatly enhance the impedance bandwidth, which is up to 37% (SWR < 1.5). The cross-polarization is less than −20 dB across the operating band. The antenna has symmetric co-polar radiation patterns in both E- and H-planes, and the antenna has a gain of 9 dBi.

The geometry of the meandering-probe fed stacked patch antenna[33] is shown in Figure 6.4. The center frequency of the proposed antenna is chosen at 1.975 GHz, and the antenna dimensions are selected after

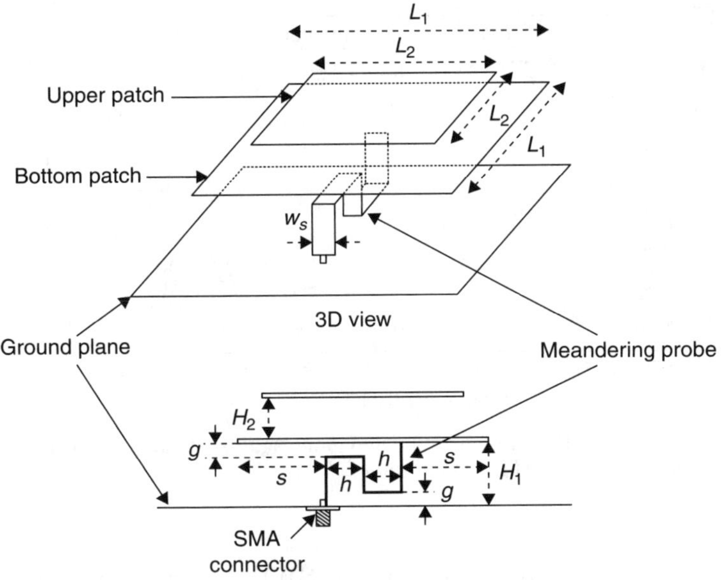

Parameters	L_1	L_2	H_1	H_2	s	h	g	w_s
Value/mm	61 $(0.4\lambda_0)$	49 $(0.32\lambda_0)$	15.5 $(0.1\lambda_0)$	10.5 $(0.066\lambda_0)$	21 $(0.138\lambda_0)$	8.75 $(0.058\lambda_0)$	1.5 $(0.01\lambda_0)$	10 $(0.066\lambda_0)$

Figure 6.4 Geometry of the meandering-probe fed stacked patch antenna: 3D view and side view[33]

a detailed parametric study of the antenna's performance. The upper and lower aluminum patches, where both have thicknesses of 1 mm, are square shaped and have lengths $L_2 = 49$ mm $(0.32\lambda_0)$ and $L_1 = 61$ mm $(0.4\lambda_0)$, respectively. The upper patch is a parasitic element, whereas a meandering probe feeds the lower patch. The height of the lower patch H_1 is 15.5 mm $(0.1\lambda_0)$, and the separation between the upper patch and the lower patch H_2 is 10.5 mm $(0.066\lambda_0)$. The patches are supported by foam spacers for testing. In real applications, they can be supported by plastic posts. The meandering probe, which has a rectangular cross-section with a thickness of 0.5 mm and a width $w_s = 10$ mm $(0.066\lambda_0)$, has one end connected to a 50-ohm SMA connector and the other end connected to the lower patch with a plastic screw. The case of the meandering probe is aligned to the center of the patches and the ground plane. Referring to Figure 6.4, the dimensions of the meandering probe are $g = 1.5$ mm $(0.01\lambda_0)$, $h = 8.75$ mm $(0.058\lambda_0)$, and $s = 21$ mm $(0.138\lambda_0)$. The aluminum ground plane dimensions are $G_W \times G_L = 200 \times 300$ mm² $(1.32\lambda_0 \times 1.97\lambda_0)$, where G_W is the width of the ground plane and G_L is the length of the ground plane. The antenna is excited in TM_{01} mode.

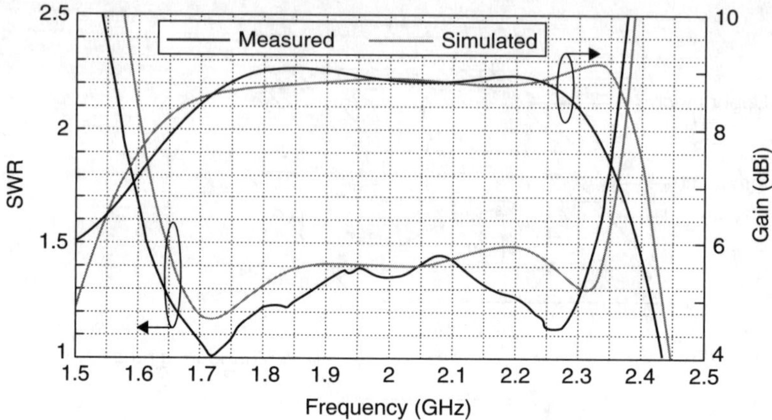

Figure 6.5 Measured and simulated SWRs and gains against frequency of the meandering-probe fed stacked patch antenna[33]

Measured and simulated results of SWR and gain are shown in Figure 6.5. The antenna operates from 1.575 to 2.36 GHz with a measured bandwidth of 40% (SWR < 2). If the commercial standard of SWR < 1.5 is used as a reference, the impedance bandwidth of the antenna is 37% (from 1.6 to 2.325 GHz). The antenna has a stable gain of about 9 dBi across its operating band. The measured and simulated SWRs and gains are in good agreement. Figure 6.6 shows the simulated and measured radiation patterns of the antenna at 1.6 GHz, 1.9 GHz, and 2.3 GHz. The radiation pattern of the antenna was measured by a near-field antenna measurement system with ORBIT/FR MiDAS Cylindrical Near-Field Antenna Measurement and Analysis System. The far-field radiation patterns are transformed from the measured near-field patterns by the FFT available from the measurement system. It is observed that both the simulated and measured cross-polarization levels are less than −20 dB. The E-plane co-polar radiation pattern of the proposed antenna is also symmetric. The backlobe radiation is less than −20 dB within the operating band. The measured 3-dB beamwidths are 66° in the E-plane and 76° in the H-plane at the center frequency, which is in good agreement with the simulated 3-dB beamwidths of 64° in the E-plane and 78° in the H-plane. This antenna is particularly suitable for wideband antenna array design.

6.2.3 Differential-Plate Fed Patch Antenna

In the interest of realizing wideband antennas, *differential-fed patch antennas (DFPA)* have been studied and developed.[48–49] The mutual capacitance between the two adjacent feeding probes, together with the

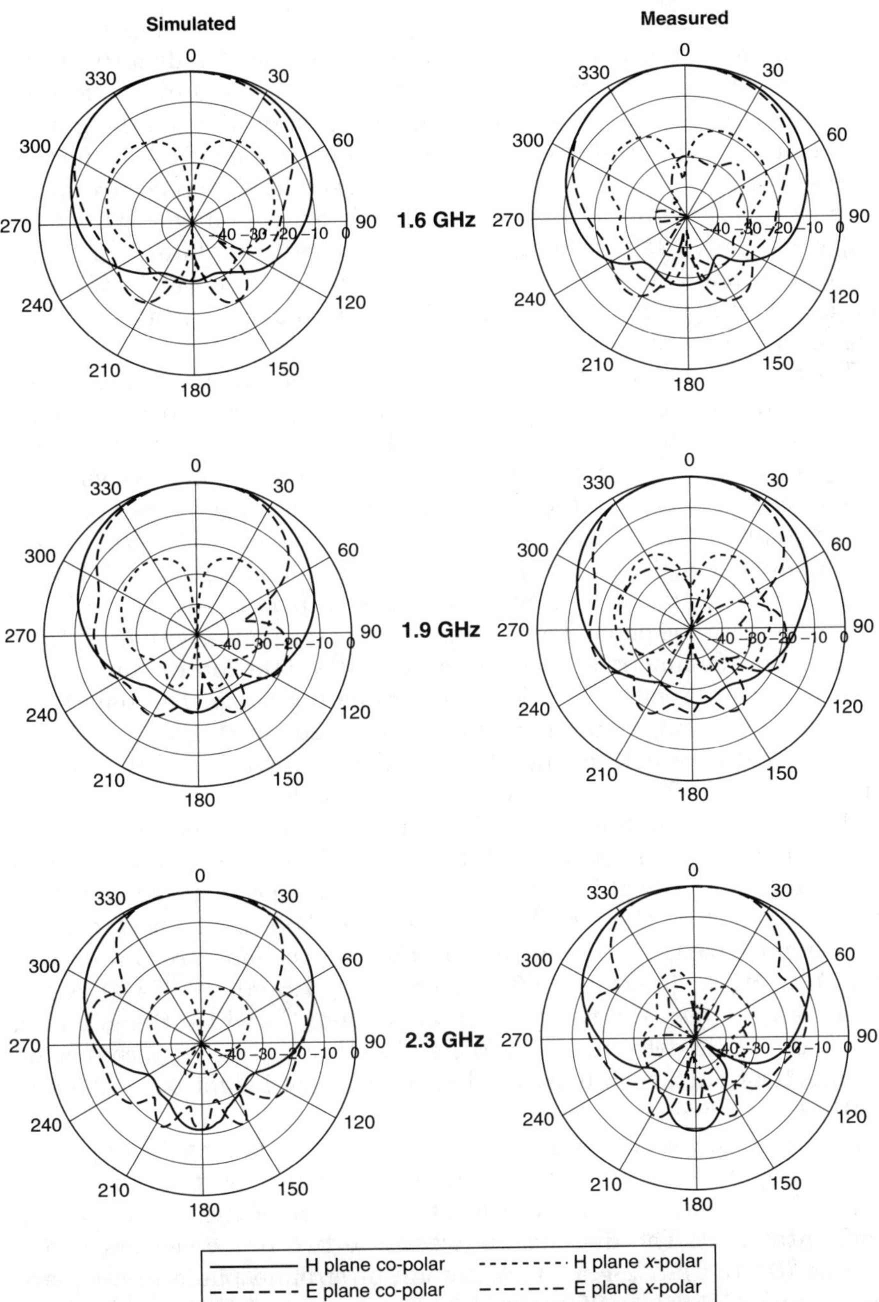

Figure 6.6 Simulated and measured radiation patterns of the meandering-probe fed stacked patch antenna[33]

capacitance induced by the patch slots,[48] can effectually compensate for the probe inductance, resulting in wide impedance bandwidth. In the meantime, the unwanted cross-polar radiation excited by the feeding probes is suppressed due to the anti-phase current flow on the probes. Subsequently, as a refinement of the concept of the DFPA for further bandwidth broadening, a new approach designated as *a vertical-plate pair as a differential feeding scheme* has been presented.[48] This technique is to modify the feeding probe into the vertical plate. As a result, impedance bandwidth is significantly enhanced. Moreover, by virtue of the differential feed, excellent radiation performance is attained within the whole operating band.

The basic geometry of the broadband patch antenna with a vertical plate pair[50] is illustrated in Figure 6.7. The proposed antenna is designed with the center frequency chosen at $f_o = 2.17$ GHz ($\lambda_o = 138.25$ mm). It consists of a planar patch, a ground plane, and a pair of vertical plates connected to the feeding probes of the 50 Ω SMA launchers. The copper patch is rectangular in shape with $P_L = 68$ mm ($0.492\lambda_o$) and $P_W = 60$ mm ($0.434\lambda_o$). It is located over the ground plane at a height of $h_p = 16$ mm ($0.116\lambda_o$). The patch is directly driven by a pair of planar plates that are vertically placed beneath the patch. The feeding edges of the plates, with $L = 43$ mm ($0.311\lambda_o$), are symmetrically located on two sides of the central line (y-axis) along the nonresonant direction of the radiating patch. The separation between the feeding edges is selected to be $s = 26$ mm ($0.188\lambda_o$).

Slightly different from the aforementioned vertical-plate fed patch antenna, this DFPA employs feeding plates folded at the bottom. Each folded plate is composed of a vertical part and a horizontal part. The capacitance due to each horizontal part and the ground plane offers an additional design flexibility to accomplish good impedance matching. By properly selecting the lengths and widths of both parts, performance can be optimized. The dimensions of the two sections are $L = 43$ mm ($0.311\lambda_o$), $W_V = 13.5$ mm ($0.098\lambda_o$), and $W_H = 7.5$ mm ($0.054\lambda_o$). A gap is formed between the two horizontal portions, in which the gap size, denoted as g, is determined by W_H and s. Having one edge connected to a vertical part, each horizontal part, with the height of $t = 2.5$ mm ($0.018\lambda_o$), is driven by a probe.

Resorting to the use of a differential feed, two probes, having radii of 0.5 mm, are directly connected to the horizontal parts of the folded plates. Each contact point locates at the center of the corresponding horizontal part. The distance between the two probe centers is $d = 20$ mm ($0.145\lambda_o$). A 2-mm thick aluminum ground plane in a square shape with $G_L = G_W = 250$ mm ($1.808\lambda_o$) is utilized. The center of the antenna is aligned with that of the ground plane, resulting in the symmetric configuration with reference to both the x-axis and y-axis. Detailed dimensions are depicted in Figure 6.7.

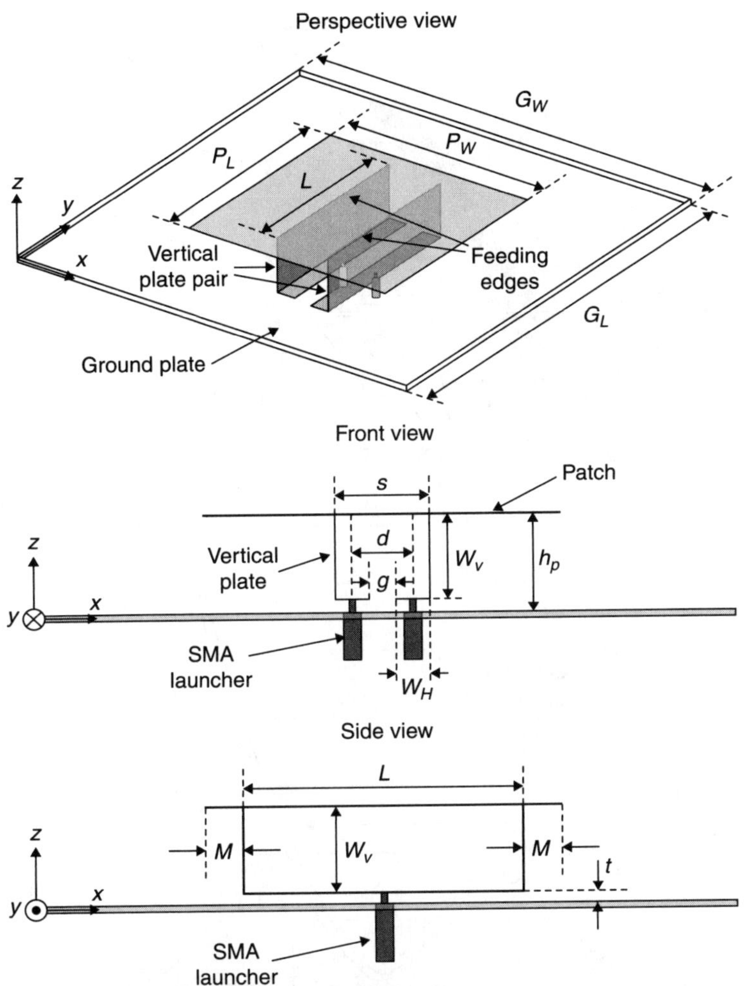

P_W	P_L	$G_L = G_W$	h_p	L	M	W_H	W_V	S	g	d	t
60.0 $(0.43\lambda_0)$	68.0 $(0.49\lambda_0)$	350 $(2.53\lambda_0)$	16.0 $(0.12\lambda_0)$	43.0 $(0.31\lambda_0)$	12.5 $(0.09\lambda_0)$	7.5 $(0.05\lambda_0)$	13.5 $(0.10\lambda_0)$	26.0 $(0.19\lambda_0)$	11.0 $(0.08\lambda_0)$	20.0 $(0.14\lambda_0)$	2.5 $(0.02\lambda_0)$

Figure 6.7 Geometry of the proposed DFPA driven by a vertical-plate pair[50] (© 2007 IEEE)

For the sake of feeding the antenna with a differential signal, a wideband 180° power divider, which was introduced in Itoh and Cheng,[51] was utilized to transform a single-ended signal to two signals, in which one travels out-of-phase with the other. As described in Figure 6.8, the two output ports of the divider are designated as Port 2 and Port 3. Port 2 and Port 3 are then connected to the antenna via coaxial cables.

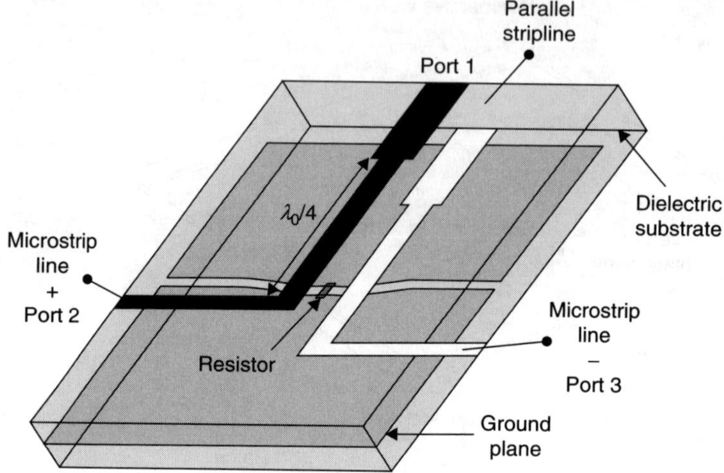

Figure 6.8 Geometry of the wideband 180° out-of-phase power divider[51]

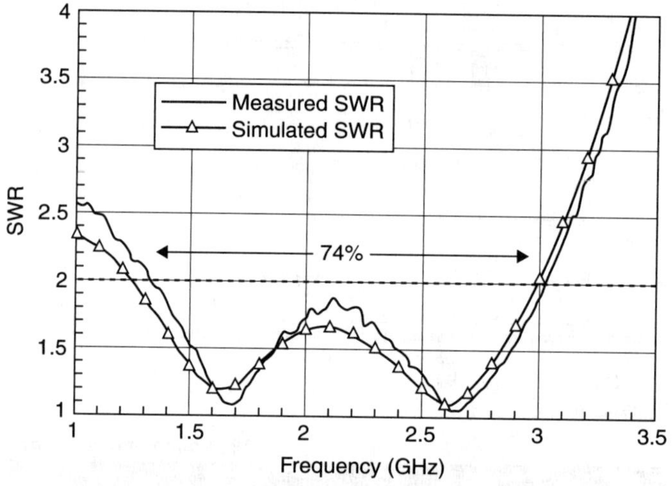

Figure 6.9 Simulated and measured SWRs of the proposed antenna[50]
(© 2007 IEEE)

Figure 6.9 reveals that the proposed antenna can operate from 1.37 to 2.97 GHz with a measured impedance bandwidth (SWR ≤ 2) of 74%. In Figure 6.10, the resonances of two modes are also obtained for the proposed DFPA. Over the operating band, ranging from 1.37 GHz to 2.97 GHz, the input resistance varies from 30 Ω to 63 Ω, while the input reactance changes interiorly between −10 Ω and 27 Ω. In Figure 6.11, the antenna gain is stable across the operating band and has an average value of 8.5 dBi. The performance of high and stable antenna gain is mostly owing to the high polarization purity.

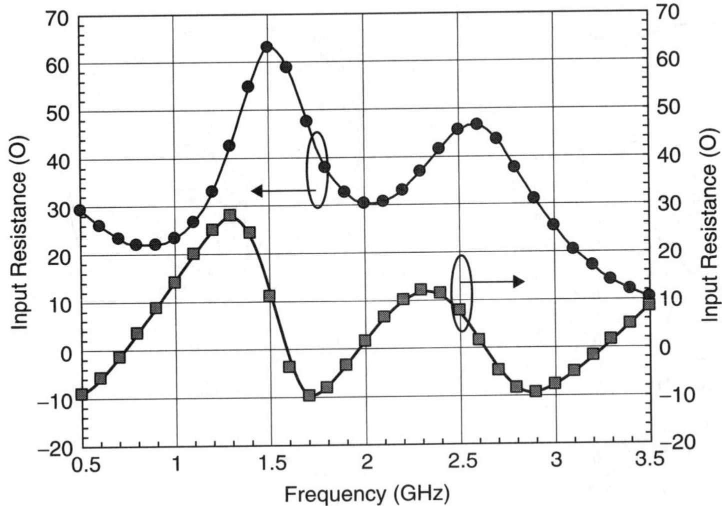

Figure 6.10 Simulated input impedance of the designed DFPA[50] (© 2007 IEEE)

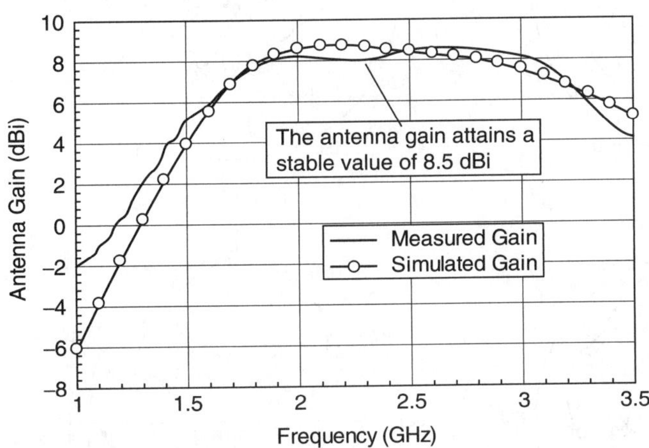

Figure 6.11 Simulated and measured antenna gains of the devised DFPA[50] (© 2007 IEEE)

It is well understood that for the vast majority of broadband patch antennas, radiation performance is inevitably degenerated, especially in the E-plane, at higher operating frequencies. By making use of the differential plate feeding scheme, the instability issue is evidently ameliorated. Simulated and measured radiation patterns of the proposed antenna in both E-plane (x-z plane, $\phi = 0°$) and H-plane (y-z plane, $\phi = 90°$) at 1.37 GHz, 2.17 GHz, and 2.97 GHz are illustrated in Figure 6.12, showing

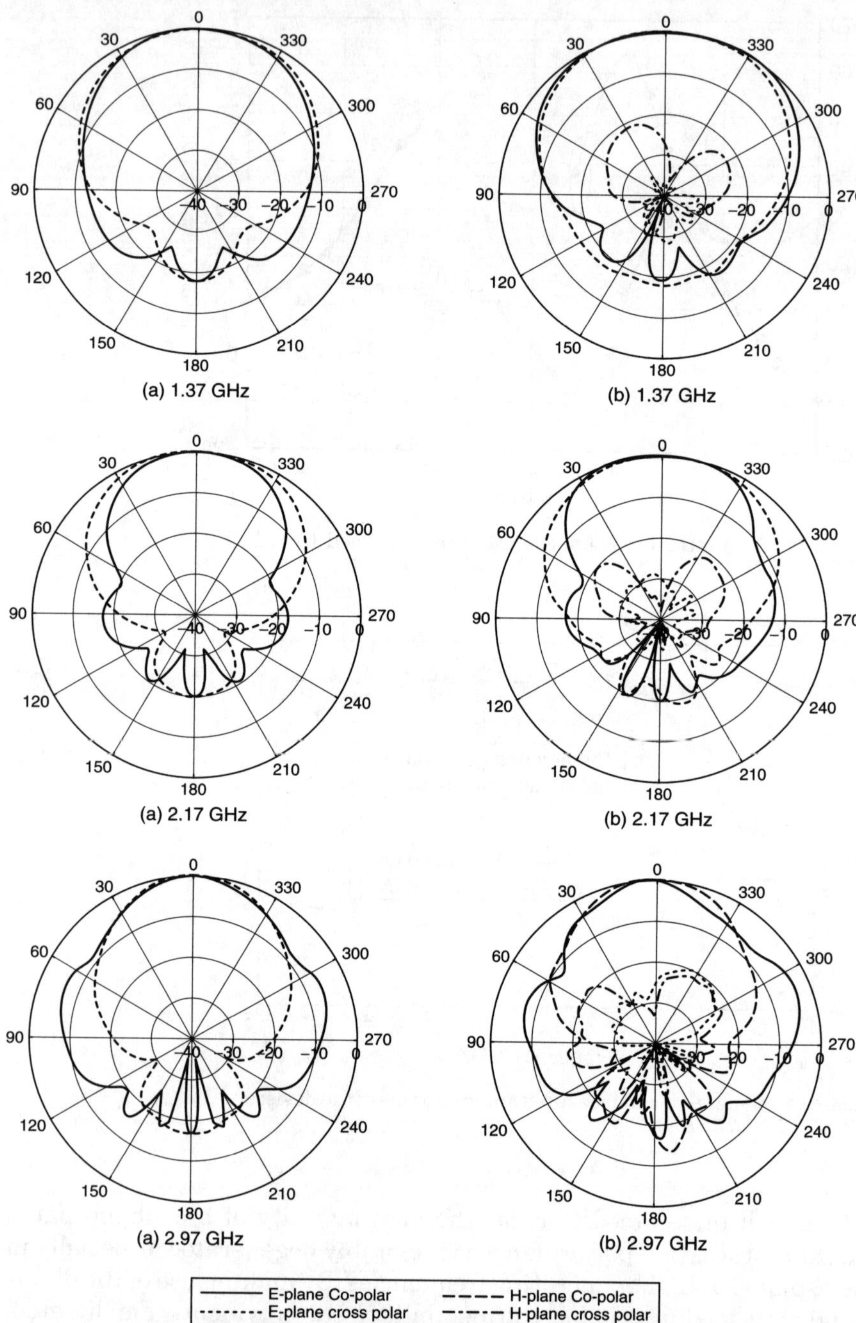

Figure 6.12 (a) Simulated and (b) measured radiation patterns of the patch antenna fed by a vertical plate pair at different frequencies[50] (© 2007 IEEE)

that the radiation characteristics are stable across the band. Obviously, the measured cross-polarization levels are around 20 dB less than the co-polarization levels. The simulated cross-polarization in both E-plane and H-plane is vanishingly small across the operating band under the ideal conditions and is thereby not able to be observed. In addition, taking advantage of the structural symmetry, the co-polarized radiation patterns in both E-plane and H-plane are symmetric with respect to the broadside direction within the band. Noteworthy, the measured maximum power is always in the broadside direction, which offers benefits to many wireless communications systems, in which stable radiation patterns and the purity of wave polarization within operating bandwidths are required. In addition, the backlobe radiation level is less than -15 dB.

6.3 Complementary Antennas Composed of an Electric Dipole and a Magnetic Dipole

After several decades of developing wideband antennas, many simple and wideband antenna elements have flourished and contributed to real applications in many wireless communication systems. Among many categories of antennas, dipole and patch antennas are the two popular types. They have been diversified into many wideband antenna designs. The basic structures of these two types of wideband antennas are low in profile, easily fabricated, and convenient for having directional radiation patterns. Nonetheless, they have the disadvantages of large variations in gain and beamwidth over the operating bandwidth, different beamwidths in the E- and H-planes, and strong radiation in the backlobe. In this section, the design of wideband unidirectional antennas with low cross-polarization, low back radiation, symmetric E- and H-plane patterns, and stable gain is introduced. To accomplish this mission, a wideband complementary antenna consisting of an electric dipole and a magnetic dipole has been developed. This concept of exciting two complementary sources was first initiated in 1954 for achieving equal E- and H-plane patterns. To realize this idea, much subsequent research has been done and has resulted in similar antenna performances. Most of these suggest using a combination of either slot-dipole or slot-monopole to form complementary antennas with a directional pattern that is equal in the E- and H-planes. However, they are still complicated in structure or are narrow in bandwidth. To overcome these problems, a new wideband antenna element that is formed by combining a planar dipole antenna and a shorted patch antenna is acquainted. This new wideband element achieves excellent performance in all electrical parameters. In particular, its low back radiation characteristic makes it highly attractive for developing various kinds of indoor and outdoor base station antennas for modern cellular communications. This is because the interference between different cells operating at the same frequency can be reduced substantially. Due to their wideband characteristics and desirable radiation patterns, they can easily find conceivable applications for the current wireless communication systems like GSM1800/1900, 3G, WiFi, WiMax, ZigBee, and so on.

6.3.1 Basic Principle

To provide more detailed understanding of a complementary antenna consisting of an electric dipole and a magnetic dipole for achieving a unidirectional symmetric radiation pattern, this section begins with a review of the characteristics of several previous works available in the literature. The approach of exciting an electric dipole and a magnetic dipole simultaneously for achieving an identical E- and H-plane radiation pattern was first revealed by Clavin in 1954.[35] Figure 6.13 shows his proposed idea about two sources that have complementary types of radiation characteristics oriented at right angles to one another. These two sources can be realized by using an electric dipole and the open end of a waveguide, which is mentioned in Clavin.[35] Figure 6.13 describes schematically how the field patterns of two complementary sources would appear. An electric dipole and a magnetic dipole are separated by a specific distance, which can be used to control the proper amplitude and phase of two complementary sources so they can perform with equal E- and H-plane radiation patterns. The electric dipole has a figure-8 radiation pattern in its E-plane and a figure-O pattern in the H-plane, but the magnetic dipole has a figure-O pattern in the E-plane and a figure-8 in the H-plane. As a result, the advantages of combining two

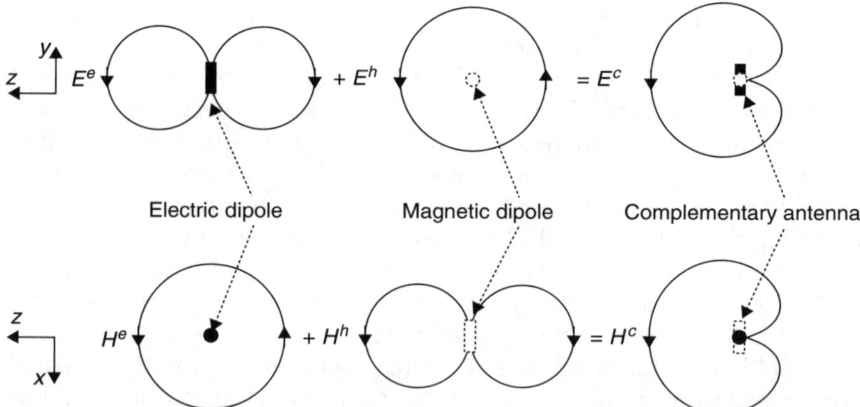

Figure 6.13 Basic principle of a complementary antenna consisting of an electric dipole and a magnetic dipole

complementary sources—(1) equal E- and H-plane patterns, (2) low back radiation, (3) low cross-polarization radiation, and (4) stable gain across the operating bandwidth obtained—are accomplished.

6.3.2 Complementary Antennas Composed of Slot Antenna and Parasitic Wires

Apart from Clavin's method,[35] some studies[36–40] were conducted for demonstrating the complementary antenna approach. King[37] deliberated modifying a slot antenna by placing a passive dipole in front of the slot. Then Gabriel[39] and Wilkinson[40] subsequently used a slot-dipole antenna to implement the complementary antenna. These antennas[36–40] are all based on combining a slot antenna and a dipole to achieve the goal.

Figure 6.14 shows one of the antennas presented by Clavin again in 1974.[36] Two parasitic inverted-L wires are placed beside a rectangular slot antenna. This arrangement is equivalent to providing a magnetic dipole from the slot and an electric dipole from the inverted-L wires. The configuration of this structure is very simple. The achievable bandwidth is about 10–20%.

6.3.3 Complementary Antennas with a Slot Antenna and a Conical Monopole

In addition to the slot-dipole antenna, some have proposed slot-monopole antennas[52–53] for realizing similar E- and H-plane patterns. Itoh[52] utilized a crossed slot and a monopole element to obtain steerable cardioid patterns. Mayes[53] employed a conical monopole with a slot for studying a broadband unidirectional antenna. Figure 6.15 depicts the antenna presented by Mayes.[53] The conical monopole is placed with an offset distance from the center of the slot, such that the proper amplitude and phase of these two radiating elements can be adjusted to provide a desirable directional pattern. This antenna has a very wide impedance bandwidth that is about 60%.

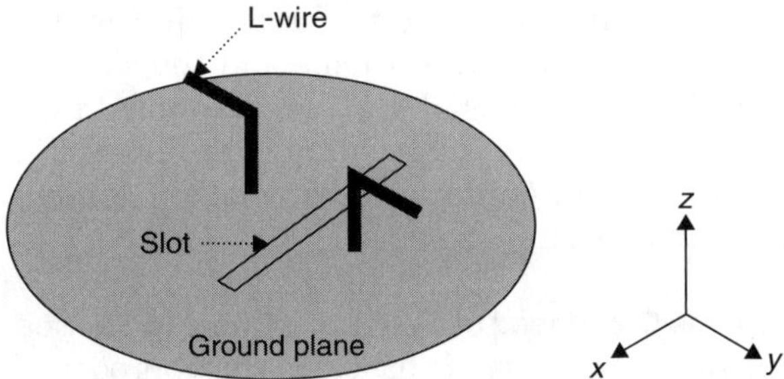

Figure 6.14 Configuration of a complementary antenna with a slot and inverted-L wires[36]

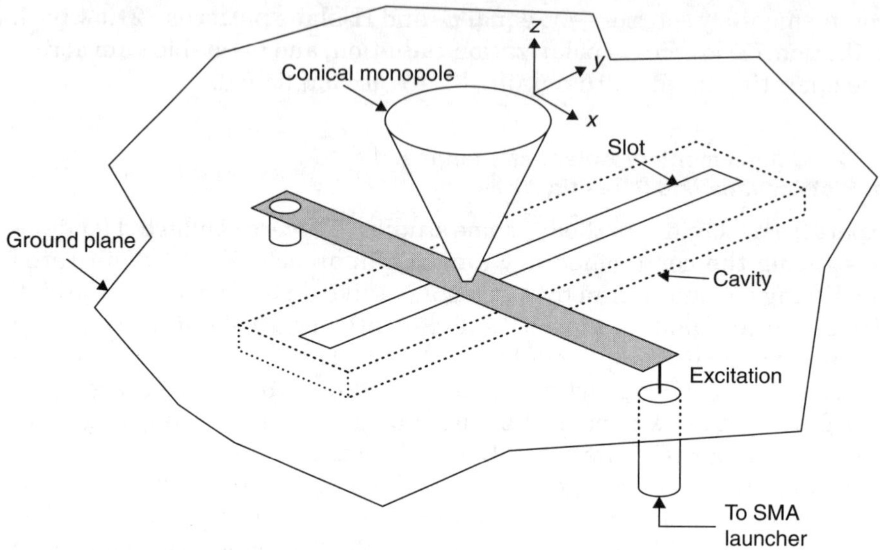

Figure 6.15   Configuration of a monopole-slot antenna[53]

6.3.4   New Wideband Unidirectional Antenna Element

A novel wideband antenna with unidirectional radiation pattern is presented in this section. This newly designed antenna has many advantages, including a simple structure, wide bandwidth, low cross-polarization, a symmetrical radiation pattern, and in particular, extremely low back radiation. The gain and beamwidth of the antenna are almost constant over the operating frequency band. The antenna can be simply developed by using two metallic L-shaped plates with a Γ-shaped feed. Such architecture offers an advantage of forming a vertically oriented shorted patch antenna and a planar dipole. With the presence of the shorted patch, a magnetic dipole can be implemented, while an electric dipole can be realized together through the planar dipole. In this design, the antenna has a wide impedance bandwidth due to the Γ-strip coupled line and the double resonance from the planar dipole and the shorted patch antenna. Because the Γ-feed consists of an L-strip coupled line, which is a well-known bandwidth-broadening technique for microstrip antennas and short-circuited patch antennas, achieving a wide impedance bandwidth for the proposed antenna is easy.

6.3.4.1   Complementary Antennas Composed of a Vertically Oriented Shorted Patch and Planar Dipole   The proposed design is based on the approach of combining an electric dipole antenna and a magnetic dipole antenna. From among many choices of electric dipoles, a planar dipole antenna has been chosen, as shown in Figure 6.16a, whereas a wideband short-circuited

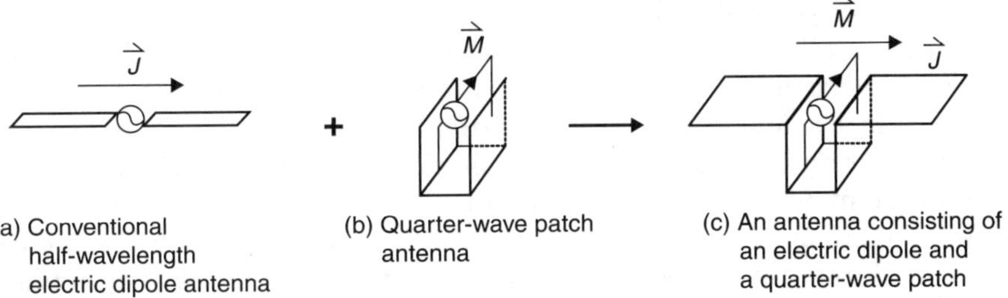

(a) Conventional half-wavelength electric dipole antenna

(b) Quarter-wave patch antenna

(c) An antenna consisting of an electric dipole and a quarter-wave patch

Figure 6.16 Principle of a wideband complementary antenna element for unidirectional radiation[41]

patch antenna has been selected as the magnetic dipole depicted in Figure 6.16b. To combine these two antennas, the short-circuited patch is placed vertically and is connected to the planar dipole, as illustrated in Figure 6.16c. Based on this approach, a new wideband antenna has been developed and its geometry is shown in Figure 6.17. This antenna operates at the center frequency of 2.5 GHz. Each side of the planar dipole has a width $W = 60$ mm (0.5λ) and a length $L = 30$ mm (0.25λ). The shorted patch antenna has a length $H = 30$ mm (also close to 0.25λ). For wideband operation, the separation of the two vertical plates, $S = 17$ mm, of the shorted patch antenna should be close to 0.14λ and the width of the dipole and the patch W should be around 0.5λ. The size of the ground plane can be used to adjust the back radiation. The optimum dimensions of the ground plane are 160 mm × 160 mm (1.3λ by 1.3λ).

To excite the antenna, an Γ-shaped probe feed is employed. This feed consists of three portions, which are made by folding a straight metallic strip of rectangular cross-section into a Γ-shape. The first portion, which is vertically oriented, has one end connected to a coaxial launcher mounted below the grounded plane. This portion acts as a

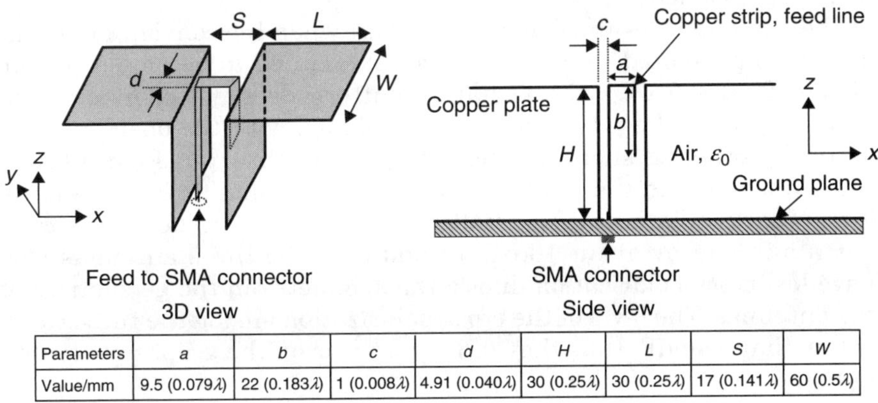

Parameters	a	b	c	d	H	L	S	W
Value/mm	9.5 (0.079λ)	22 (0.183λ)	1 (0.008λ)	4.91 (0.040λ)	30 (0.25λ)	30 (0.25λ)	17 (0.141λ)	60 (0.5λ)

Figure 6.17 Configuration of a wideband complementary antenna with a planar dipole and a vertically oriented shorted patch[41]

50 Ω air microstrip line for transmitting electrical signal from the coaxial launcher to the second portion of the feed. The second portion, which is located horizontally, is responsible for coupling the electrical signal to the planar dipole and the shorted patch antenna. The input resistance of the antenna is controlled by the length of this portion. This portion is equivalent to an inductive reactance, which causes the antenna to be totally mismatched. The third portion incorporated with the second vertical plate forms an open circuited transmission line. The equivalent circuit of this line is a capacitor. By selecting the appropriate length for this portion, its capacitive reactance can be used to compensate for the inductive reactance caused by the second portion.

In the conventional designs of complementary antennas,[35–40,52,53] both electric dipoles and magnetic dipoles are excited and separated with a specific distance (around 0.25λ) for controlling the proper amplitude and phase of the two radiating sources. On the contrary, in our proposed design, two dipoles are presented with a zero separation, as shown in Figure 6.16. The unique structure of this proposed antenna results in forming an inverted-U equivalent magnetic current from the aperture of the vertically oriented patch antenna. Therefore, the width of the antenna, W, is one of the key parameters for adjusting the proper amplitude and phase of the two current elements for achieving equal E- and H-plane radiation patterns.

A comparison of the simulated radiation patterns of a thin dipole, a planar dipole, and a wideband complementary antenna is depicted in Figure 6.18. The three antennas have the same ground plane size (160 mm × 160 mm) as well as the same antenna height (30 mm, 0.25λ). The length of the dipole, L, for the three cases is chosen to be 60 mm, which is equal to 0.5λ at the operating frequency of 2.5 GHz. For cases Figure 6.18a and b, the antennas are excited by a conventional coaxial cable with a balun; for Figure 6.18c, the antenna is excited by a Γ-shaped strip feed.

The simulated results demonstrate that when the conventional thin dipole, as presented in Figure 6.18a, is modified to become a planar dipole (Figure 6.18b), the radiation pattern does not change much. However, when the planar dipole is combined with the open end of a vertically oriented shorted patch antenna, as shown in Figure 6.18c, the beamwidths in the E- and H-planes become similar. Moreover, the level of back radiation is also smaller than the cases without the shorted patch antenna by about 10 dB. In addition, the three antennas also have low cross-polarization due to the symmetry in the geometries of the antennas. The level of the cross-polarization among the three cases is less than −40 dB, thus they did not appear on the graphs presented in Figure 6.18.

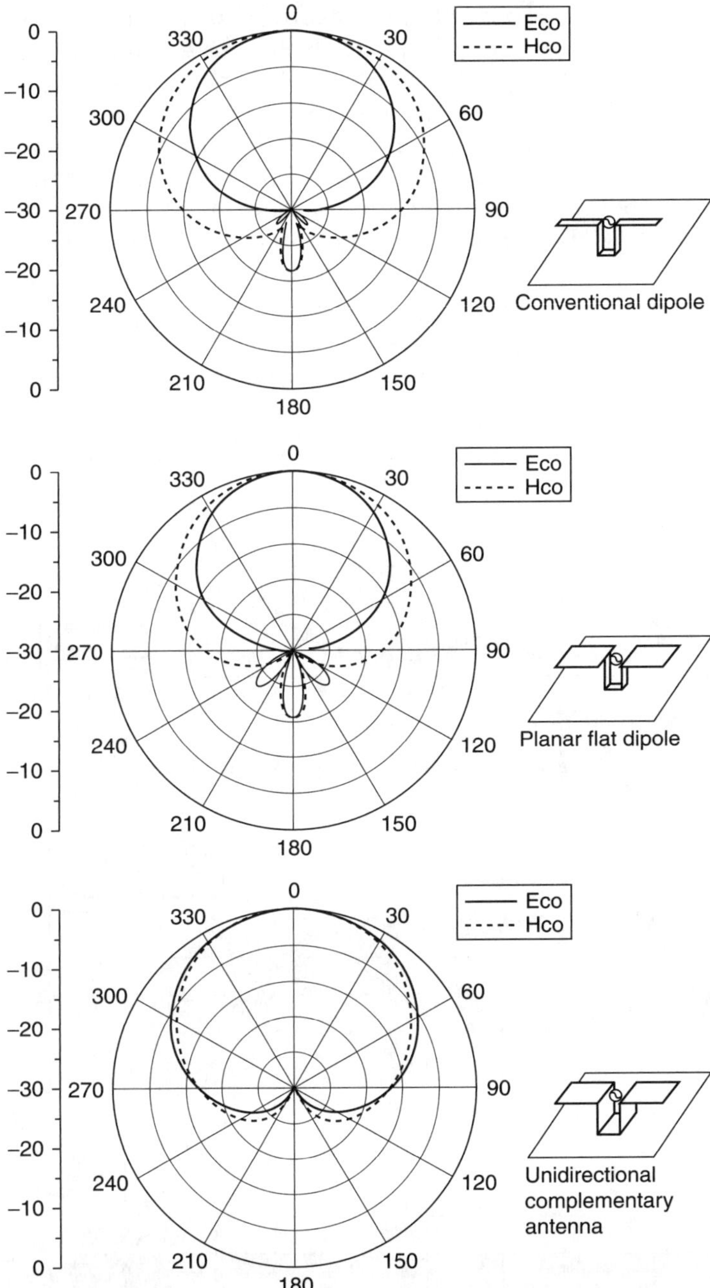

Figure 6.18 Radiation patterns for conventional dipole, planar dipole, and unidirectional complementary antenna at the center frequency of 2.5 GHz[41]

Figure 6.19 shows a comparison of the measured and simulated SWRs of the proposed antenna. As seen from the SWR curves, the antenna has a wide impedance bandwidth of 52% (SWR $\leq$ 2) from 1.75 to 3.0 GHz. Figure 6.20 illustrates the measured and simulated gain curves of the antenna. It can be observed that the proposed antenna has an average gain of approximately 8 dBi, varying from 7.5 dBi to 8.2 dBi across the operating bandwidth. Radiation patterns at frequencies of 1.75, 2.5, and 3 GHz were measured and are shown in Figure 6.21.

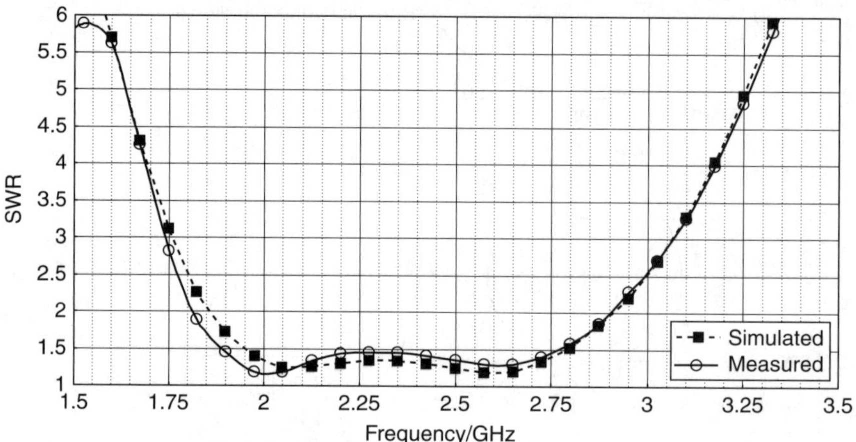

Figure 6.19 SWR against frequency for a wideband complementary antenna[41]

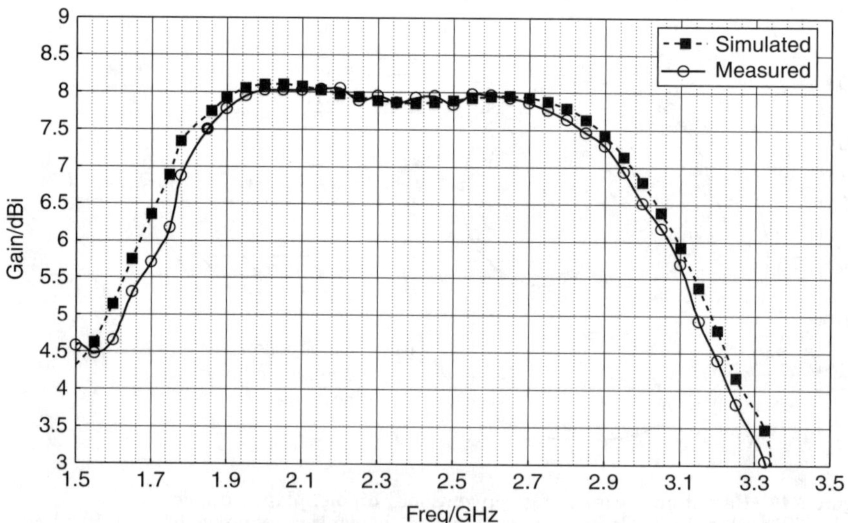

Figure 6.20 Antenna gain against frequency for a wideband complementary antenna[41]

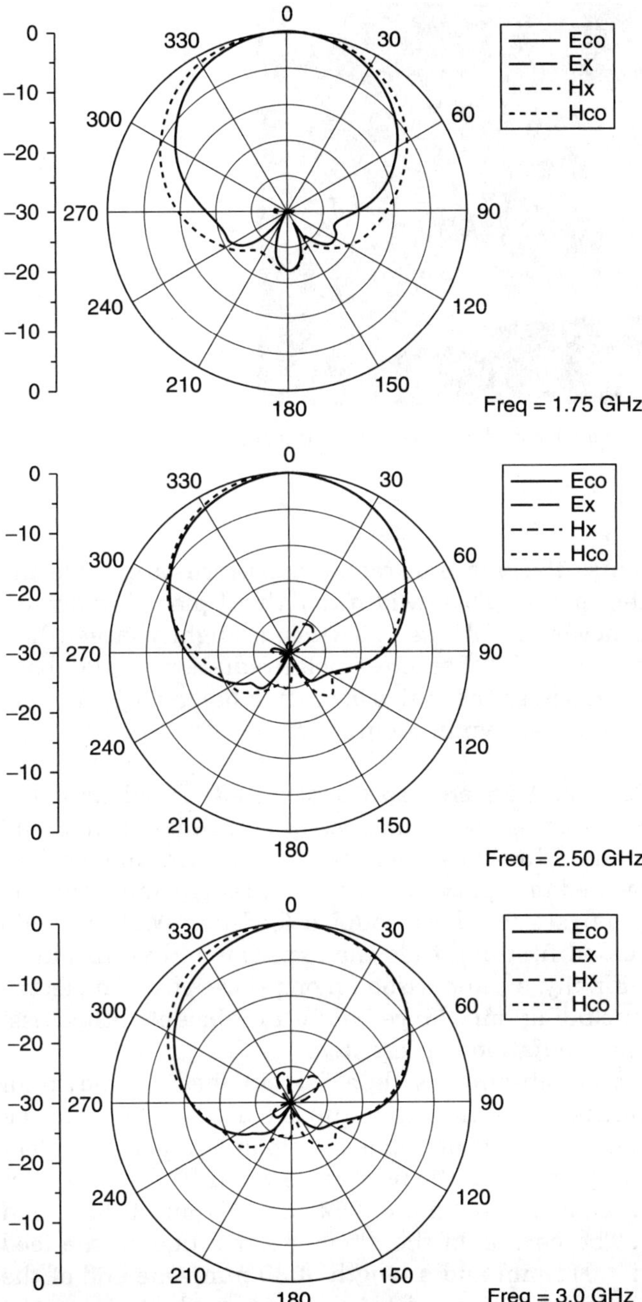

Figure 6.21 Measured radiation pattern at frequencies of 1.75, 2.5, and 3.0 GHz[41]

Figure 6.22 Photograph of a wideband complementary antenna with a Γ-shaped strip feed[41]

In both E- and H-planes, the broadside radiation patterns are stable and symmetric across the operating bandwidth, and the H-plane beamwidth at the center frequency of 2.5 GHz is 79°, which is slightly larger than that in the E-plane of about 75°. Low cross-polar radiation and low back radiation are achieved across the entire operating bandwidth. A photo of a fabricated antenna is shown in Figure 6.22.

6.3.4.2 L-Strip, T-Strip, and Square-Cap Coupled Fed Previously, the proposed Γ-feed consisted of an air microstrip transmission line and an L-strip coupled line.[54] In this section, two alternative coupled feed structures are suggested to replace the L-strip: a T-strip and a square-cap.[55] Simulation analyses for these coupled feeds in SWR and gain responses are discussed first and then the experimental verifications are demonstrated. Finally, a comparison among these cases in electrical performances, including impedance bandwidth, beamwidth, cross-polarization, and back radiation is presented.

The geometry of an antenna with a T-strip feed is shown in Figure 6.23. The dimensions of these radiating elements, consisting of a vertically oriented shorted patch antenna and a planar dipole, are identical to the antenna described in the previous section. The key parameters for the antenna are $L = 30$ mm (0.25λ), $S = 17$ mm (0.14λ), and $W = 60$ mm (0.5λ). The portion of the air microstrip line on the feed line has a width of 4.911 mm and a length of 30 mm. One end of the microstrip line is connected to the SMA connector, which is located underneath the ground plane. The other end of the microstrip line is combined with the T-strip. This T-strip has two parameters: $T_1 = 9.5$ mm (0.079λ) and $T_2 = 27$ mm (0.225λ). The width of the strip, which

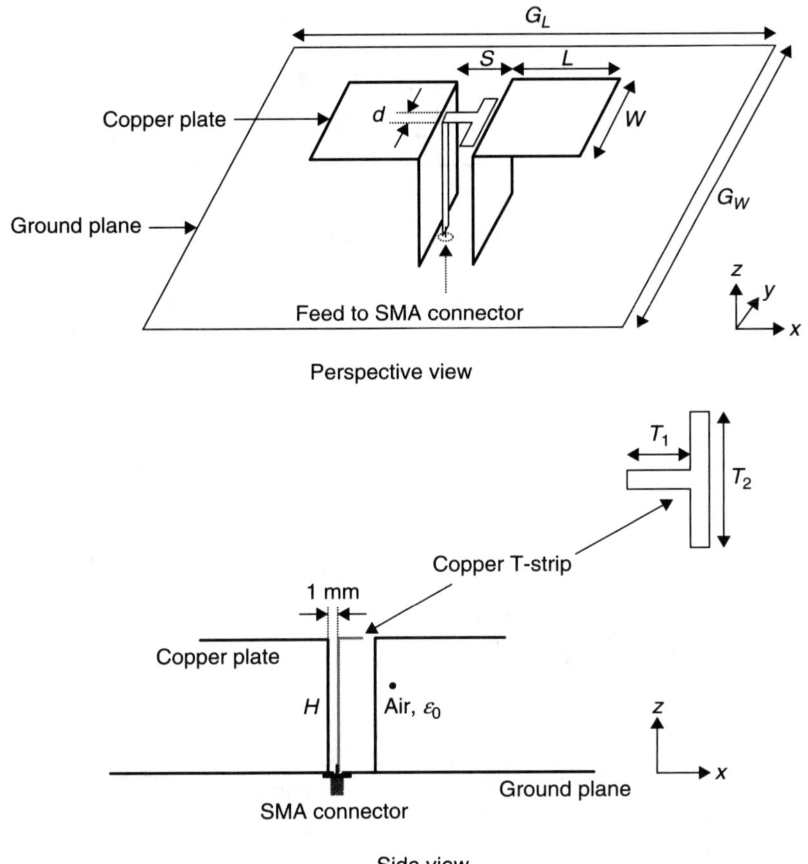

Parameters	T_1	T_2	d	H	L	S	W	G_L	G_W
Value/mm	9.5 (0.079λ)	27 (0.225λ)	4.91 (0.040λ)	30 (0.25λ)	30 (0.25λ)	17 (0.141λ)	60 (0.5λ)	160 (1.3λ)	160 (1.3λ)

Figure 6.23   A wideband unidirectional antenna with a T-strip coupled feed

is the same as the microstrip line, is 4.911 mm. The whole antenna is made of 0.3-mm-thick copper plates and its ground plane is made of an aluminum plate 160 mm × 160 mm with a 2-mm thickness.

Similarly, the configuration of the proposed antenna fed by a square-cap is illustrated in Figure 6.24. For this antenna, the microstrip line has a vertical length of 24 mm and a horizontal length of 12.5 mm. The end of the vertical line is connected to the SMA connector, while the horizontal portion is associated with the metallic square-cap, which has a length of 10 mm.

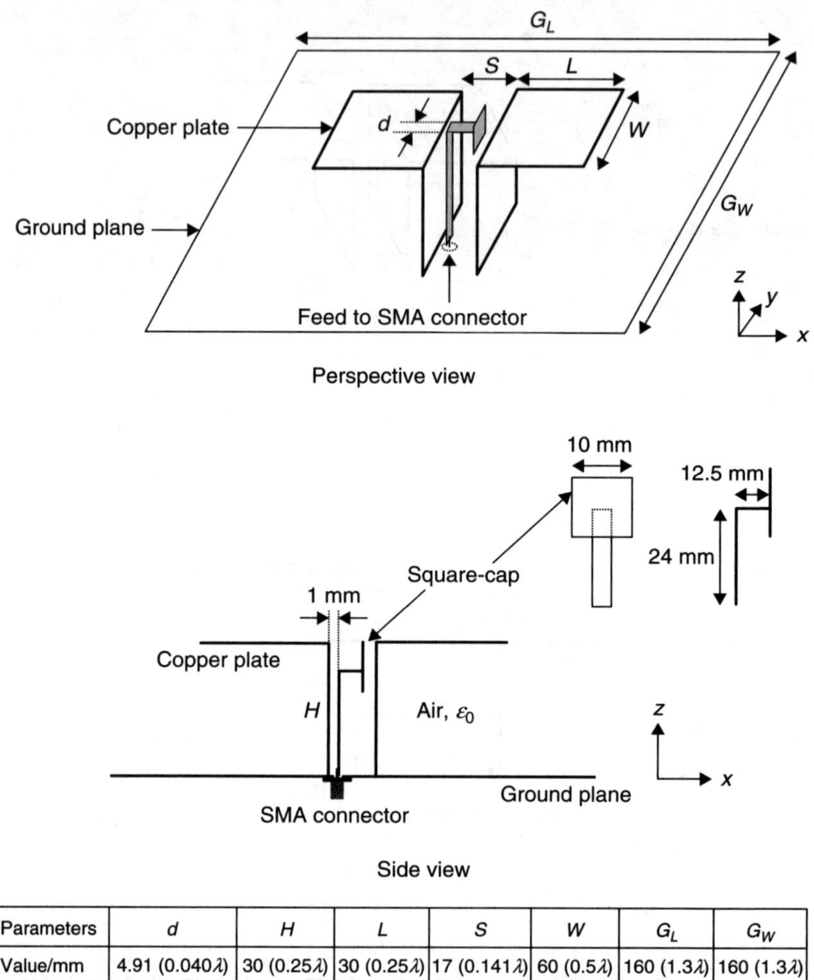

Parameters	d	H	L	S	W	G_L	G_W
Value/mm	4.91 (0.040λ)	30 (0.25λ)	30 (0.25λ)	17 (0.141λ)	60 (0.5λ)	160 (1.3λ)	160 (1.3λ)

Figure 6.24   A wideband complementary antenna with a square-cap coupled feed[55]

Figure 6.25 shows the simulated SWR curves for the three different feeding structures. The impedance bandwidth for the case using an L-strip is 44.5% (1.85 GHz – 2.91 GHz) for an SWR ≤ 2. The bandwidths for the T-strip and square-cap cases are 65.6% (1.68 GHz – 3.32 GHz) and 65.2% (1.84 GHz – 3.62 GHz), respectively.

Figure 6.26 depicts the simulated gain against frequency for the three different feeding techniques. The antenna gains are around 8 dBi for the three cases when the frequencies range between 1.8 GHz to 2.9 GHz. Because the impedance bandwidths for the T-strip and the square-cap cases are greater than the L-strip coupled, these two cases have wider gain bandwidths. However, the antenna gains are only 6.5 dBi at a

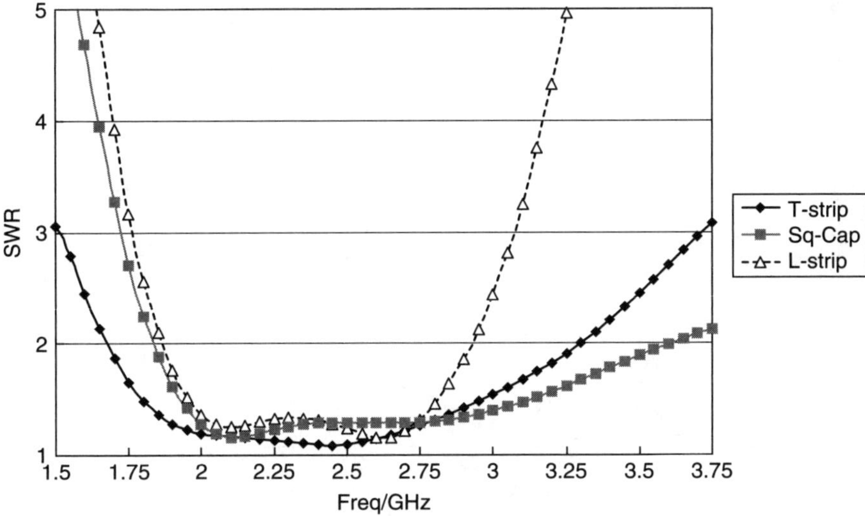

Figure 6.25   SWRs for three different feeding structures: T-strip, L-strip, and square-cap

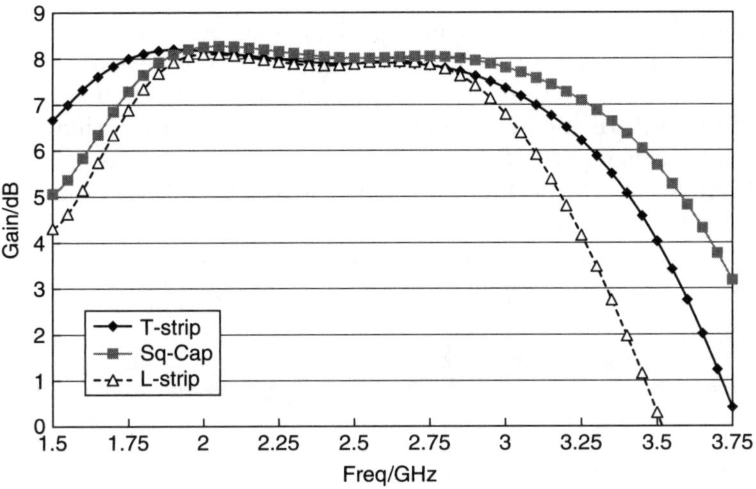

Figure 6.26   Gain curves for three different feeding structures: T-strip, L-strip, and square-cap

higher frequency within the operating bandwidths in these two cases. As a result, we can conclude the antenna gain is very stable across the operating bandwidth for the L-strip coupled feed. Additionally, there is an approximately 1.5-dB variation in the antenna gains for the two cases of T-strip and square-cap coupled feeds.

Figure 6.27 shows the measured and simulated SWRs of the T-strip case. As seen from the SWR curves, the antenna has a wide impedance

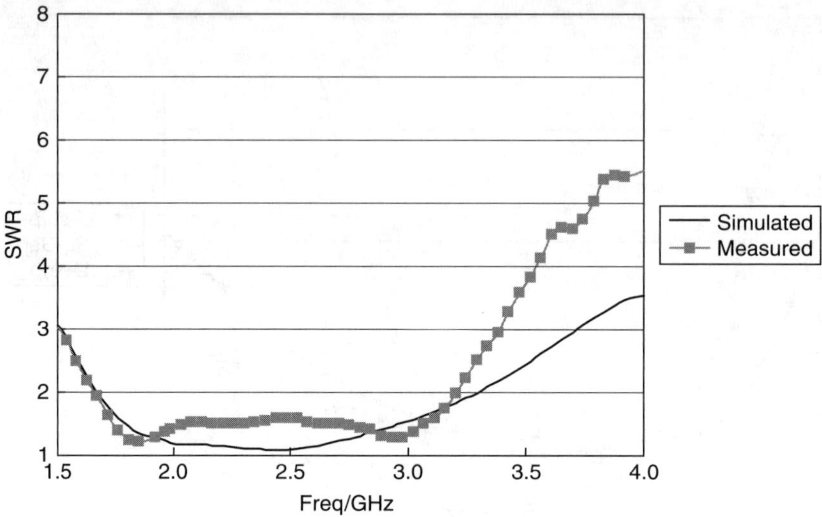

Figure 6.27 Measured and simulated SWR for a T-strip fed wideband unidirectional antenna

bandwidth of 63% (SWR ≤ 2) from 1.66 to 3.20 GHz. Figure 6.28 illustrates the measured and simulated gain curves of the antenna. From this, we can observe that the proposed antenna has a gain varying from 6.8 dBi to 8.2 dBi across the operating band. The radiation patterns at frequencies of 1.66, 2.5, and 3.2 GHz were measured and are shown in Figure 6.29. For both E- and H-planes, the broadside radiation

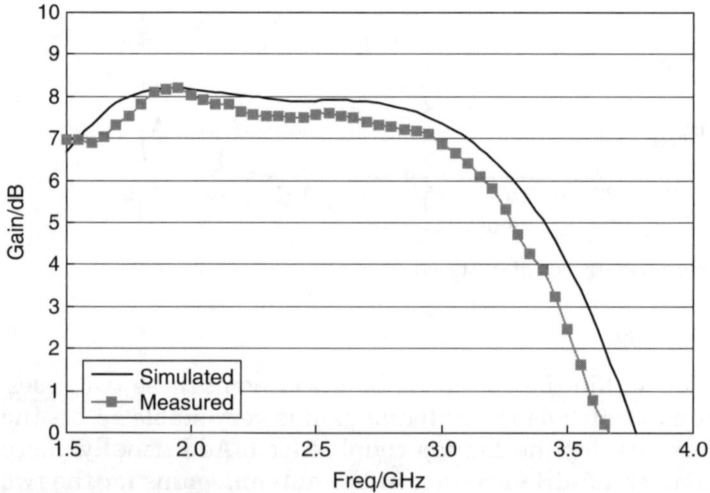

Figure 6.28 Measured and simulated gain for a T-strip fed wideband unidirectional antenna

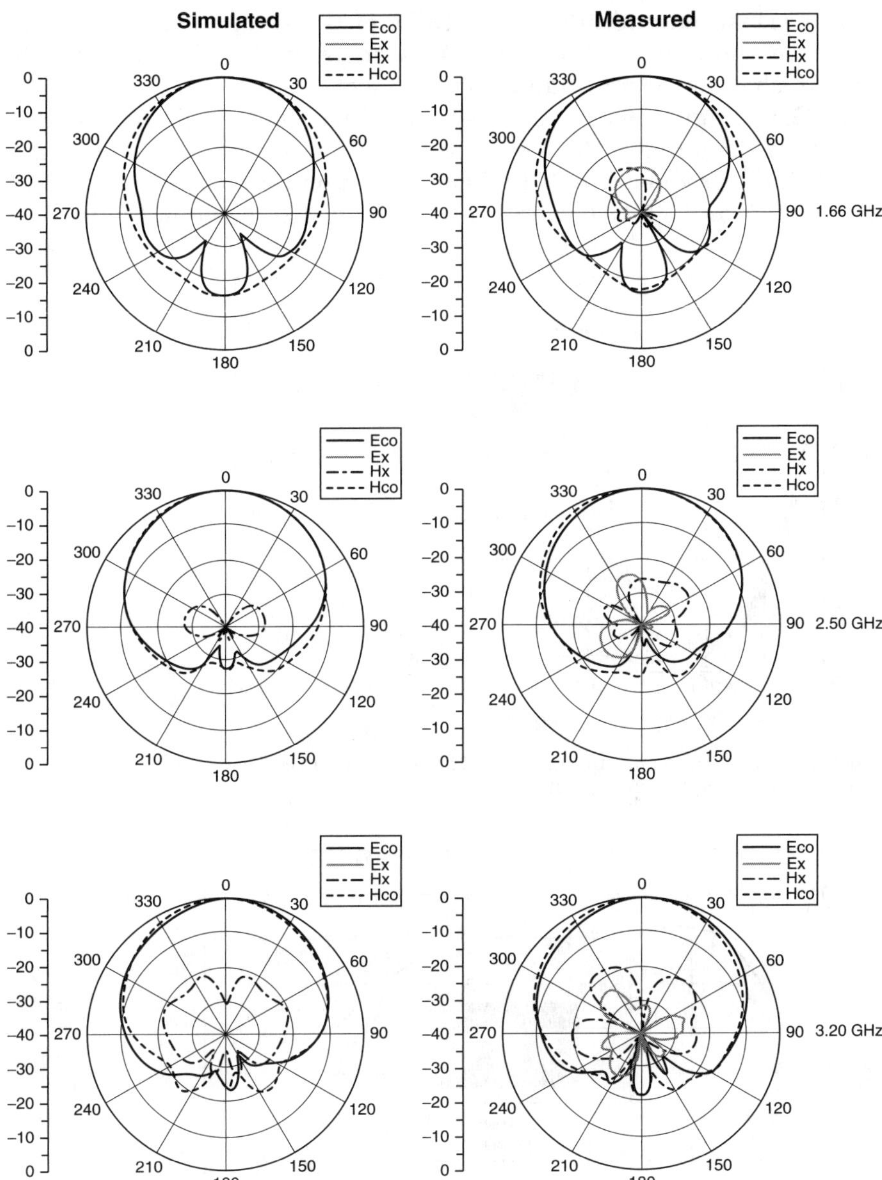

Figure 6.29 Radiation pattern of a T-strip coupled fed wideband unidirectional antenna

patterns are stable and symmetric across the operating band, and the beamwidth at the center frequency of 2.5 GHz in the H-plane is 78°, which is slightly larger than the beamwidth in the E-plane, which is about 71°. Low cross-polar radiation and low back radiation are achieved across the entire operating frequency band.

The measured and simulated SWRs of the square-cap coupled fed case are shown in Figure 6.30. As seen from the SWR curves, the antenna has a wide impedance bandwidth of 62% (SWR ≤ 2) from 1.83 to 3.50 GHz. Figure 6.31 displays the measured and simulated gain curves of the antenna. The proposed antenna has a maximum gain of 8.2 dBi at a frequency of 2.15 GHz. Radiation patterns at frequencies of 1.83, 2.5, and 3.5 GHz were measured and are shown in Figure 6.32.

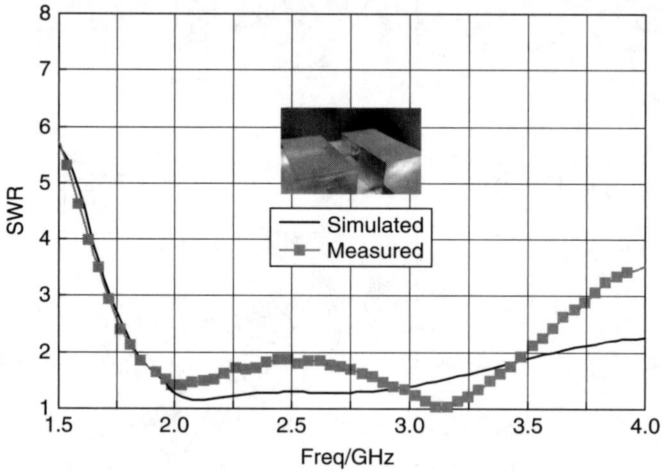

Figure 6.30   Measured and simulated SWR for a square-cap coupled fed wideband complementary antenna[55]

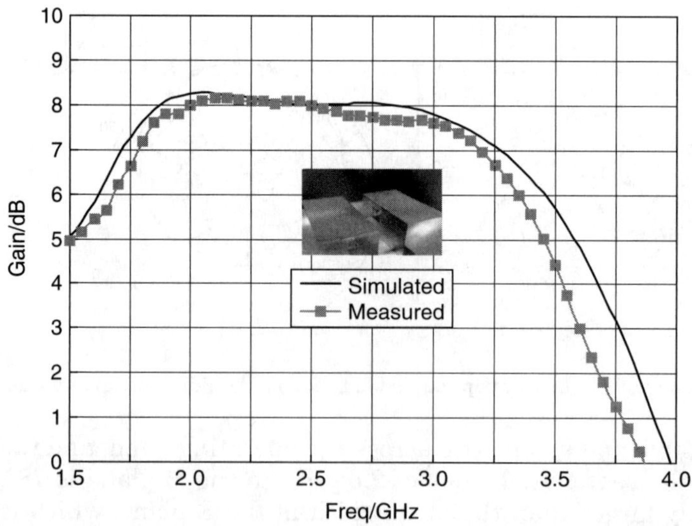

Figure 6.31   Simulated and measured gain for a square-cap coupled fed wideband complementary antenna[55]

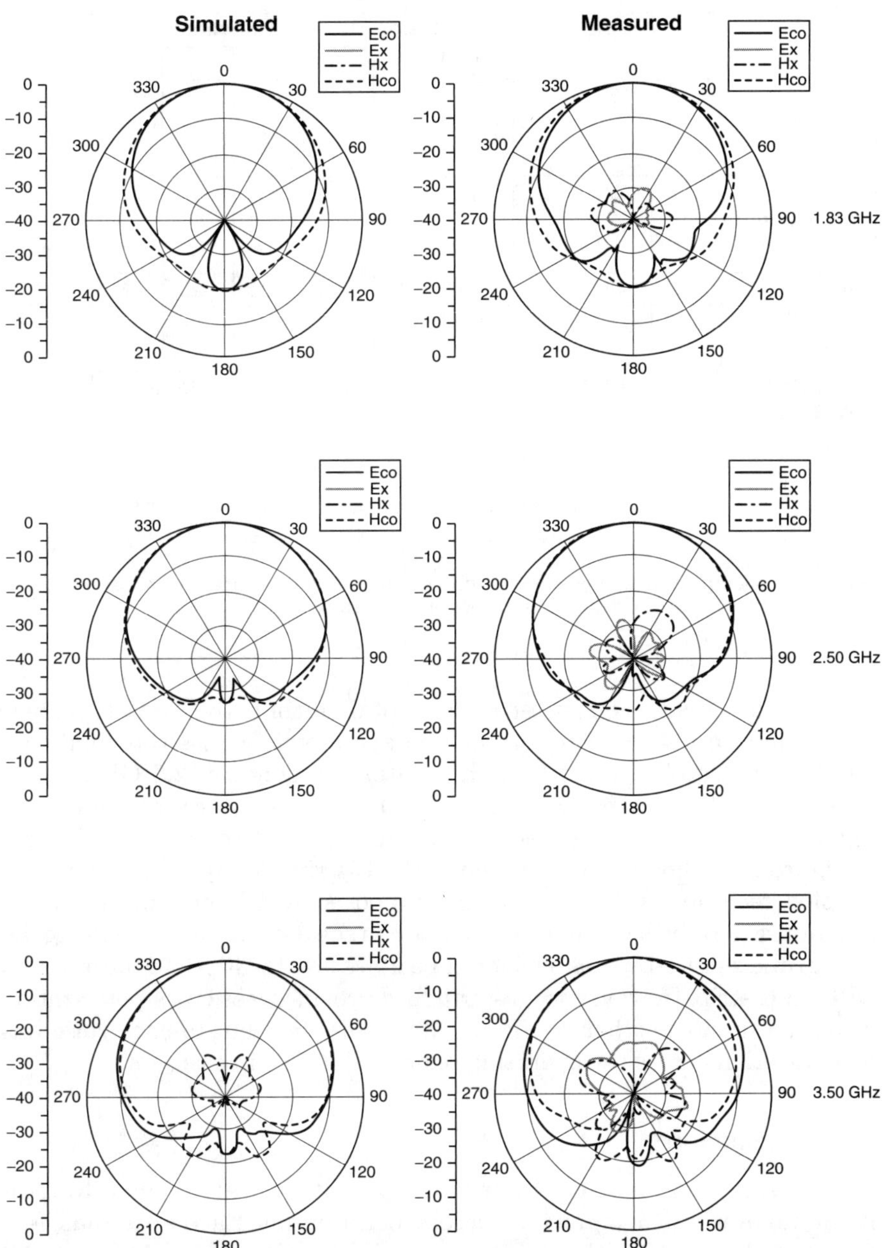

Figure 6.32 Radiation pattern of a square-cap coupled fed wideband unidirectional antenna

	L-strip			T-strip			Square-cap		
	f_L	f_o	f_H	f_L	f_o	f_H	f_L	f_o	f_H
3 dB beamwidth E-plane	67°	75°	79°	68°	71°	68°	66°	72°	78°
3 dB beamwidth H-plane	81°	79°	72°	81°	78°	78°	80°	77°	70°
X-pol/dB	−32	−28	−27	−26	−25	−18	−30	−25	−21
Front-to-back ratio	20	25	22	16	25	22	20	25	19
Bandwidth (SWR ≤ 2)	1.75–3.0 GHz, 52.6%			1.66–3.20 GHz, 63.3%			1.83–3.50 GHz, 62.6%		

f_L -lowest frequency across the operating bandwidth for (SWR ≤ 2)

f_o -center frequency across the operating bandwidth for (SWR ≤ 2)

f_H -highest frequency across the operating bandwidth for (SWR ≤ 2)

Figure 6.33 Comparison of measured results for the L-strip, T-strip, and square-cap coupled feeds for a wideband unidirectional antenna

The broadside radiation patterns were obtained for both E- and H-planes. The radiation patterns are stable and symmetric across the operating bandwidth. The beamwidth at the center frequency of 2.5 GHz in the H-plane is 77° and in the E-plane is 72°. Low cross-polar radiation levels and low back radiation are achieved across the entire operating band.

Figure 6.33 shows a comparison of the L-strip, T-strip, and square-cap coupled techniques for exciting the proposed wideband unidirectional antenna. From the measured results, we can see that the antenna using the T-strip and square-cap has a wider impedance bandwidth than the antenna with an L-strip. However, the L-strip feed provides less cross-polarization radiation than the other two feeds. Moreover, the front-to-back ratio has less variation across the operating band in the L-strip case.

6.4 Conclusion

This chapter begins with the introduction of several wideband unidirectional antenna designs for microstrip antennas. Each antennas' features are discussed with proper illustrations and related references. All designs employ an electrically thick substrate with a low dielectric constant for achieving wide impedance bandwidth performance. Moreover, the antennas using the twin L-probe feed, meandering-probe feed, or differential-plate feed not only have wide impedance bandwidths, but also possess noticeable electrical characteristics such as low cross polarization, high gain, and symmetrical E-plane radiation. After that, we discussed a new type of wideband unidirectional antenna element— a complementary antenna composed of a planar electric dipole and a shorted patch antenna, which is equivalent to a magnetic dipole. This architecture possesses advantages including a stable radiation pattern with low cross-polarization, low backlobe radiation, nearly identical E- and H-plane patterns, and stable antenna gain across the entire operating bandwidth. In addition, two alternative feeding structures, the T-shaped and square-plate coupled lines, demonstrate the flexibility in antenna feed design.

6.5 Acknowledgment

The authors would like to express their sincere thanks to Dr. C.L. Mak and Dr. H.W. Lai for contributing some research works presented in this chapter.

References

1. S. Dey, P. Venugopalan, K. A. Jose, C. K. Aanandan, P. Mohanan, and K.G. Nair, "Bandwidth enhancement by flared microstrip dipole antenna," *IEEE Antennas and Propagation Society International Symposium, 1991,* vol. 1, AP-S. Digest, 24–28 (June 1991): 342–345.
2. Y. D. Lin and S. N. Tsai, "Coplanar waveguide-fed uniplanar bow-tie antenna," *Electron. Lett.,* vol. 45 (1997): 305–306.
3. E. Levine, S. Shtrikman, and D. Treves, "Double-sided printed arrays with large bandwidth," *IEE Proceedings – Part H,* vol. 135, no. 1 (1988): 54–59.
4. S. Dey, P. Venugopalan, K. A. Jose, C. K. Aanandan, P. Mohanan, and K. G. Nair, "Bandwidth enhancement by flared microstrip dipole antenna," *IEEE Antennas and Propagation Society International Symposium, 1991,* vol. 1, AP-S Digest, 24–28 (June 1991): 342–345.
5. Y. D. Lin and S. N. Tsai, "Coplanar waveguide-fed uniplanar bow-tie antenna," *Electron. Lett.,* vol. 45 (1997): 305–306.
6. K. Kiminami, A. Hirata, and T. Shiozawa, "Double-sided printed bow-tie antenna for UWB communications," *IEEE Antenna and Wireless Propagation Lett.,* vol. 3 (2004): 152–153.
7. J. I. Kim, B. M. Lee, and Y. J. Yoon, "Wideband printed dipole antenna for multiple wireless services," *IEEE Radio and Wireless Conference* (August 19–22, 2001): 153–156.
8. G. A. Evtioushkine, J. W. Kim, and K. S. Han, "Very wideband printed dipole antenna array," *Electron. Lett.,* vol. 34 (1998): 2292–2293.
9. K. M. Luk, C. L. Mak, Y. Chow, and K. F. Lee, "Broadband microstrip patch antenna," *Electron. Lett.,* vol. 34, (1998): 1442–1443.
10. C. L. Mak, K. M. Luk, K. F. Lee, and Y. L. Chow, "Experimental Study of a Microstrip Patch Antenna with an L-shaped Probe," *IEEE Transactions on Antennas and Propagation,* vol. AP-48, no. 5 (May 2000): 777–783.
11. Y. X. Guo, C. L. Mak, K. M. Luk, and K. F. Lee, "Analysis and Design of L-Probe Proximity Fed-Patch Antennas," *IEEE Transactions on Antennas and Propagation,* vol. AP-49, no. 2 (February 2001): 145–149.
12. H. Wong, K. L. Lau, and K. M. Luk, "Design of dual-polarized L-probe patch antenna arrays with high isolation," *IEEE Transaction on Antenna Propagation,* vol. 52, no. 1 (January 2004): 45–52.

13. H. Wong and K. M. Luk, "A Low-Cost L-Probe Patch Antenna Array," *Microwave and Optical Technology Letters*, vol. 29, no. 4 (May 2001): 280–282.

14. F. Croq and D. M. Pozar, "Millimeter wave design of wide-band aperture-coupled stacked microstrip antennas," *IEEE Trans. Antennas Propogat.*, vol. 39 (December 1991): 1770–1776.

15. V. Rathi, G. Kumar, and K. P. Ray, "Improved Coupling for Aperture Coupled Microstrip Antennas," *IEEE Transactions on Antennas and Propagation*, vol. AP-44, no. 8 (August 1996): 1196–1198.

16. P. L. Sullivan and D. H. Schaubert, "Analysis of an Aperture Coupled Microstrip Antenna," *IEEE Transactions on Antennas and Propagation*, vol. AP-34, no. 8 (August 1986): 977–984.

17. R. B. Waterhouse, "Design of probe-fed stacked patches," *IEEE Trans. Antennas Propogat.*, vol. 47, no. 11 (November 1999): 1780–1784.

18. R. Q. Lee and K. F. Lee, "Experimental Study of the Two-Layer Electromagnetically Coupled Rectangular Patch Antenna," *IEEE Transactions on Antennas and Propagation*, vol. AP-38, no. 8 (August 1990): 1298–1302.

19. T. M. Au and K. M. Luk, "Effect of Parasitic Element on the Characteristics of Microstrip Antennas," *IEEE Transactions on Antennas and Propagation*, vol. AP-39, no. 8 (August 1991): 1247–1251.

20. H. Legay and L. Shafai, "New Stacked Microstrip Antenna with Large Bandwidth and High Gain," *IEE Proceeding-H, Microwaves, Antennas and Propagation*, vol. 141, no. 3 (June 1994): 199–204.

21. R. B. Waterhouse, "Design and Scan Performance of Large, Probe-Fed Stacked Microstrip Patch Array," *IEEE Transactions on Antennas and Propagation*, vol. AP-50, no. 6 (June 2002): 893–895.

22. K. F. Lee, K. M. Luk, K. F. Tong, S. M. Shum, T. Huynh, and R. Q. Lee, "Experimental and simulation studies of the coaxially fed U-slot rectangular patch antenna, " *IEE Proc.-Microw. Antenna Propag.*, vol. 144, no. 5 (October 1997): 354–358.

23. T. Huynh and K. F. Lee, "Single-layer single-patch wideband microstrip antenna," *Electron. Lett.*, vol. 31, no. 16 (3 August 1995): 1310–1312.

24. K. F. Lee, K. M. Luk, K. F. Tong, Y. L. Yung, and T. Huynh, "Experimental study of a two-element array of U-slot patches," *Electron. Lett.*, vol. 32, no. 5 (29 February 1996): 418–420.

25. M. Clenet and L Shafai, "Multiple resonances and polarisation of U-slot patch antenna," *Electron. Lett.*, vol. 35, no. 2 (21 January 1999): 101–103.

26. K. F. Tong, K. M. Luk, K. F. Lee, and R. Q. Lee, "A Broad-Band U-slot Rectangular Patch Antenna on a Microwave Substrate," *IEEE Transactions on Antennas and Propagation*, vol. AP-48, no. 6 (June 2000): 954–960.

27. Y. X. Guo, K. M. Luk, K. F. Lee, and Y. L. Chow, "Double U-slot rectangular patch antenna," *Electron. Lett.*, vol. 34, no. 19 (17 September 1998): 1805–1806.

28. A. Petosa, A. Ittipiboon, and N. Gagnon, "Suppression of unwanted probe radiation in wideband probe-fed microstrip patches," *Electron. Lett.*, vol. 35, no. 5 (1999): 355–357.

29. C. L. Mak, H. Wong, and K. M. Luk, "High-gain and wide-band single-layer patch antenna for wireless communications, " *IEEE Trans. on Vehicular Technology*, vol. 54, no. 1 (January 2005).

30. H. W. Lai and K. M. Luk, "Design and study of wide-band patch antenna fed by meandering probe," *IEEE Trans. Antennas Propogat.*, vol. 54, no. 2 (February 2006).

31. H. W. Lai and K. M. Luk, "Wideband patch antenna with low cross-polarisation," *Electron. Lett.*, vol. 40 (2004): 159–160.

32. P. Li, H. W. Lai, K. M. Luk, and K. L. Lau, "A wideband patch antenna with cross-polarization suppression," *IEEE Antennas Wireless Propagat. Lett.*, vol. 3 (2004): 211–214.

33. H. W. Lai and K. M. Luk, "Wideband stacked patch antenna fed by a meandering probe," *Electron. Lett.*, vol. 41 (2005): 297–298.

34. H. W. Lai and K. M. Luk, "Wideband patch antenna fed by a modified L-shaped probe," *Microwave and Opt. Technol. Lett.*, vol. 48, no. 5 (2006): 977–979.

35. A. Clavin, "A new antenna feed having equal E- and H-plane patterns," *IRE Trans. Antennas Propogat.*, vol. AP-2 (1954): 113–119.

36. A. Clavin, D. A. Huebner, and F. J. Kilburg, "An improved element for use in array antennas," *IEEE Trans. Antennas Propogat.*, vol. AP-22, no. 4 (July 1974): 521–526.

37. R. W. P. King and G. H. Owyang, "The slot antenna with coupled dipoles, " *IRE Trans. Antennas Propogat.*, vol. AP-8 (March 1960): 136–143.

38. W. W. Black and A. Clavin, "Dipole augmented slot radiating element," U. S. Patent 3594806, July 1971.

39. W. F. Gabriel and L. R. Dod, "A complementary slot-dipole antenna for hemispherical coverage," NASA-Goddard Space Flight Center, Greenbelt, MD, NASA TM X-55681, October 1966.

40. E. J. Wilkinson, "A circularly polarized slot antenna," *Microwave J.*, vol. 4 (March 1961): 97–100.

41. Kwai Man Luk and Hang Wong, "A New Wideband Unidirectional Antenna Element," *International Journal of Microwave and Optical Technology,* vol. 1, no. 1 (June 2006): 35–44.

42. C. Wood, "Improved bandwidth of microstrip antennas using parasitic elements," *IEE proceedings – microwaves, antennas and propagation*, vol. 127 (1980): 231–234.

43. G. Kumar and K. C. Gupta, "Nonradiating edges and four edges gap-coupled multiple resonator broad-band microstrip antennas," *IEEE Trans. Antennas Propagat.*, vol. AP-33 (1985): 173–178.

44. T. M. Au, K. F. Tong, and K. M. Luk, "Characteristics of aperture-coupled coplanar microstrip subarrays," *IEE Proc. – Microw. Antennas Propagat.*, vol. 144 (1997): 137–140.

45. P. S. Hall, "Probe compensation in thick microstrip patches," *Electron. Lett.*, vol. 23 (1987): 606–607.

46. R. Q. Lee, R. Acosta, and K. F. Lee, "Radiation characteristics of microstrip arrays with parasitic elements," *Electron. Lett.*, vol. 23 (1987): 835–837.

47. R. E. Munson, "Conformal microstrip antennas and microstrip phased arrays," *IEEE Trans. Antennas Propagat.*, vol. 22 (1974): 74–78.

48. Q. Xue, X. Y. Zhang, and C. H. K. Chin, "A Novel Differential-Fed Patch Antenna," *Antennas and Wireless Propagation Letters*, vol. 5, no. 1 (December 2006): 471–474.

49. C. H. K. Chin, Q. Xue, H. Wong, and X. Y. Zhang, "Broadband Patch Antenna with Low Cross-Polarisation," *Electron. Lett.*, vol. 43, no. 3 (February 2007): 137–138.

50. C. H. K. Chin, Q. Xue, and H. Wong, "Broadband Patch Antenna with a Folded Plate Pair as a Differential Feeding Scheme," *IEEE Transactions on Antennas and Propagation*, vol. 55, no. 9 (September 2007): 2461–2467.

51. J.-X. Chen, C. H. K. Chin, K. W. Lau, and Q. Xue, "180° Out-of-Phase Power Divider Based on Double-Sided Parallel Striplines," *Electron. Lett.*, vol. 42, no. 21 (October 2006): 1229–1230.

52. K. Itoh and D. K. Cheng, "A novel slots-and-monopole antenna with a steerable cardioid pattern," *IEEE Trans. on Aerospace and Electronic Systems*, vol. AES-8, no. 2 (March 1972): 130–134.

53. P. E. Mayes, W. T. Warren, and F. M. Wiesenmeyer, "The monopole-slot: a broad-band, unidirectional antenna," *1971 G-AP Int. Symp. Dig.* (September 1971): 109–112.

54. H. Wong and K. M. Luk, "Unidirectional antenna composed of a planar dipole and a shorted patch," *Asia Pacific Microwave Conference 2006* (December 2006): 85–88.

55. H. Wong and K. M. Luk, "A new wideband antenna: Comprising planar-dipole and short-circuited-patch," *2006 European Conference on Antennas and Propagation* (November 2006): 1–4, 6–10.

Antennas for WLAN (WiFi) Applications

Zhi Ning Chen, Wee Kian Toh, Shie Ping See, Xianming Qing

Institute for Infocomm Research

7.1 Introduction

7.1.1 WLAN (WiFi)

Wireless Local Area Networks (WLANs) provide wireless network communications, in particular, between computers and other portable devices with fixed access points over a short distance, typically in the order of tens of meters. WLAN's rapidly increasing popularity in consumer electronics is primarily due to its convenience, mobility, easy deployment and expandability, and cost efficiency, as well as its ease of integration with other networks and devices. WLANs include Wireless Fidelity (WiFi, IEEE 802.11a/b/g/n)[1] and High Performance Radio LAN (HIPERLAN). A typical WLAN is connected to stations equipped with wireless network interface cards (WNICs). The wireless stations are categorized into fixed base stations, namely access points (APs) and clients, or customer premises equipment (CPE), or subscriber unit (SU). These include mobile devices such as laptops, personal digital assistants, IP phones, or fixed devices such as desktops and workstations that are equipped with a wireless network interface. WLAN communication can be peer-to-peer, bridge, or via wireless distribution. Peer-to-peer communication is achieved through an ad-hoc network without a base station and permission to talk. In other words, peer-to-peer communication allows wireless devices to talk directly to each other without any access points. Typically, two computers can be connected to form a network. A bridge in the WLAN is used to achieve the connections among different networks. For instance, a wireless Ethernet bridge connects the devices in a wired Ethernet network to a wireless LAN. The APs in a WLAN can be set up as repeaters when connecting all of the access points within a network through wires is difficult. Therefore, the WLAN can serve as a wireless distribution system if required.

The WLANs operate in the unlicensed Industrial, Scientific, and Medical (ISM) bands, namely, the 2.4-GHz band with a frequency range of 2.4–2.485 GHz and the 5-GHz bands with frequency ranges of 5.150–5.350 GHz, 5.470–5.725 GHz, and 5.725–5.850 GHz. The channel bandwidth within each band varies from 5 MHz to 20 MHz. Generally, the data rate of a wireless communication link increases with a larger bandwidth, but decreases with mobility. Table 7.1 shows the various IEEE 802.11 standards. In general, the actual data rate in a typical office environment is about half of its maximum. Furthermore, the data of a conventional wireless communication link is much less than that of the multiple-input-multiple-output (MIMO) setup based on IEEE 802.11n. The data rate of 11-Mbps for IEEE 802.11b remains relatively constant for a distance of up to 50 meters, whereas the data rate of 54-Mbps for IEEE 802.11a declines linearly to about 11 Mbps at a distance of 50 meters. One of the main challenges in deploying and operating WLAN stems from the fact that other wireless systems, such as Wireless Personal Area Networks (WPANs), Bluetooth (IEEE 802.15.1), ZigBee (IEEE 802.15.4),[2] and microwave ovens share the same unlicensed frequency bands, which can lead to interference among electrical devices and systems.

In addition, the standardization bodies in individual countries, for instance, the Federal Communications Commission (FCC)[3] in the United States and the European Telecommunications Standards Institute (ETSI)[4] in the European Union, regulate electromagnetic radiation. The effective isotropic radiated power (EIRP) defined by FCC Rules part 15.247[5] for an omnidirectional antenna of less than 6-dBi gain is 1 W. The IEEE Standard C95.1-1991 recommends that the power density for human exposure to radio frequency (RF) and microwave electromagnetic field emissions be 1–10 mW/cm^2 from 1–10 GHz. As the lower frequency is more penetrative in nature, a lower radiation restriction is enforced.

TABLE 7.1   IEEE 802.11 Standard Family

IEEE 802.11	Frequency bands, GHz	Maximum data rate, Mbps	Modulation	Radius indoor, m	Radius outdoor, m
a	5	54	OFDM	< 35	< 120
b	2.4	11	DSSS	< 38	< 140
g	2.4	54	OFDM	< 38	< 140
n	2.4, 5	600	OFDM	< 70	< 250

Antennas play an important role in the design of WLANs. The antennas for the client devices have critical constraints in terms of cost and size, which severely limits the antenna performance.[6] The antennas for base stations in point-to-point (P2P) and/or point-to-multipoint (P2MP) systems face different challenges such as the performance, cost, size, and integration of multiple functions into one antenna design, as well as the integration of antennas into radios.

7.1.2 MIMO in WLANs

Over the past decade, a wide range of applications has spurred the need to provide a reliable high-speed wireless communication link. This is especially challenging in an indoor wireless environment, where transmitted signals are received through multiple paths, which may add up destructively at the receiver, resulting in serious degradation in the overall system performance. This phenomenon is known as *multipath fading*. Multipath is the arrival of the transmitted signal at an intended receiver through different angles and/or different time delays and/or different frequency shifts (i.e., Doppler effect). Consequently, the received signal fluctuation can be larger than 10dB within one wavelength due to angle spread and/or frequency due to delay spread and/or time due to Doppler spread through the random superposition of the impinging multipath components. The scarcity of available bandwidth, transmit power constraints, hardware complexity, and signal interference are some of the other challenges that high-speed wireless communication faces.

With the steady increase in the number of new wireless applications and the expansion of existing ones, the limited frequency spectrum problem can be alleviated effectively by using diversity techniques, which provide the receiver with independently faded copies of the transmitted signal. This increases the probability of reception at the receiver. There are many diversity techniques based on frequency, time, angle, space, polarization, and spatial diversity. Usually, the MIMO system uses multiple (two or more) antennas in both its transmitter and receiver sides. A significant advantage of MIMO technology is that it provides a substantial increase in channel capacity, resulting in higher data throughputs with a low bit error rate, i.e., enhanced data transmission reliability. The wireless channel's impairments can also be overcome by channel coding. Examples include convolution and block coded modulation, trellis coded modulation, bit-interleaved coded modulation, and turbo and low-density parity check codes. The combination of various diversity techniques, such as space-time coding in MIMO systems, offers high data rates with increased reliability. To maximize the transmission reliability, transmit diversity schemes should be adopted. In this case, the data rate is the same as for a single-input-single-output system since all the degrees of freedom in the MIMO channel are used to improve transmission reliability (or decrease the frame-error rate, or FER). A MIMO channel with N_T transmitting antennas and N_R receiving antennas potentially offers $N_T N_R$ independent fading links.

On the other hand, the data rate can be maximized through spatial multiplexing where multiple independent signals are transmitted but at the expense of an increase in the FER. In general, the number of data streams that can be reliably supported by a MIMO channel is equal to the minimum number of transmitting and receiving antennas, i.e., $\min\{N_T, N_R\}$. Capacity is enhanced by a multiplicative factor equal to the number of data streams. Therefore, a tradeoff has to be made between the data rate and reliability.[7] In a MIMO context, diversity gain (d) is often associated with reducing the FER while multiplexing gain (r) is associated with increasing the data rate. The maximum multiplexing gain $r_{\max}$, as shown in Eq. 7.1, is given by the slope of the outage capacity (for a fixed FER) plotted as a function of the SNR (γ) on a log-linear scale

$$r_{\max} = \lim_{\gamma \to \infty} \frac{C_{\text{out},p}(\gamma)}{\log_2 \gamma} \tag{7.1}$$

where $C_{\text{out},p}(\gamma)$ is the outage capacity defined as the data rate that can be supported by $(100 - p)\%$ of the fading realizations of the channel.[8] For a fixed FER, the transmission rate can be increased by $\min\{N_T, N_R\}$ bps/Hz for every 3-dB increase in the SNR.

The maximum diversity gain $d_{\max}$, shown in Eq. 7.2, that can be achieved is given by the negative of the asymptotic slope of FER for a fixed data rate, plotted as a function of SNR on a log-linear scale. P_e denotes the probability that the frame will be decoded incorrectly. With every 3-dB increase in the SNR for a fixed transmission rate, the FER decreases by a factor of $2^{-N_T N_R}$.

$$d_{\max} = -\lim_{\gamma \to \infty} \frac{\log_2 P_e(\gamma, R)}{\log_2 \gamma} \tag{7.2}$$

It may be desirable that the increase in the SNR be a combination of an increase in transmission rate and decrease in FER. The optimal tradeoff curve for the $\boldsymbol{H}_w$ MIMO channel, $d(r)$, is piecewise linear such that

$$d(r) = (N_R - r)(N_T - r) \tag{7.3}$$

This equation implies that if the data rate is increased by r bps/Hz over a 3-dB increase in SNR, the corresponding reduction in the FER will be $2^{-d(r)}$.

Coherent combining of the wireless signals at the receiver through spatial processing at the receiving antenna array and/or preprocessing at the transmitting antenna array is able to achieve an increase in the signal-to-noise ratio (SNR) known as *array gain*. Array gain improves resistance to noise, thereby improving the coverage of a wireless network. Also, the signal-to-noise-plus-interference ratio (SINR) will be improved, and as a result, interference in wireless networks can be mitigated. Reducing interference by increasing the separation between users and directing the energy toward intended users improves the coverage and range of a wireless network.

MIMO technology is found in several standards for future wireless communication systems, especially WLANs and cellular networks. The standardization of MIMO technology is currently under development. The IEEE 802.11n, which has yet to be finalized, supports MIMO communications with peak data rates of 600 Mbps. The IEEE 802.16 standard has been developed for the world interoperability for microwave access (WiMAX), which is intended to deliver high data rates over long distances. MIMO communication has been incorporated in the IEEE 802.16e version of this standard, where 2×1 and 4×4 MIMO configurations are considered. The 3GPP technology, also known as *wideband code division multiple access (W-CDMA),* is used for 3G cellular networks, and MIMO has been incorporated into this standard, particularly in Releases 7 and 8. In Release 7, 2×1 and 4×2 configurations employing space-time block coding were used, whereas in Release 8 (TSG-R1(04)0336(2004)), the 2×2 and 4×4 configurations were employed. In addition, MIMO is also considered in IEEE 802.20 and IEEE 802.22 standards. The former is used to enable the worldwide deployment of multivendor interoperable mobile broadband wireless access networks, whereas the latter is aimed at constructing wireless regional area networks (WRANs), utilizing channels that are not employed within the already allocated television frequency spectrum.

7.2 Design Considerations for Antennas

Generally, the major considerations for WLAN antenna design include electrical properties such as frequency range/bandwidth in terms of gain, impedance matching, and polarization, as well as other factors such as size, cost effectiveness, and mechanical robustness. Fulfilling the electrical properties of the antenna is a priority, which stems from system requirements. In addition to an antenna's electrical performance, other factors are also vital for a successful design. For example, a low-cost antenna product is always preferable for the commercial market. The relevant cost factors include the material used, manufacturing, mechanical tolerance, installation, switching from one standard to another, and integration with the radio. In addition, the form factor of the antenna is also an important consideration. Indoor antennas are generally designed to be compact and less protrusive.

7.2.1 Materials, Fabrication Process, Time to Market, Deployment, and Installation

7.2.1.1 Material

A. Dielectric Substrate   Low-cost dielectric substrates, such as Flame Retardant-4 (FR-4) with an $\varepsilon_r \approx 4.4$ and a $\tan\delta \approx 0.02$, are widely employed for printed circuit board (PCB) antenna designs usually up to 6 GHz. At higher frequencies, the dielectric loss incurred becomes greater. The thickness of the FR-4 material ranges from 5–60 mils, and it normally comes with copper cladding with a thickness ranging from 0.5–2.0 oz. The dielectric substrates that can be used for high frequencies, such as Rogers 4003 ($\varepsilon_r \approx 3.38$, $\tan\delta \approx 0.002$) and RT/Duroid 5880 ($\varepsilon_r \approx 2.2$, $\tan\delta \approx 0.0004$), are generally not selected for mass production as they are relatively more expensive. Instead, the air substrate is commonly adopted for low-cost and broadband solutions.

B. Conductor   Conductors with good conductivity, such as copper $(5.8 \times 10^7 \text{ S/m})$ and brass, are commonly employed as radiators and feeding networks in antenna design. Aluminum and galvanized steel with appropriate surface treatments are normally used for the antenna ground plane and casing. Aluminum varies from the 1xxx–7xxx series. Its oxide and magnesium contents make aluminum difficult to process by using conventional low-temperature (225–490°C) soft-soldering. Mounting an RF connector on an aluminum ground or casing is carried out by riveting instead of soldering.

C. Radome   A wide variety of antenna radome materials are available in the market. They include fiber reinforced plastics (FRPs), glass reinforced plastics (GRPs), polypropylene (PP), acrylonitrile butadiene styrene (ABS), and a variety of polycarbonates. The radome should be electromagnetically transparent with respect to the operating frequency of the antenna. In addition, it should be thin, lightweight, homogeneous, and uniform in thickness. The radome used for outdoor applications has to be weatherproof and resistant to vandalism. Polypropylene is commonly used in injection molding, but is susceptible to marring or chalking after prolonged UV irradiation.

D. Supporting Structures   In order to lower the Q-value of the antenna, and provide ample support, Styrofoam^{TM} material ($\varepsilon_r \approx 1.1$), nylon ($\varepsilon_r \approx 3.5$) stubs, rods, and screws are commonly used as supporting materials. Metallic screws are also used as shorting pins or as support at the noncritical portions of the antenna, e.g., the center of a patch antenna.

E. Cables and Connectors   Low-cost coaxial cables such as the RG-58/U (50 Ω), RG-178/U (50 Ω), and RG-59/U (75 Ω) are commonly used to connect the antenna and the radio. The cable length is often kept as short as possible in order to minimize the loss and phase shifting of signals, so that the integrity of the received signal can be preserved. Connectors include the SMA, N-type, U.FL/IPEX, and MMCX. They provide both RF shielding and mechanical transition between the cable and the feed of the antenna.

7.2.1.2 Fabrication Process   Fabrication processes usually include machine tooling, injection molding, assembling, soldering, quality control, and testing. The Six Sigma (6σ) strategy is used by companies to reduce defects and errors in the manufacturing process. Each process has various degrees of manufacturing tolerances. For example, a coplanar waveguide feeding structure with a ground plane (CPWG) having an eight mil gap between the strip and ground plane is easily fabricated on a PCB. However, it is difficult to fabricate suspended metallic structures with high precision using low-cost turret punching and millings. Therefore, costly laser cutting and wire electrical discharge machining (EDM) are needed for such high-precision metal cutting.

7.2.1.3 Time to Market   Time to market (TTM) is also a critical factor for new products and solutions. The appropriate design tools, pragmatic estimations, and swift prototyping will shorten the overall time required. For example, a common RF board that supports multiple mini-PCI express radio cards, as well as several different versions of broadband antenna designs, will be more cost effective.

7.2.1.4 Deployment and Installation   Figure 7.1 depicts the various radio deployment scenarios for (*a*) base station (BS), (*b*) subscriber station (SS), and (*c*) access point (AP) antennas. Despite the differences in radiated power, they have many similarities in their antenna design considerations in terms of impedance, polarization, gain, and radiation pattern.

These antennas can be linearly (vertically or horizontally) polarized (LP). Circularly polarized (CP) antennas are also used for P2P communication links, so as to reduce the effects of multipath and depolarization of the radiated fields. Dual-polarized (DP) antennas are also often employed, which make use of the orthogonal property of electromagnetic waves to provide channel isolation by transmitting both the horizontally and vertically polarized waves simultaneously.

BS P2P antennas typically have high gain (> 12 dBi) and narrow beamwidth (< 40°) in both the E- and H-planes. Low-gain antennas such as half-wavelength dipole and single microstrip patch antennas are not

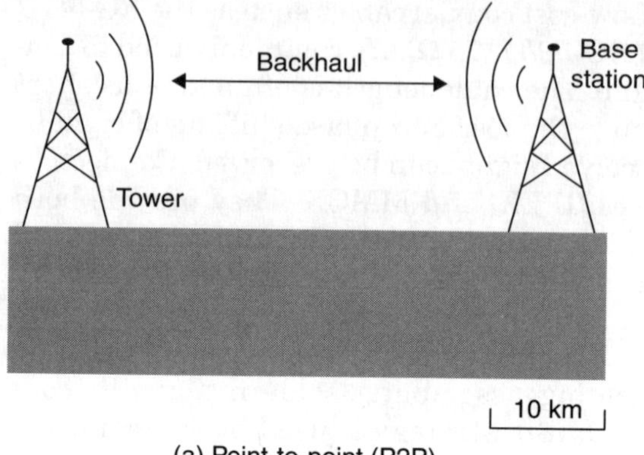

(a) Point-to-point (P2P)

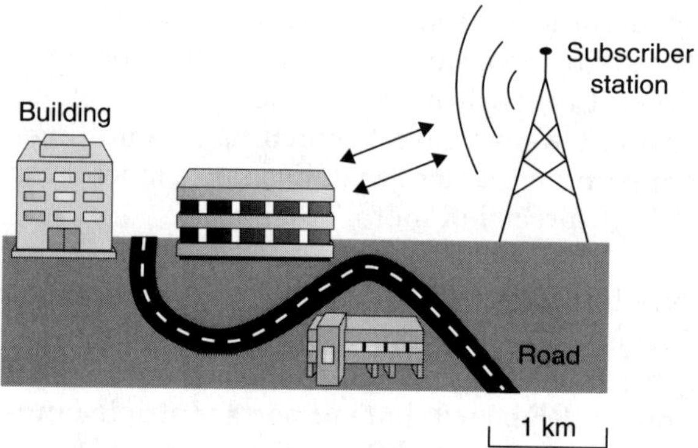

(b) Outdoor point-to-multiple points (P2MP)

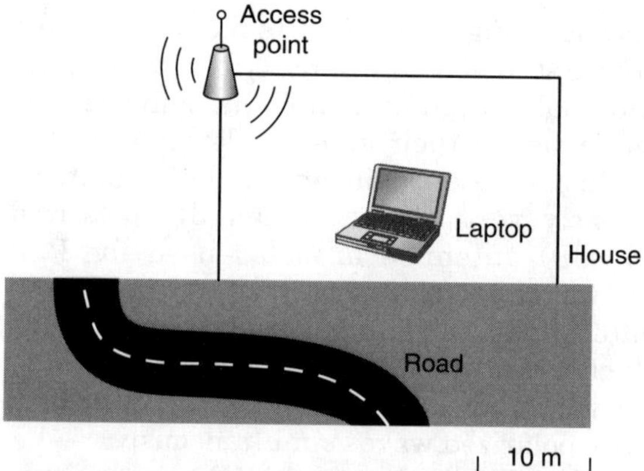

(c) Indoor point-to-multiple points (P2MP)

Figure 7.1 Radio deployment scenarios

adequate for such applications; instead, patch antenna arrays or reflector antennas are preferred. The planar patch antenna array has the advantage of having a low profile and high gain, typically to 12–18 dBi. Reflector antennas (such as parabolic dishes) and horn antennas are used when the required gain exceeds 18 dBi, despite having a larger profile and volume.

SS and AP antennas usually provide P2MP links. These antennas require broad radiation patterns in the plane where there are multiple terminals for achieving wide coverage. An array of radiating elements is required to manipulate the radiation patterns. The radiation patterns of the antenna array can be tailored to achieve a sectored or omnidirectional coverage based on the systems' requirements.

7.2.2 MIMO Antenna System Design Considerations

In a MIMO system, the effect of the antenna on system performance will be to some degree distinct from conventional WLANs. Such an effect stems from MIMO system requirements. Therefore, understanding the features of MIMO systems will be conducive to designing the antennas.

7.2.2.1 MIMO Communication System
Figure 7.2 shows the generic MIMO system as a reference for the MIMO communication channel. A stream of $Q \times 1$ vector symbols $\boldsymbol{b}^{(k)}$, where k is a time index, are fed into a space-time encoder to generate a stream of $N_T \times 1$ complex vectors $\boldsymbol{x}^{(k)}$, where N_T refers to the number of transmit antennas. The pulse shaping filters transform each element of the vector to create a $N_T \times 1$ time-domain signal vector $\boldsymbol{x}(t)$, which is up-converted to a suitable transmission carrier. The resulting signal vector $\boldsymbol{x}_A(t)$ drives the transmit transducer array, which in turn radiates energy into the propagation environment. The impulse response h_P relates the field radiated by the transmit antenna array to the field incident on the receive antenna array. The time-variant impulse response is due to the motion of

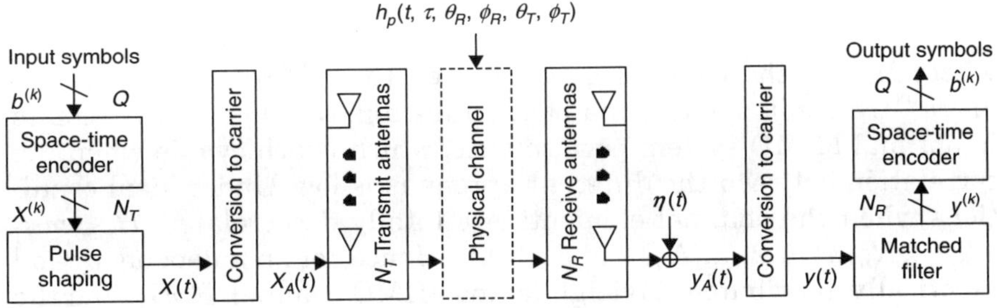

Figure 7.2 A MIMO communication system

scatterers within the propagation environment or the motion of the transmitter and/or receiver. The time delay relative to the excitation time t is represented by τ. It is assumed that the input response is finite, i.e., $h_P = 0$ for $\tau > \tau_0$ and that h_P remains constant over a time interval τ_0 so that the physical channel can be treated as a linear, time-invariant system over a single transmission. The input signal $x_A(t)$ creates the field $x_P(t, \theta_T, \phi_T)$ radiated from the transmit array, where (θ_T, ϕ_T) denote the elevation and azimuth angles. At the receive array, the field distribution is expressed as the convolution:

$$y_p(t,\theta_R,\phi_R) = \int_0^{2\pi} \int_0^{\pi} \int_{-\infty}^{\infty} h_p(t,\tau,\theta_R,\phi_R,\theta_T,\phi_T) x_p(t-\tau,\theta_T,\phi_T) \sin(\theta_T) d\tau d\theta_T d\phi_T \quad (7.4)$$

The element in the receive array samples this field and generates $N_R \times 1$ signal vector, $y'_A(t)$, at the array terminals. The noise from the propagation channel and receiver front-end electronics (thermal noise) is lumped as a $N_R \times 1$ vector $\eta(t)$ and is injected at the receive antenna terminals. The resulting signal-plus-noise vector, $y_A(t)$, is then downconverted to produce $N_R \times 1$ baseband output vector $y(t)$, which is finally passed through a matched filter whose output is sampled once per symbol to produce $y^{(k)}$, after which the space-time decoder produces estimates $\hat{b}^{(k)}$ of the originally transmitted symbols.

The nature of the MIMO channel is important in the design of efficient communication algorithms and understanding its performance limits. For a system with N_T transmitting antennas and N_R receiving antennas, and assuming frequency-flat fading over the bandwidth of interest, the MIMO channel at a given time instant can be written as

$$\boldsymbol{H} = \begin{bmatrix} H_{1,1} & H_{1,2} & \cdots & H_{1,N_T} \\ H_{2,1} & H_{2,2} & \cdots & H_{2,N_T} \\ H_{N_R,1} & H_{N_R,2} & \cdots & H_{N_R,N_T} \end{bmatrix} \quad (7.5)$$

where $H_{m,n}$ is the channel gain between the mth receiving antenna and n^{th} transmitting antenna pair. A full-rank transfer matrix results in optimal MIMO system performance, which is achievable when the correlation between the different antennas is low. Under ideal conditions when the channel elements are totally decorrelated, $H_{m,n}(m = 1,2,....,N_R, n = 1,2,....,N_T) \sim$ i.i.d. N(0,1). Hence, for an independent and identically distributed Rayleigh fading MIMO channel, $\boldsymbol{H}=\boldsymbol{H_w}$ and the spatial diversity order is equal to $N_T N_R$. However, with the increasing bandwidth and/or delay spread, the channel becomes frequency dependent. The correlation properties in the frequency domain are a function of the power delay profile. The coherence bandwidth, which is inversely proportional to the delay spread of the channel, is defined as the minimum bandwidth separation in order to achieve decorrelation. Furthermore, due to the motion of scatterers or the transmitter and/or receiver, the channel realizations also vary with time. The coherence time, which is inversely proportional to the Doppler spread, is defined as the minimum time separation that is required for the decorrelation of the time-varying channel. In the real world, due to antenna spacing and scattering, H may differ significantly from $H_{\mathbf{w}}$. Also, the presence of a line-of-sight (LOS) component will result in Ricean fading. The MIMO channel can then be modeled as the sum of a fixed and fading component:

$$H = \sqrt{\frac{K}{1+K}}\,\bar{H} + \sqrt{\frac{1}{1+K}}\,H_w \tag{7.6}$$

where $\sqrt{\frac{K}{1+K}}\,\bar{H}$ is the LOS component and $\sqrt{\frac{1}{1+K}}\,H_w$ is the uncorrelated fading component. K (≥ 0) is the Ricean factor of the channel and is defined as the ratio of the power in the LOS component to the power of the fading component. $K = 0$ corresponds to a Rayleigh channel and $K = \infty$ corresponds to a nonfading channel.

7.2.2.2 MIMO System Capacity

Capacity (bits/s/Hz) of a communication system can be defined as the maximum rate at which reliable communication is possible, which can be characterized in terms of the mutual information between the input and the output of the channel. For a time-invariant additive white Gaussian noise (AWGN) channel with bandwidth B and received SNR γ, the Shannon's channel capacity can be given as[9]

$$C = B \log_2(1 + \gamma) \tag{7.7}$$

In the case where there are N channels and the transmit power is equally divided among them, the capacity becomes

$$C = \sum_{n=1}^{N} \log_2\left(1 + \frac{\gamma}{N}\right) = N \log_2\left(1 + \frac{\gamma}{N}\right) \tag{7.8}$$

For deterministic MIMO channels, the channel gain matrix H is fixed. It is assumed that the transmitter does not have any channel state information (CSI) and hence cannot optimize the power allocation among the antennas. The input vectors are independent complex Gaussian-distributed random variables with equal variance σ^2. Therefore, the channel capacity can be expressed as

$$C = \max_{\{R_x : \mathrm{Tr}(R_x) \leq P_T\}} \log_2 \left| I + \frac{HR_x H^H}{\sigma^2} \right| \tag{7.9}$$

The diagonal elements of the transmit covariance matrix represent the transmit power from each antenna. The off-diagonal elements of R_x represent the correlation between the transmitted signal streams. An increased correlation will result in a decreased capacity. Also, the term $HR_x H^H$ represents the covariance of the received signal in the absence of noise, such that the λ_{ii} eigenvalue represents the received signal level in the ith eigenchannel.

When the transmitter has no CSI, it divides its power equally among the transmit antennas to form N_T independent streams, or $R_x = (P_T/N_T)I$. The capacity can be given by

$$C = \log_2 \left| I + \frac{P_T}{N_T \sigma^2} HH^H \right| \tag{7.10}$$

When the transmitter has CSI, the use of equal power allocation is suboptimal. The optimal solution can be obtained by applying the water-filling principle. In order to maximize Eq. 7.11, $R'_{x,ii}$ must be diagonal.[10] Capacity is given by

$$C = \max_{\left\{ R'_x : \sum_i R'_{x,ii} \leq P_T \right\}} \sum_{i=1}^{N_R} \log_2 \left(1 + \frac{S_{ii}^2 R'_{x,ii}}{\sigma^2} \right) \tag{7.11}$$

where $R'_{x,ii}$ represents the optimal transmit power on the i_{th} unencoded stream and S_{ii}^2 is the power gain of the i_{th} eigenchannel. The values of $R'_{x,ii}$ that maximize Eq. 7.11 can be determined using Lagrange multipliers to obtain the water-filling solution.[9–12] This method allocates power to the high-gain channels and generally does not use weaker channels.

7.2.2.3 Antenna Effects on MIMO Capacity

Antenna properties, such as impedance matching, pattern, polarization, array configuration, and mutual coupling affect the correlation. Angle (pattern) diversity occurs when the antennas have distinct radiation patterns. Large capacity gains are possible when the element patterns are appropriately designed to minimize the correlation. Also, by directing the majority of the radiation in the direction where most of the multipath components are concentrated, higher capacity can be achieved.[13] The correlation can be calculated from the S-parameters according to Eq. 7.12[14] or

Eq. 7.13 if the electric fields for both polarizations over the entire three-dimensional space are known.

$$\rho = \frac{|\,S_{11}^{*}S_{12} + S_{21}^{*}S_{22}\,|^{2}}{(1 - (\,|\,S_{11}\,|^{2} + |\,S_{21}\,|^{2}\,))(1 - (\,|\,S_{22}\,|^{2} + |\,S_{12}\,|^{2}\,))} \tag{7.12}$$

$$\rho_{e,ij} = \frac{\left|\,\iint_{4\pi}\left[\vec{F}_{i}(\theta,\phi) \bullet \vec{F}_{j}(\theta,\phi)\right] d\Omega\,\right|^{2}}{\iint_{4\pi}\left|\vec{F}_{i}(\theta,\phi)\right|^{2} d\Omega \iint_{4\pi}\left|\vec{F}_{j}(\theta,\phi)\right|^{2} d\Omega} \tag{7.13}$$

From Eq. 7.12, the effects of mutual coupling and impedance matching on the correlation can be studied directly. Figure 7.3 shows that as the return loss ($|\,S_{11}\,|$) increases, the mutual coupling S_{21} has to be reduced in order to achieve low correlation. The mutual coupling requirement can be significantly relaxed when the antenna is well-matched. For instance, when the return loss is −10 dB, the mutual coupling can be around −4 dB and yet achieve a correlation of 0.7, assuming that the S_{11} and S_{21} are both in phase.

However, if they are not in phase, the correlation will be sensitive to the phase difference if the mutual coupling ($|\,S_{21}\,|$) is too high, as shown in Figure 7.4. The effects of the phase difference on the correlation becomes insignificant when $|\,S_{21}\,| < -10$ dB. Hence, if the mutual coupling and return loss are kept lower than −10 dB, a very low correlation of less than 0.1 can be achieved.

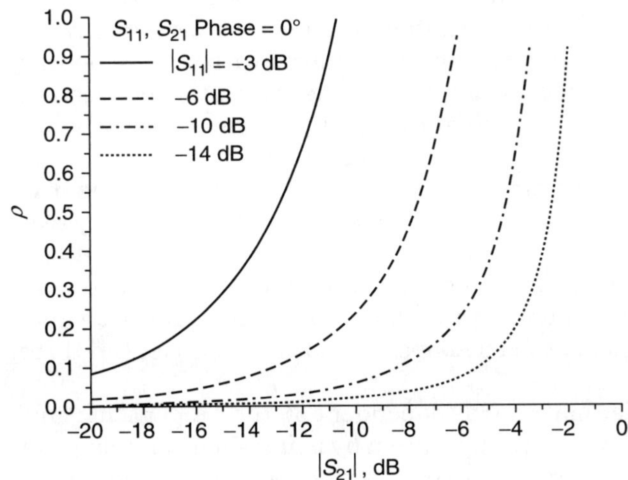

Figure 7.3 Effects of $|\,S_{11}\,|$ and $|\,S_{21}\,|$ on correlation

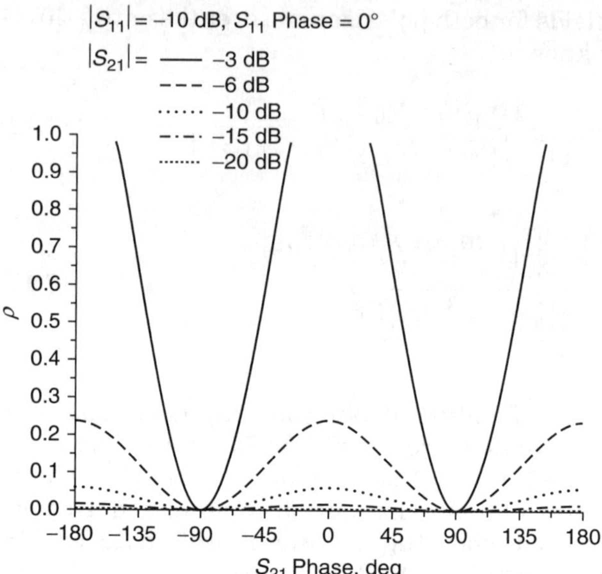

Figure 7.4   Effects of phase difference on correlation

From Eq. 7.13, the information about the pattern and polarization diversity can be directly derived, which will be conducive to the optimization of the antenna system configuration. First, if the radiating elements are of orthogonal polarization, i.e., $\bar{F}_i(\theta,\phi) \bullet \bar{F}_j(\theta,\phi) = 0$, ρ_e will be zero. Second, $\rho_e = 0$ can also be achieved with pattern diversity, which means that the antenna elements can have the same polarization but their radiation patterns do not overlap each other in space, i.e., $|\iint_{4\pi} [\bar{F}_i(\theta,\phi) \bullet \bar{F}_j(\theta,\phi)]\, d\Omega| = 0$. In other words, the average spatial fidelity of the radiation patterns of the two antenna elements is zero. As a result, ρ_e can be minimized by having radiation patterns with complementary coverage in space. Therefore, by minimizing the correlation between the antenna elements, both the polarization and pattern diversity will be able to enhance the capacity of the MIMO system.

In addition, the envelope correlation coefficient computed by the truncated Laplacian model encompasses the effect of all three diversity techniques (spatial, polarization, and pattern) on the signal correlation.[15] However, this two-dimensional method can only evaluate the average correlation coefficient over a particular plane and does not provide information on the variation of the envelope correlation coefficient in three-dimensional space.

The channel transfer matrix $\boldsymbol{H}$ is dependent on the propagation environment as well as the array configuration by using single or dual polarization and multibeam array structures at the base station and dipole arrays at the mobile.[16] Studies have shown that the average capacity is relatively insensitive to the array configuration. However, it is possible to enhance the capacity by having an adaptive system that selectively connects a subset of available antennas to the electronic modules.[17–20] Furthermore, in a rich multipath environment, six uncorrelated signals at the receiver can be achieved from sensing the three Cartesian vector components of the electric and magnetic fields. A large multipath angle spread is able to offer a higher number (maximum of six) independent communication channels.[21–23] However, constructing half wavelength dipoles or full wavelength loops to achieve this is difficult due to mutual coupling, and nonideal radiation pattern characteristics, which reduces the number of effective (independent) channels. Typically, when two polarizations are used, scattering leads to a 4–10-dB higher co-polarized signal as compared to the cross-polarized signal.[24] As a result, the transfer matrix exhibits low correlation coupled with weak channel gain between the two orthogonally polarized channels.[25–28] A capacity gain of around 10–20% is achieved for dual-polarization over single polarized spatially separated elements in an indoor environment.[25] Regardless of the environment, with the use of dual polarization, at least two channels will be enabled. Mutual coupling is also one of the key issues in MIMO systems.[29–30] Studies have shown that when the power collection capability is enhanced through mutual coupling, the capacity when two dipoles are closely coupled to each other through proper termination can be increased.[31–32] For a fixed length array, a strong mutual coupling will lead to an upper bound in the capacity performance, especially when the spacing between the elements is less than $\lambda/2$.[33]

Therefore, in the antenna design for MIMO systems, the mutual coupling between the elements has a critical impact on system performance although the overall system performance is influenced by many factors, in particular, the channels through which the RF signals propagate. The optimized design of antennas with low mutual coupling will be able to exploit the full benefits of diversity in MIMO systems. As a result, antennas with low mutual coupling will be conducive to the performance of MIMO systems. Furthermore, MIMO systems will simultaneously benefit from all types of possible diversity, for instance, spatial, pattern, and polarization diversity.

7.3   State-of-the-Art Designs

There are many types of commercial antennas that meet the demands of the fast growing market (Table 7.2). A series of practical antenna designs that are commercially available are discussed in the following sections.

7.3.1   Outdoor Point-to-Point Antennas

Outdoor antennas have to be weatherproof (IEC IP66),[34] i.e., operate under direct and indirect sunlight, resistant to wind, rain, snow, and provide lightning surge protection. Common outdoor P2P antennas include

TABLE 7.2   Various Types of WLAN Antennas in the Market

Antennas	Gain, dBi			
	0–6	6–8	8–18	18–30
Single monopole	✔			
Helix (normal mode)	✔			
Single dipole	✔			
Slot		✔		
Log-periodic		✔		
Dipole, slot, patch (arrays)			✔	✔
Yagi-Uda			✔	✔
Horn			✔	✔
Reflector, dish				✔

patch arrays, Yagi-Uda arrays, log-periodic arrays, helix antennas, cavity-backed slot antennas, waveguide slot arrays, horn antennas, and reflector antennas. The beamwidth of these antennas is usually between 20°–60° based on the system requirements. A beamwidth of less than 10° requires additional antenna alignment procedures. Also, an antenna with a high directivity has a lower probability of encountering unpredictable interferences.

Figure 7.5 shows a 2.4–2.5-GHz P2P antenna array that has a gain of 14 dBi. It uses weatherproof galvanized sheet metal with surface treatment for the ground plane. The single-layered radiators with a feeding network are cut from a 0.5-mm brass sheet and suspended above the

Frequency	2.4–2.5 GHz
Gain	14 dBi
Return loss	>10 dB
Polarization	Linear
H-plane 3-dB beamwidth	28°
E-plane 3-dB beamwidth	28°
Connectors	N-type
Size	200×200×5 mm

(a)   (b)

Figure 7.5   A 14-dBi P2P antenna: (*a*) photo of antenna, (*b*) antenna specifications, (*c*) schematic diagram, (*d*) return loss, and (*e*) radiation patterns at 2.4 GHz (Photo courtesy of Compex Systems Pte Ltd.)

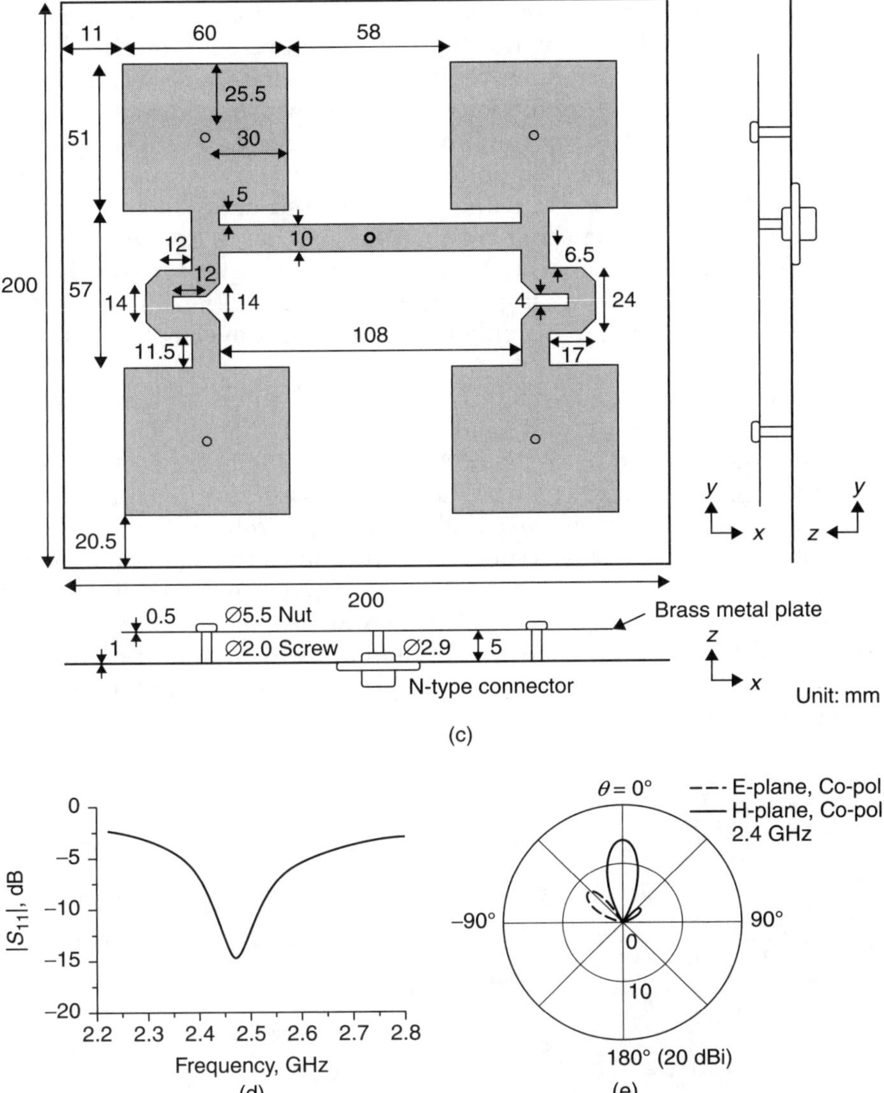

Figure 7.5 A 14-dBi P2P antenna: (a) photo of antenna, (b) antenna specifications, (c) schematic diagram, (d) return loss, and (e) radiation patterns at 2.4 GHz (Photo courtesy of Compex Systems Pte Ltd.) (*Continued*)

ground plane for enhancing the operating bandwidth. The radiators are supported by ripped polycarbonate material—commonly used for plastic bottles—with riveted screws at the middle of the radiators. The folded ground plane at the edges allows the radome to be secured onto the antenna easily. In order to reduce the length of the feeding lines, the array is fed in the middle using an N-type connector bolted to the ground plane.

The meandered feeding lines after the T-junctions are designed to ensure that the signals at the top radiators are equi-amplitude but 180° out-of-phase with respect to the bottom radiators. As a thin meandered line results in a reduction of the bandwidth, whereas a closely coupled line increases the coupling loss, optimizing the feeding lines is, therefore, necessary. The patches are separated by approximately half a wavelength so as to achieve the desired array factor. Grating lobes occur when the spacing between the patches is greater than one wavelength, while a separation of less than one quarter wavelength increases the mutual coupling and reduces the radiation efficiency. The height of this patch array is only 5 mm (4% of a wavelength); hence, it is narrowband.

Figure 7.6 shows a 5.4–5.9 GHz, 16 dBi P2P antenna comprising 12 brass radiator elements and an aluminum ground plane. This antenna is designed to be fed near the bottom edge of the ground plane so as to fit into a low-profile casing. The feeding structure is simplified to ensure a low fabrication cost and mechanical robustness. However, the long series-fed array structure limits the usable operating bandwidth because a non-squinting radiation pattern is required. It is difficult to ensure an equi-amplitude and in-phase signal distribution to all the radiators across a broad bandwidth. The first sidelobe level can be reduced by altering the phase or spatial distribution of the radiators.

Frequency	5.4–5.9 GHz
Gain	16 dBi
Return loss	>10 dB
Polarization	Linear
H-plane 3-dB beamwidth	16°
E-plane 3-dB beamwidth	16°
Connectors	SMA
Size	155×213×5 mm

(a)   (b)

Figure 7.6 A 16-dBi P2P antenna: (*a*) photo of antenna, (*b*) antenna specifications, (*c*) schematic diagram, (*d*) return loss, (*e*) radiation patterns at 5.6 GHz, and (*f*) radiation patterns at 5.8 GHz

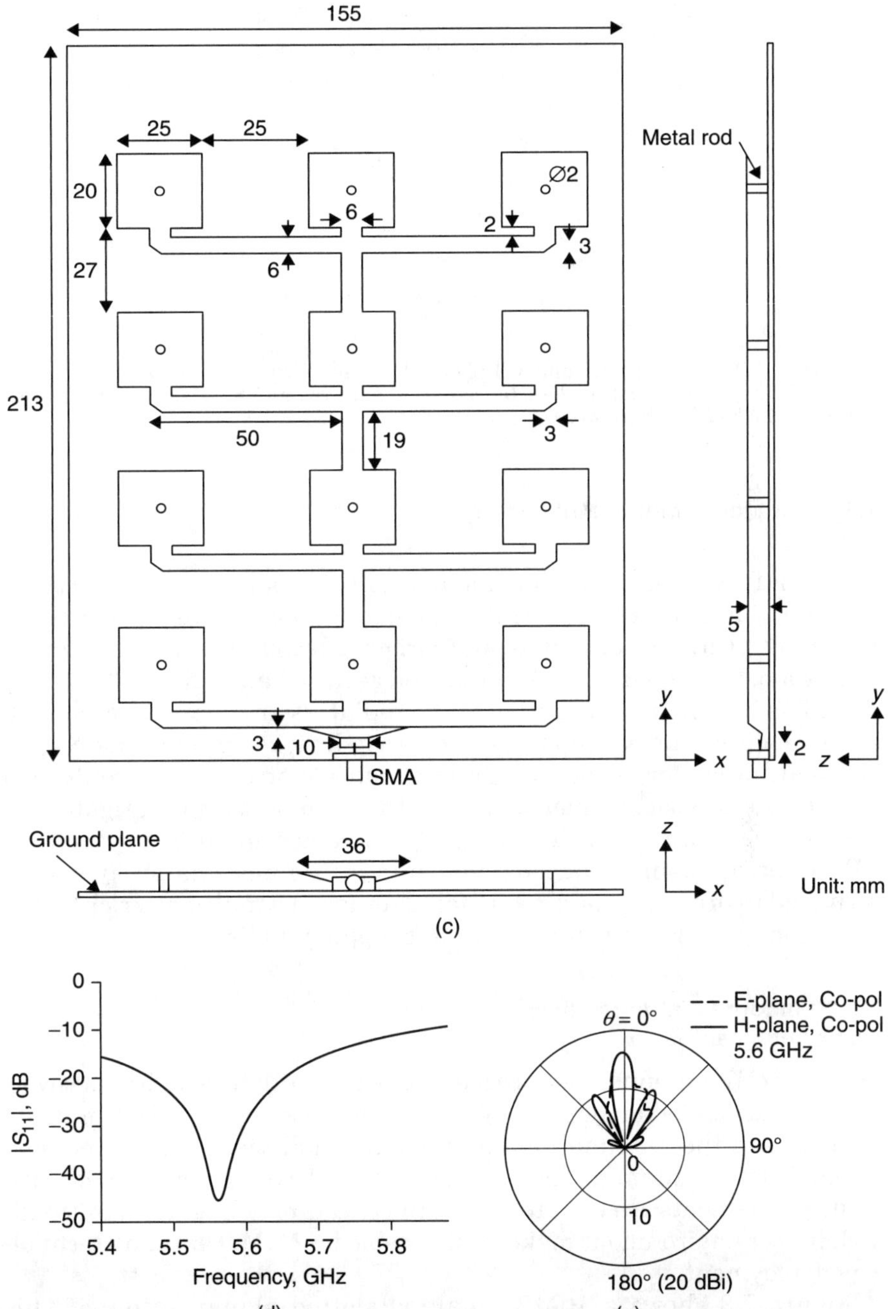

Figure 7.6 A 16-dBi P2P antenna: (*a*) photo of antenna, (*b*) antenna specifications, (*c*) schematic diagram, (*d*) return loss, (*e*) radiation patterns at 5.6 GHz, and (*f*) radiation patterns at 5.8 GHz (*Continued*)

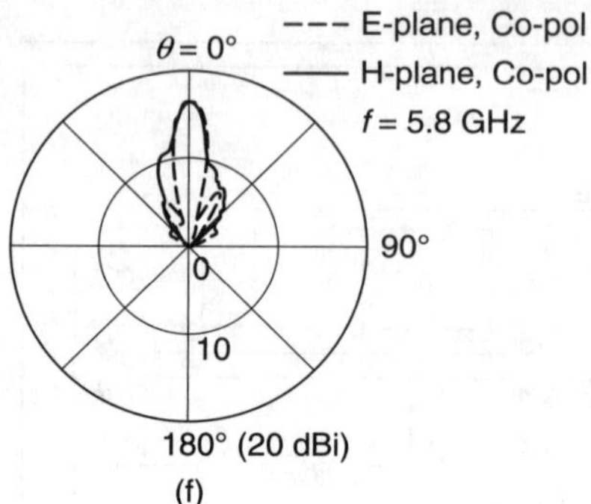

Figure 7.6 A 16-dBi P2P antenna: (*a*) photo of antenna, (*b*) antenna specifications, (*c*) schematic diagram, (*d*) return loss, (*e*) radiation patterns at 5.6 GHz, and (*f*) radiation patterns at 5.8 GHz (*Continued*)

7.3.2   Outdoor Point-to-Multiple-Point Antennas

Outdoor P2MP antennas include omnidirectional antennas, sectored antennas, and arrays, such as the patch antennas, sleeve dipoles, colinear dipoles, turnstile antennas, and corner reflector antennas. Figure 7.7 shows a 5.4–5.9 GHz, 17-dBi center-fed sectored antenna with a series feed. The length and separation of the radiators are approximately half a wavelength. The width of the microstrip line, soldered to the N-type probe at the center, is widened for impedance matching. This center-fed structure is almost symmetrical along the E-plane, thus it mitigates the undesirable beam-squinting effect. By implementing it in a cylindrical MIMO array fashion, various elements in the antenna array can be switched to provide omnidirectional coverage in a multipath rich urban environment, e.g., on campuses or in shopping malls.

7.3.3   Indoor Point-to-Multiple Point Antennas

Indoor P2MP antennas require a broad beamwidth for maximum coverage. A base loaded monopole, e.g., a rubber duck antenna, is commonly used due to the antenna's omnidirectional radiation. Broadband suspended patch antennas are also employed. Diversity antennas with multiple elements are used to reduce the effects of fading. The multipath rich indoor environment makes it favorable for MIMO antenna technology deployment.

Figure 7.8 shows a 10-dBi dual-fed slotted planar antenna. This antenna operates from 4.9–6.0 GHz, which covers the IEEE 802.11j (Japan) band, public safety band (U.S.), and the IEEE 802.11a band. The radiator is single-layered and easy to manufacture. The dual-fed

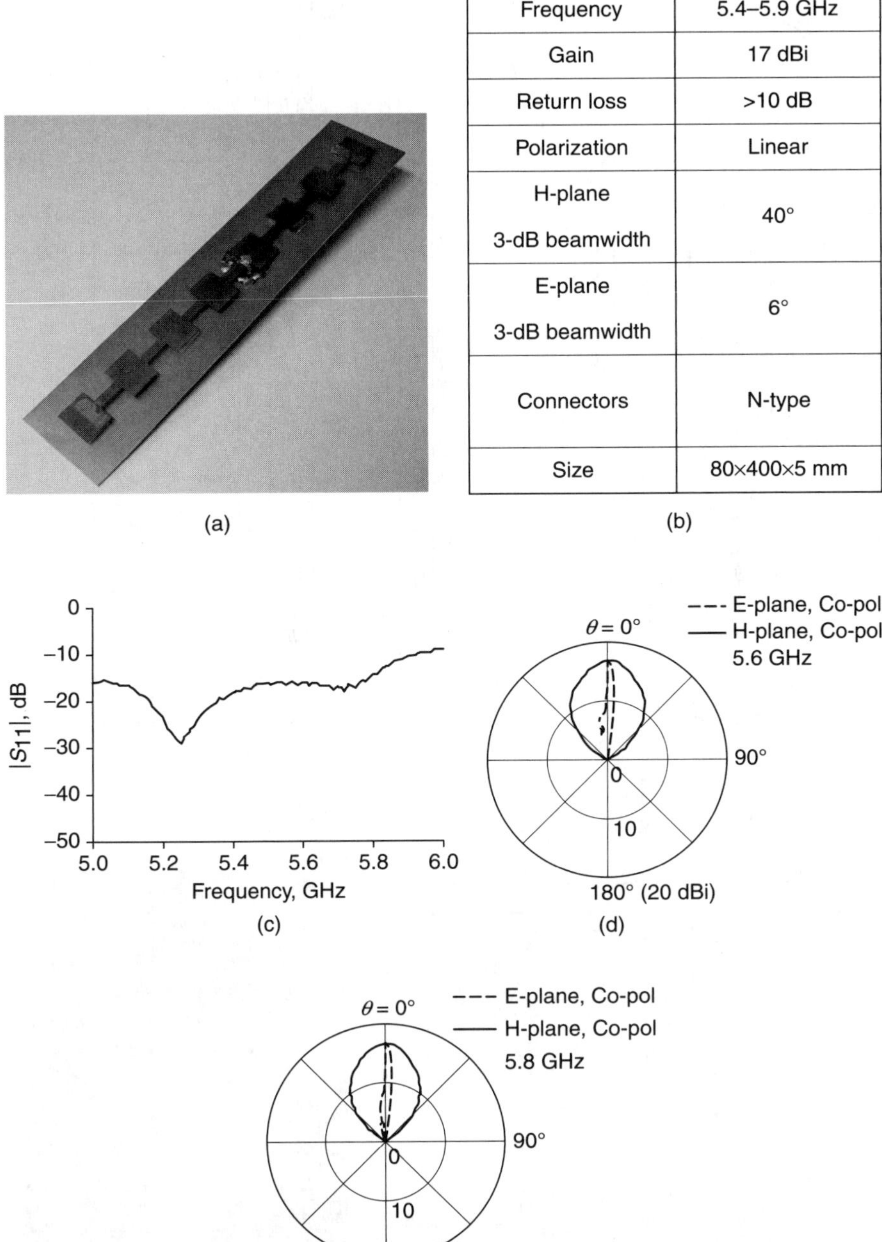

Frequency	5.4–5.9 GHz
Gain	17 dBi
Return loss	>10 dB
Polarization	Linear
H-plane 3-dB beamwidth	40°
E-plane 3-dB beamwidth	6°
Connectors	N-type
Size	80×400×5 mm

Figure 7.7 A 17-dBi P2MP antenna: (*a*) photo of antenna, (*b*) antenna specifications, (*c*) return loss, (*d*) radiation patterns at 5.6 GHz, (*e*) radiation patterns at 5.8 GHz, and (*f*) MIMO sectored array

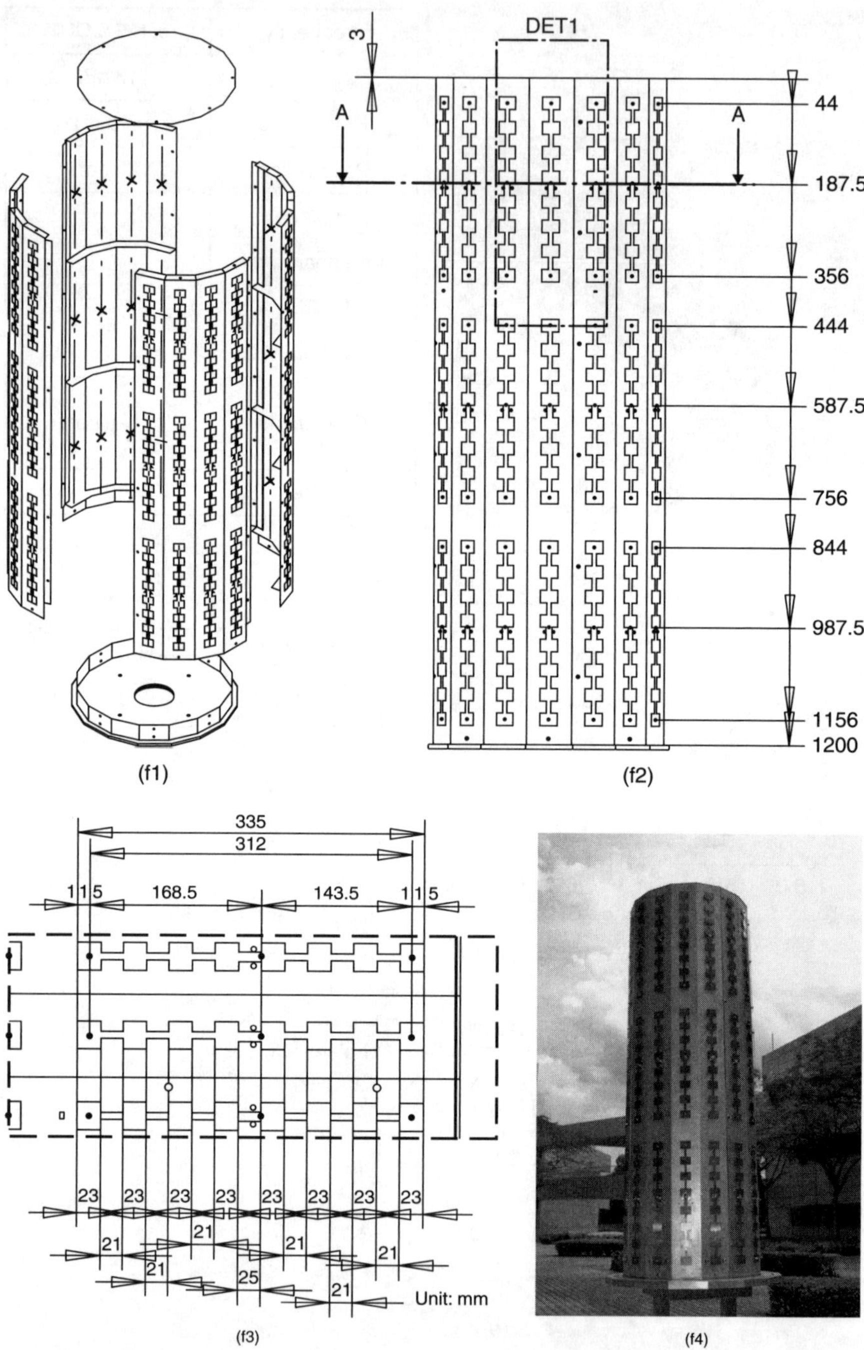

Figure 7.7 A 17-dBi P2MP antenna: (*a*) photo of antenna, (*b*) antenna specifications, (*c*) return loss, (*d*) radiation patterns at 5.6 GHz, (*e*) radiation patterns at 5.8 GHz, and (*f*) MIMO sectored array (*Continued*)

(a)

Frequency	4.9–6.0 GHz
Gain	10 dBi
Return loss	>10 dB
Polarization	Linear
H-plane 3-dB beamwidth	45°
E-plane 3-dB beamwidth	45°
Connectors	N-type
Size	100×100×5 mm

(b)

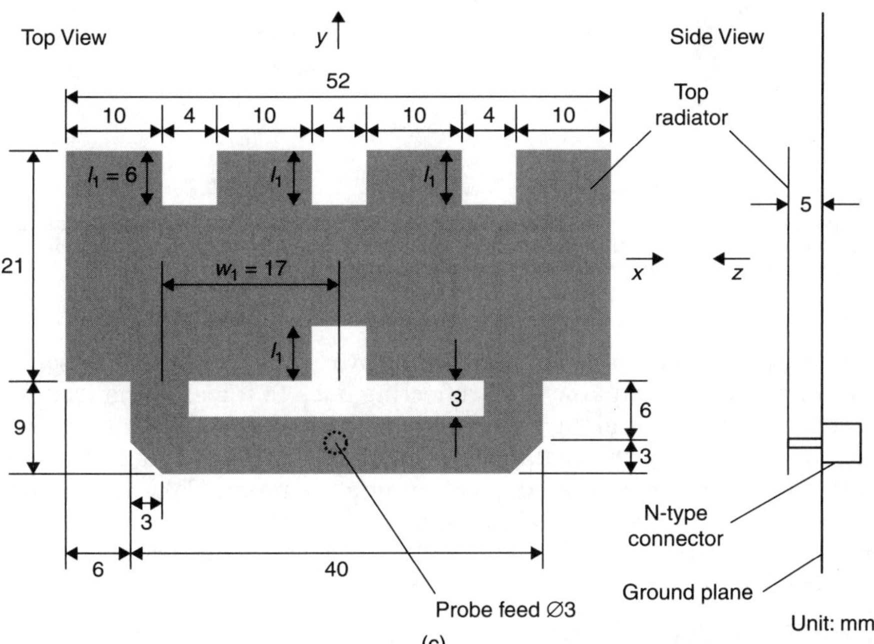

(c)

Figure 7.8 A 10-dBi dual-fed slotted planar antenna: (*a*) photo of antenna, (*b*) antenna specifications, (*c*) schematic diagram, (*d*) return loss, (*e*) gain profile, (*f*) H-plane radiation patterns, and (*g*) E-plane radiation patterns

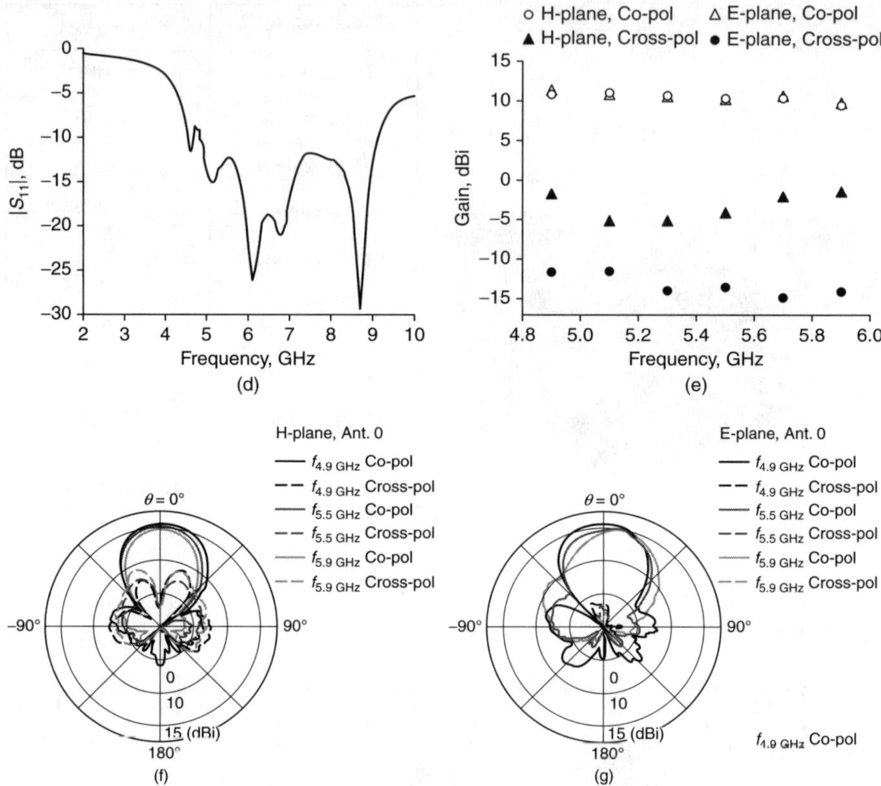

Figure 7.8 A 10-dBi dual-fed slotted planar antenna: (*a*) photo of antenna, (*b*) antenna specifications, (*c*) schematic diagram, (*d*) return loss, (*e*) gain profile, (*f*) H-plane radiation patterns, and (*g*) E-plane radiation patterns (*Continued*)

structure improves impedance matching by reducing the strong resonating waves on the radiator from reflecting back to the antenna feed. By feeding strategically along the radiating edge and introducing reactance loading by slotting the radiator, the higher order modes, e.g., TM_{10} and TM_{20} modes, are suppressed, sustaining a dominant TM_{01} mode resonant across a wide band of frequencies. Therefore, a consistent gain profile and 3-dB beamwidth, as well as low cross-polarization, can be achieved across the broad bandwidth.

7.4 Case Studies

Based on the discussion of various specifications and antenna design considerations in WLAN systems, several antenna designs are discussed in this section from an engineering perspective. The practical issues in these designs are highlighted.

7.4.1 Indoor P2MP Embedded Antenna

Figure 7.9 shows a compact WLAN antenna design with high manufacturability. The center slot and the shorting pin that is located opposite to the feed are able to reduce the size of the patch and lower the cross-polarization radiation by suppressing the higher order modes.[35–37]

The geometry of the suspended slot-loaded plate antenna is shown in Figure 7.10a. A rectangular copper plate (25 mm × 22 mm × 0.1 mm) is placed at a height of 4 mm parallel to a ground plane of dimensions 155 mm × 108 mm. The medium between the radiator and the ground plane is air and can be supported by a foam layer of $\varepsilon_r \approx 1.06$. A narrow rectangular slot of dimensions 1 mm × 13 mm,

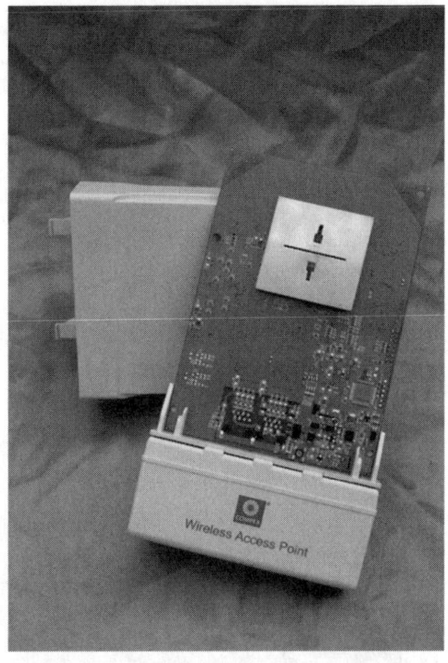

Figure 7.9 A 2.4-GHz suspended plate antenna with shorting pins and slots (Photo courtesy of Compex Systems Pte Ltd.)

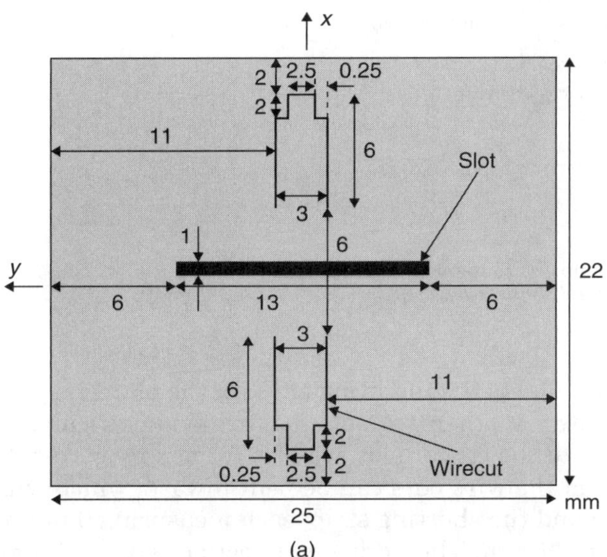

Figure 7.10 Embedded antenna: (a) radiator design, (b) side view of antenna, and (c) feed design

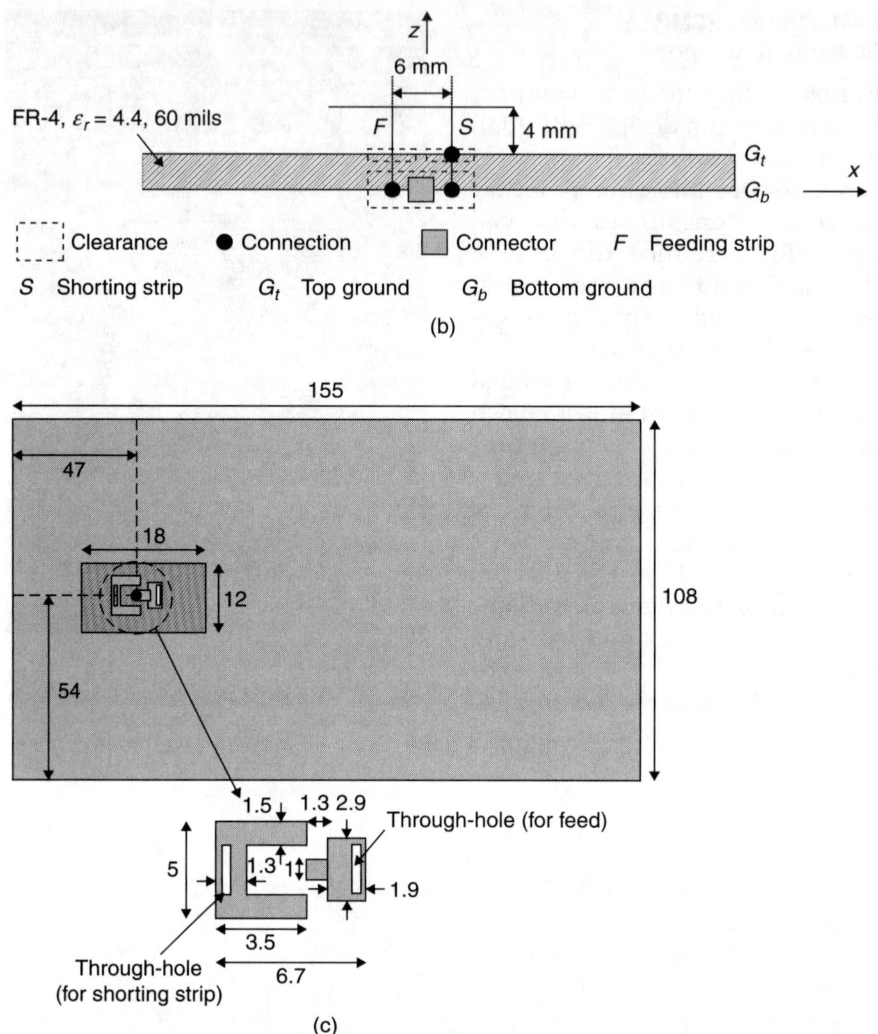

Figure 7.10 Embedded antenna: (*a*) radiator design, (*b*) side view of antenna, and (*c*) feed design (*Continued*)

with the longer sides parallel to the radiating edges of the plate, is symmetrically cut at the center. Another two slots that are symmetrical and orthogonal with respect to the center slot are also formed by wire-cut. The copper sections from the wire-cuts can be bent inward, which will act as the feeding strip and the shorting strip, each measuring 3 mm × 4 mm and separated by 6 mm. The additional section with a 2-mm width that is attached to the feeding strip will go through the hole of the same size on the ground plane in order to ensure that the height is controlled precisely at 4 mm. The feed point is located on the bottom layer of the dielectric substrate, as shown in Figure 7.10b. An area of the component ground plane measuring 18 mm × 12 mm is isolated as shown in Figure 7.10c, where the feeding and shorting strips are soldered onto the two copper pads. The IPEX connector is placed between the two copper pads.

From Figure 7.11, it can be seen that the antenna is able to achieve good impedance matching from 5.1–6 GHz and maintain stable radiation patterns and a peak gain of 6 dBi across the bandwidth. For brevity, only the radiation patterns at 5.5 GHz are given.

Figure 7.12 shows a tunable dual-band (2.26–2.54 GHz / 5.14–6.10 GHz) antenna. Dual-band antennas are preferred over broadband antennas, when it is difficult to simultaneously achieve the required gain and radiation performances across a very broad frequency range. A dual-band antenna also provides better out-of-band isolation, which reduces the filtering requirement. This antenna design consists of two patch radiators suspended above a ground plane. This allows each frequency band to be tuned independently while retaining impedance matching and radiation performance at both frequency bands.

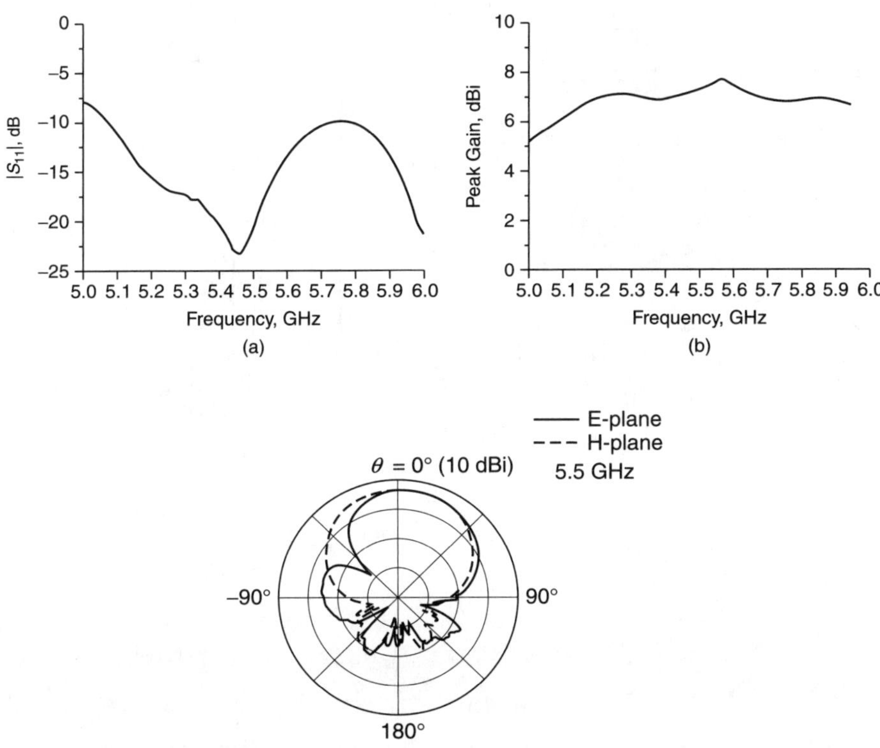

Figure 7.11 Measured results: (a) return loss, (b) gain profile, and (c) radiation patterns at 5.5 GHz

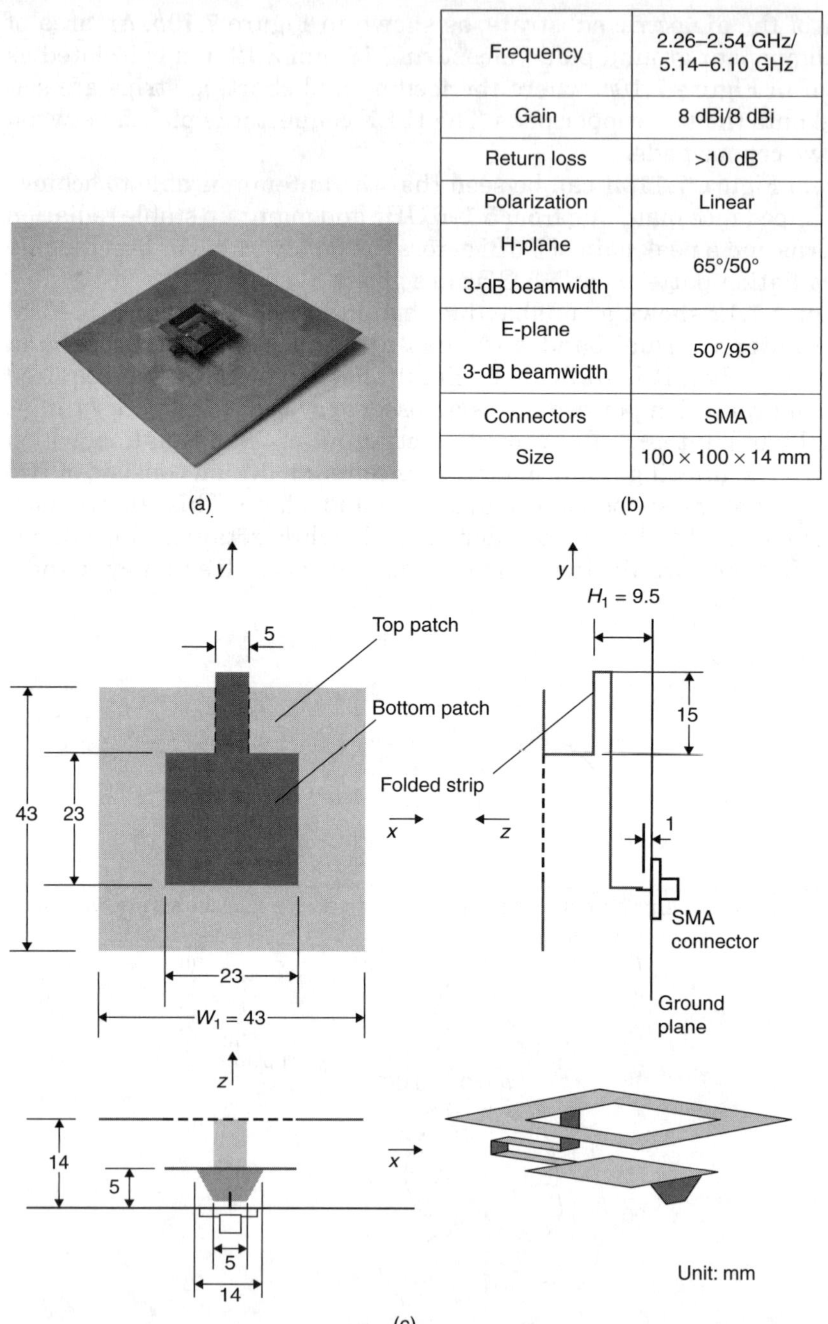

Frequency	2.26–2.54 GHz/ 5.14–6.10 GHz
Gain	8 dBi/8 dBi
Return loss	>10 dB
Polarization	Linear
H-plane 3-dB beamwidth	65°/50°
E-plane 3-dB beamwidth	50°/95°
Connectors	SMA
Size	100 × 100 × 14 mm

Figure 7.12 Tunable dual-band antenna: (a) photo of antenna, (b) antenna specifications, (c) schematic diagram, (d) gain profile, (e) effects of W_1 on the lower band, (f) effects of H_1 on the upper band, (g) radiation patterns at 2.4 GHz, and (h) radiation patterns at 5.6 GHz

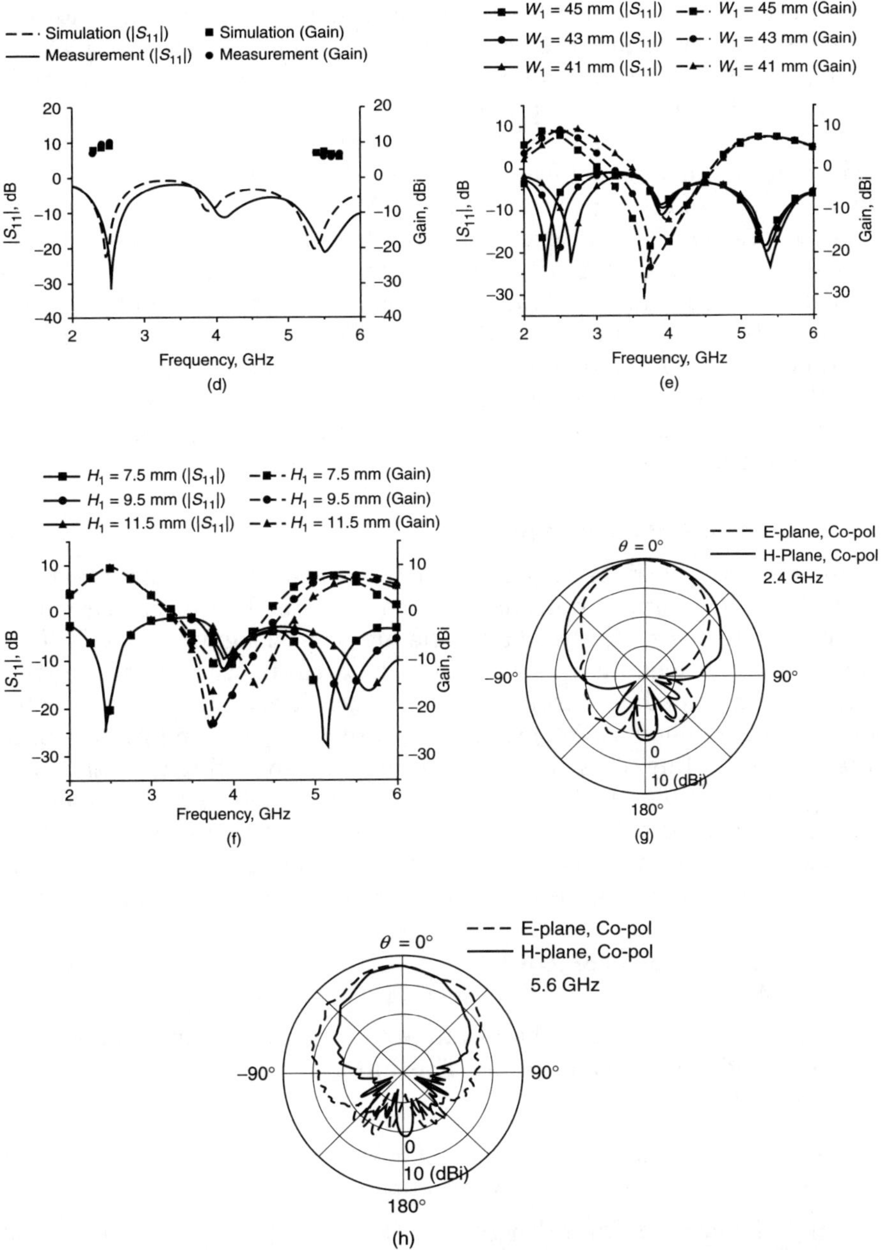

Figure 7.12 Tunable dual-band antenna: (*a*) photo of antenna, (*b*) antenna specifications, (*c*) schematic diagram, (*d*) gain profile, (*e*) effects of W_1 on the lower band, (*f*) effects of H_1 on the upper band, (*g*) radiation patterns at 2.4 GHz, and (*h*) radiation patterns at 5.6 GHz (*Continued*)

The lower and upper bands can be controlled by varying the width of the top radiator W_1 and the height of the folded feed H_1, respectively. In doing so, the reactance loading on different sections of the antenna is altered with minimum coupling.[38]

Figure 7.13 shows a broadband suspended plate antenna. A consistent gain of 8 dBi is achieved over a bandwidth of 44% (2.3–3.6 GHz) by using a folded two-layer structure.[39] The folded structure suppresses the occurrence of higher order modes when the height of the antenna is elevated beyond 10% of the free space wavelength λ_0. Besides the broadband impedance matching and consistent gain, this antenna also features broad beamwidth in the azimuth plane and low cross-polarization. The broadband characteristics are partially attributed to the coupling between the top and bottom radiators. The retarded field from the bottom radiator adds up in-phase to that of the top, resulting in a non-squinting radiation pattern.

7.4.2 Outdoor P2P Antenna Array

Antenna array designs require a feeding network with low insertion loss. The phase and amplitude of each radiator has to be regulated. Figure 7.14 shows a tri-band antenna array with a gain of 17 dBi. This single-layered feeding structure does not require the usual quarter-wavelength impedance transformer at each junction. The bandwidth of the feeding network is optimized by properly selecting the width of the microstrip line. This design maintains non-squinting radiation patterns and constant gain profile across the entire 5–6 GHz (18.2%) band. A relatively evenly distributed power is achieved by using this hybrid series- and corporate-feeding network, which ensures that most of the elements are radiating in-phase across a broad frequency band.

7.4.3 Dual-Band Outdoor P2P Antenna Array

Figure 7.15 shows a dual-band antenna array operating at both the 2.4 GHz and 5.4–5.9 GHz bands. This antenna is able to overcome the following design challenges:

- Encompassing two sets of array elements, operating at two different frequency bands within a compact space
- Single-layered radiator design
- Good impedance matching and radiation efficiency at both frequency bands
- Single port feeding
- Relatively stable radiation patterns and constant gain at both frequency bands

(a)

Frequency	2.3–3.6 GHz
Gain	8 dBi
Return loss	>10 dB
Polarization	Linear
H-plane 3-dB beamwidth	42°
E-plane 3-dB beamwidth	30°
Connectors	SMA
Size	100×100×16 mm

(b)

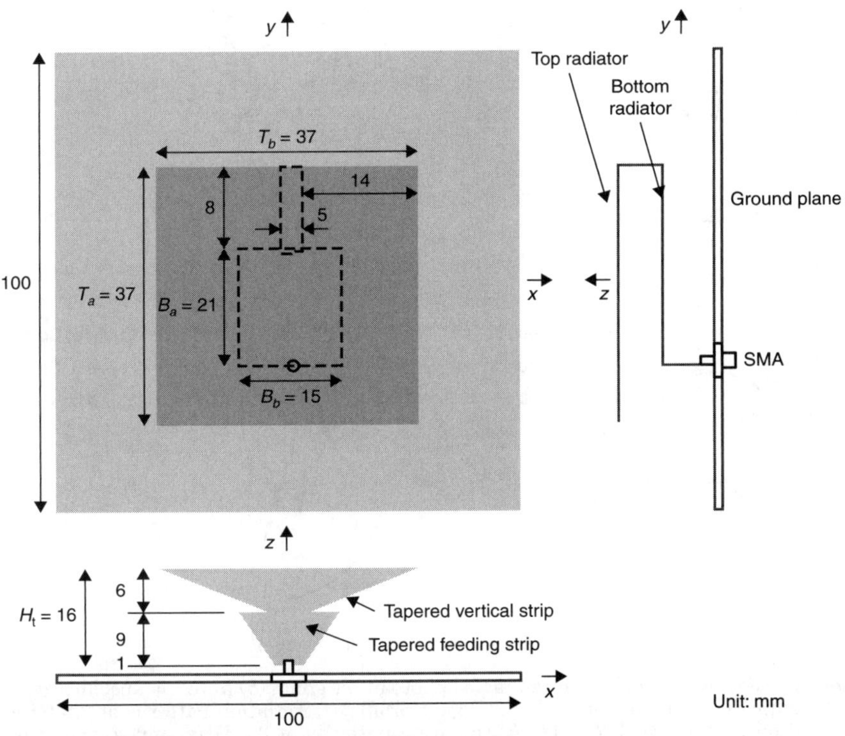

(c)

Figure 7.13 Folded broadband antenna: (*a*) photo of antenna, (*b*) antenna specifications, (*c*) schematic diagram, (*d*) return loss, (*e*) gain profile, (*f*) radiation patterns at 2.3 GHz, (*g*) radiation patterns at 2.7 GHz, (*h*) radiation patterns at 3.3 GHz, and (*i*) radiation patterns at 3.6 GHz

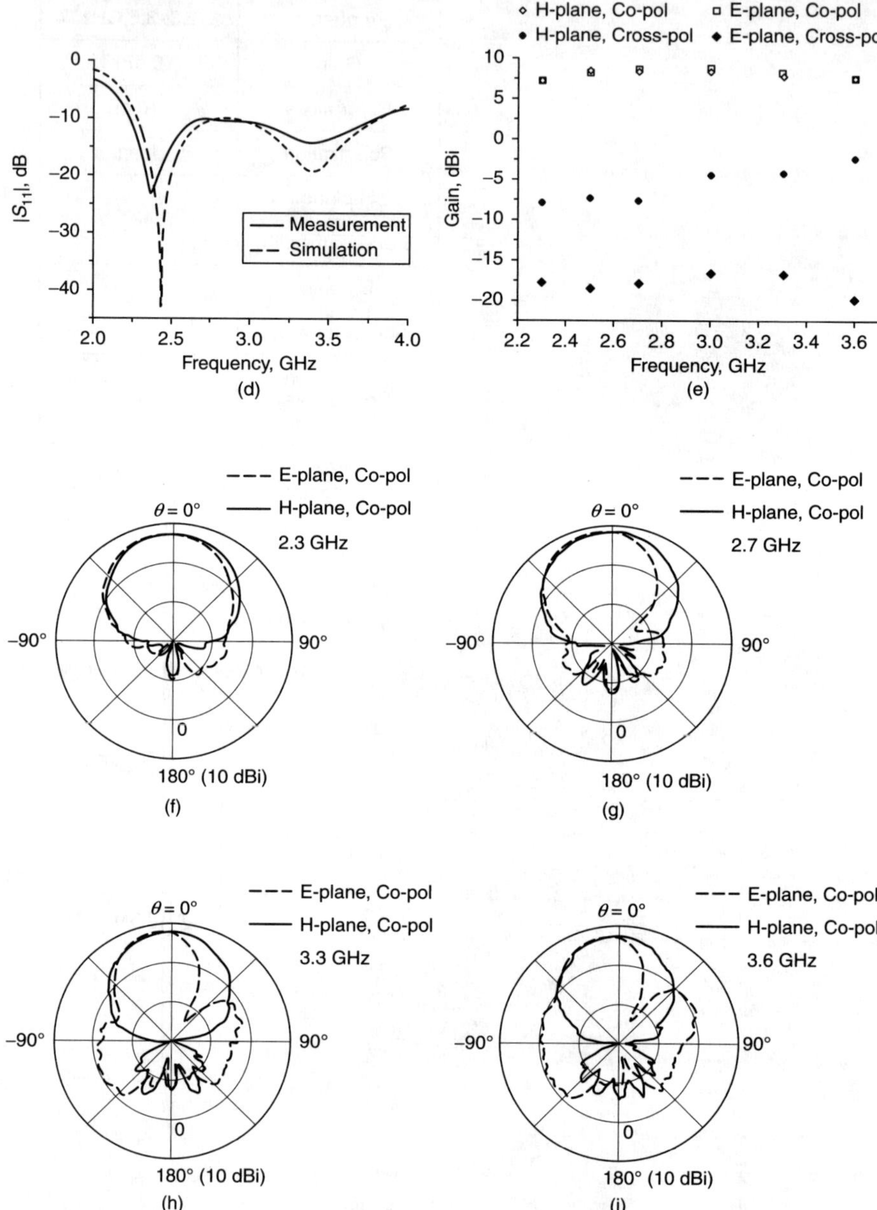

Figure 7.13 Folded broadband antenna: (a) photo of antenna, (b) antenna specifications, (c) schematic diagram, (d) return loss, (e) gain profile, (f) radiation patterns at 2.3 GHz, (g) radiation patterns at 2.7 GHz, (h) radiation patterns at 3.3 GHz, and (i) radiation patterns at 3.6 GHz (*Continued*)

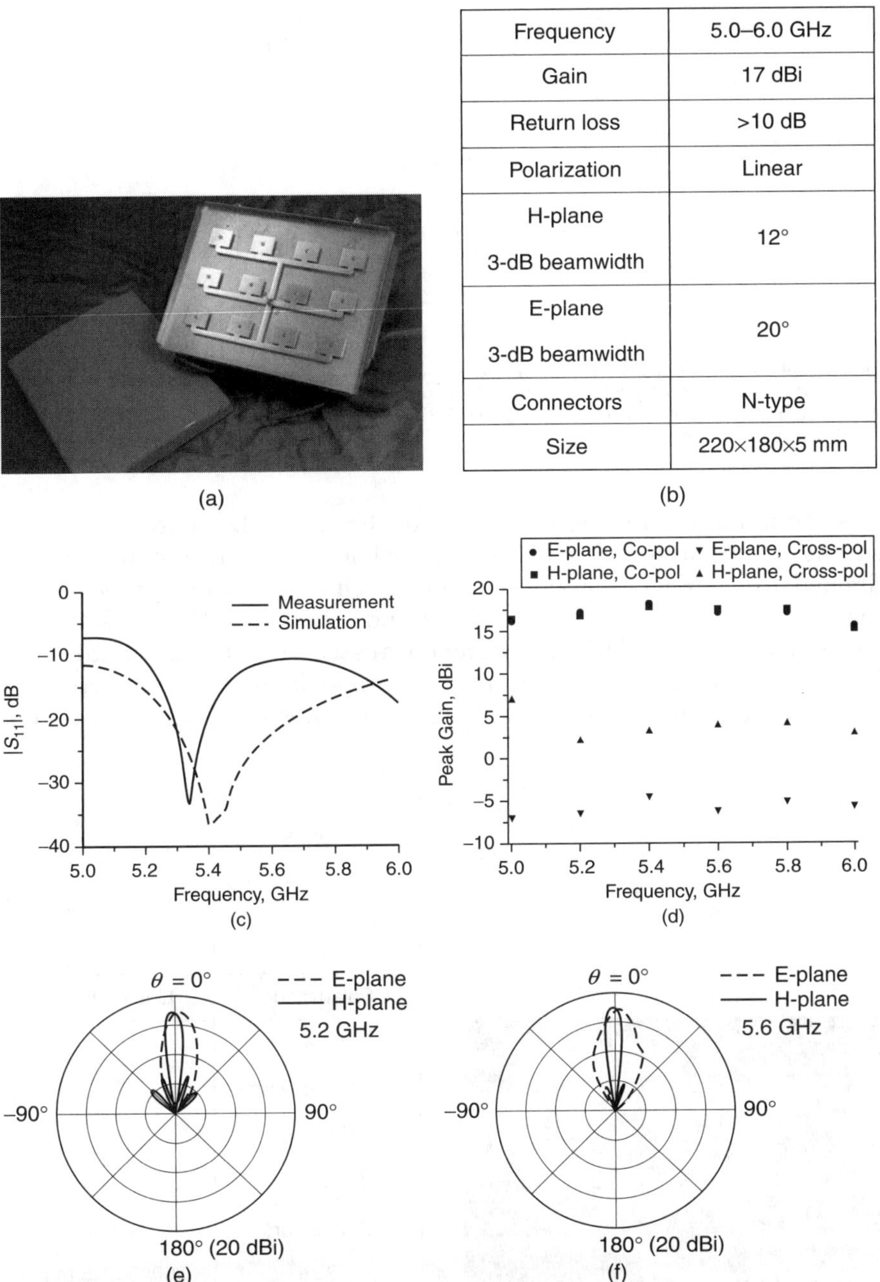

Figure 7.14 Tri-band antenna array: (*a*) photo of antenna, (*b*) antenna specifications, (*c*) return loss, (*d*) gain profile, (*e*) radiation patterns at 5.2 GHz, (*f*) radiation patterns at 5.6 GHz, and (*g*) radiation patterns at 5.8 GHz (Photo courtesy of Compex Systems Pte Ltd.)

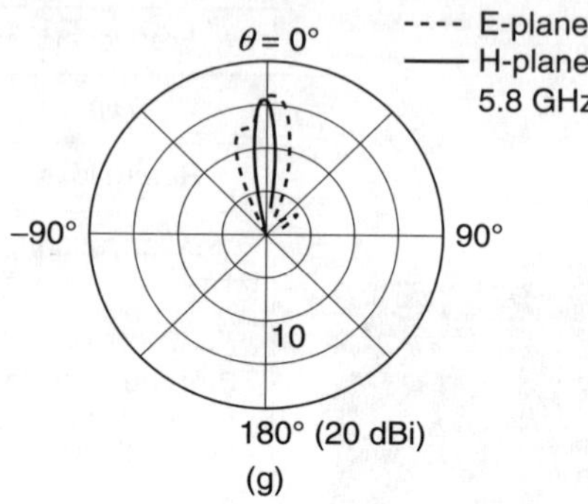

Figure 7.14 Tri-band antenna array: (*a*) photo of antenna, (*b*) antenna specifications, (*c*) return loss, (*d*) gain profile, (*e*) radiation patterns at 5.2 GHz, (*f*) radiation patterns at 5.6 GHz, and (*g*) radiation patterns at 5.8 GHz (Photo courtesy of Compex Systems Pte Ltd.) (*Continued*)

Although the radiation pattern at the 2.4-GHz band is slightly squinted, its gain remains >10 dBi at the boresight. This antenna can be divided into four sets of array elements. Each set consists of three series-fed radiators: two 5.4-GHz radiators and a large 2.4-GHz radiator at the end. Given its relatively symmetrical structure, the polarization is kept linear and cross-polarization is suppressed. The vertical microstrip section at the center provides a 90° and 180° phase shift for the 2.4-GHz

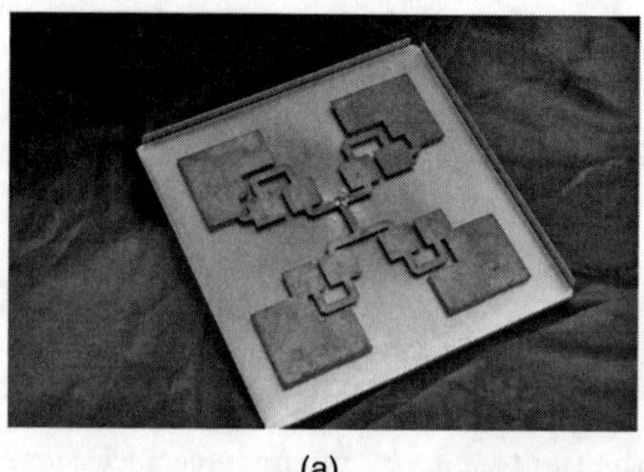

Frequency	2.4–2.5 GHz/ 5.4–5.85 GHz
Gain	10 dBi/11 dBi
Return loss	>10 dB
Polarization	Linear/Linear
H-plane 3-dB beamwidth	25°/20°
E-plane 3-dB beamwidth	25°/35°
Connectors	N-type
Size	300×300×5 mm

(a)   (b)

Figure 7.15 Dual-band antenna array: (*a*) photo of antenna, (*b*) return loss, (*c*) gain profile, (*d*) radiation patterns at 2.4 GHz, (*e*) radiation patterns at 5.5 GHz, (*g*) average current distribution at 2.4 GHz, and (*h*) average current distribution at 5.6 GHz

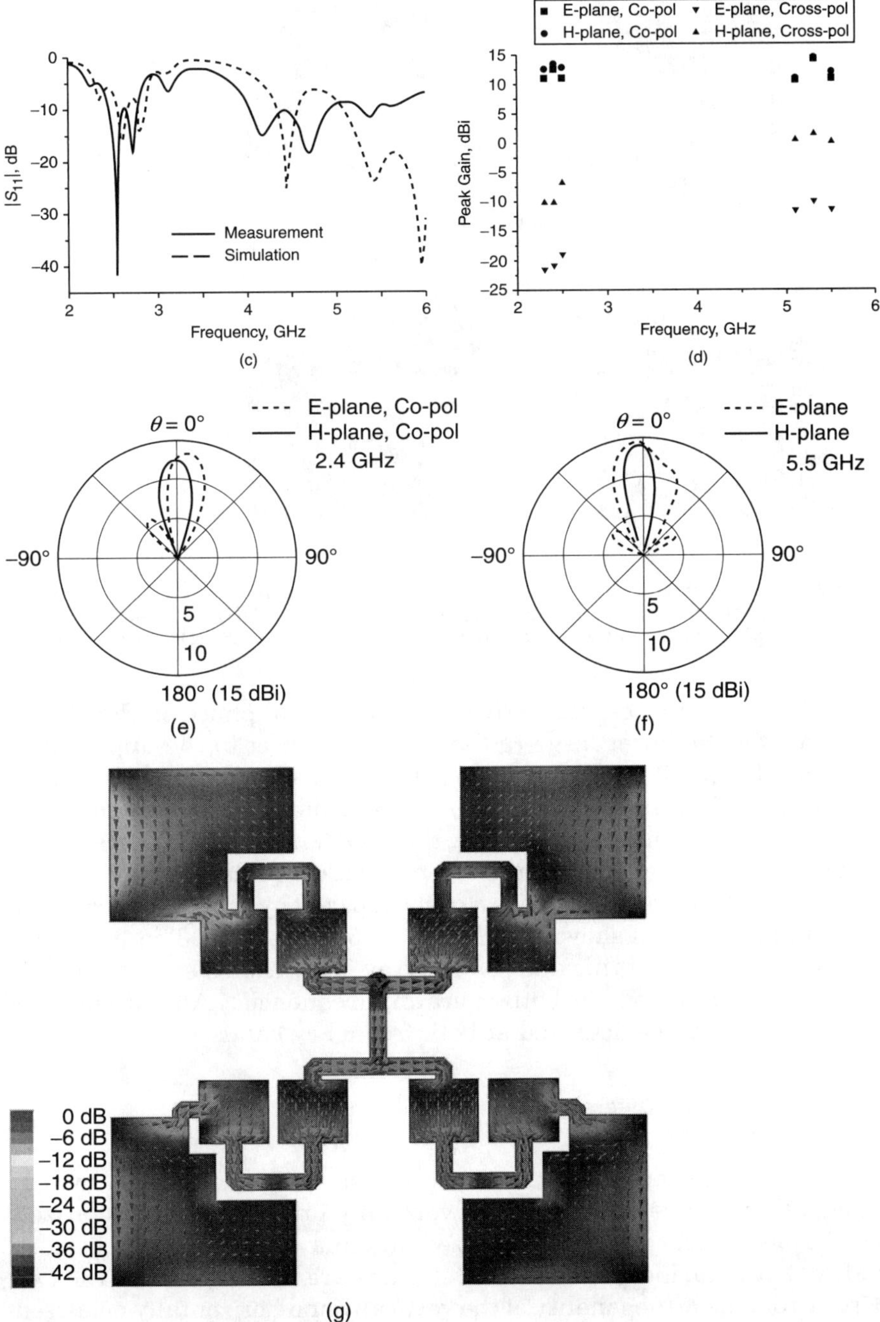

Figure 7.15 Dual-band antenna array: (*a*) photo of antenna, (*b*) return loss, (*c*) gain profile, (*d*) radiation patterns at 2.4 GHz, (*e*) radiation patterns at 5.5 GHz, (*g*) average current distribution at 2.4 GHz, and (*h*) average current distribution at 5.6 GHz (*Continued*)

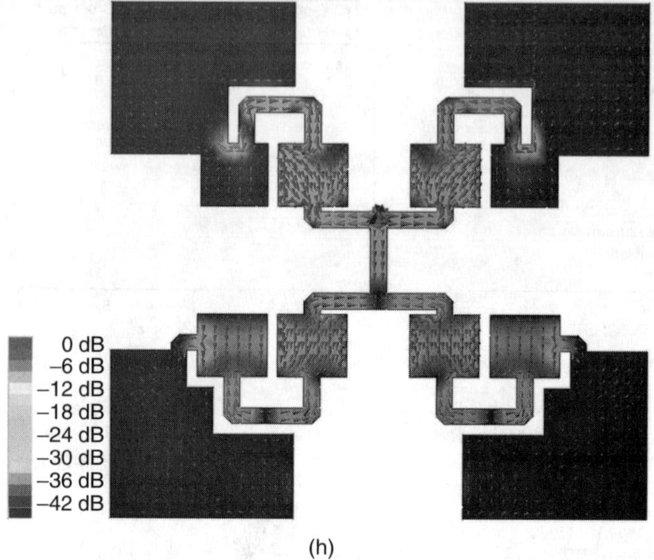

(h)

Figure 7.15 Dual-band antenna array: (*a*) photo of antenna, (*b*) return loss, (*c*) gain profile, (*d*) radiation patterns at 2.4 GHz, (*e*) radiation patterns at 5.5 GHz, (*g*) average current distribution at 2.4 GHz, and (*h*) average current distribution at 5.6 GHz (*Continued*)

and 5-GHz resonances, respectively. An additional phase shift of 90° is allocated for the lower large radiators with respect to the upper large radiators for the 2.4-GHz operation. Using this meandered structure, the larger radiators resonate at 2.4 GHz while the smaller radiators operate as a transmission line. At the 5-GHz resonance, the smaller radiators resonate while radiation from the large radiators is effectively suppressed. This phenomenon is clearly depicted by the average current distribution plots, as shown in Figures 7.15*g* and 7.15*h*. Therefore, the radiation aperture of this dual-band array is controlled so as to ensure a constant gain profile at both operating frequencies. Also, a low first sidelobe level can be achieved at both frequency bands.

7.4.4 Outdoor P2P Diversity Grid Antenna Array

Most of the P2P or P2MP antennas consist of only one port, which transmits and receives either the vertically or horizontally polarized wave. However, for the dual-polarized antenna element that is used in a MIMO system, independent data streams are fed into each of the two ports. From the orthogonality of the vertically and horizontally polarized fields, two uncorrelated channels are created for transmission so that the overall data throughput is increased. The orthogonally polarized radiation can be achieved without any increase in size. Figure 7.16 shows

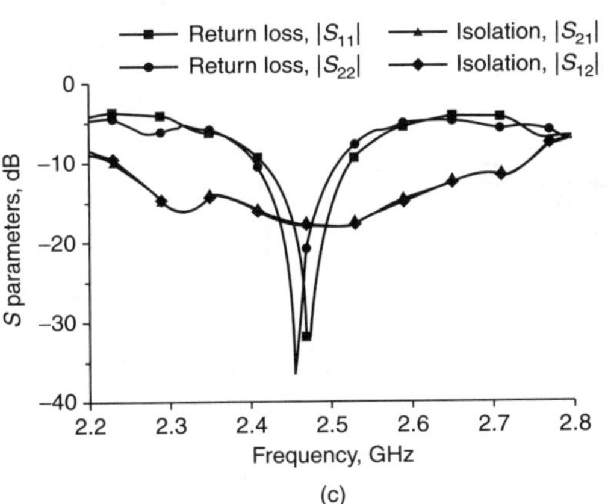

Frequency	2.4–2.5 GHz
Gain	14 dBi
Return loss	>10 dB
Polarization	Dual linear
Isolation	>10 dB
H-plane 3-dB beamwidth	30°
E-plane 3-dB beamwidth	30°
Connectors	SMA
Size	250×250×5 mm

(a) (b)

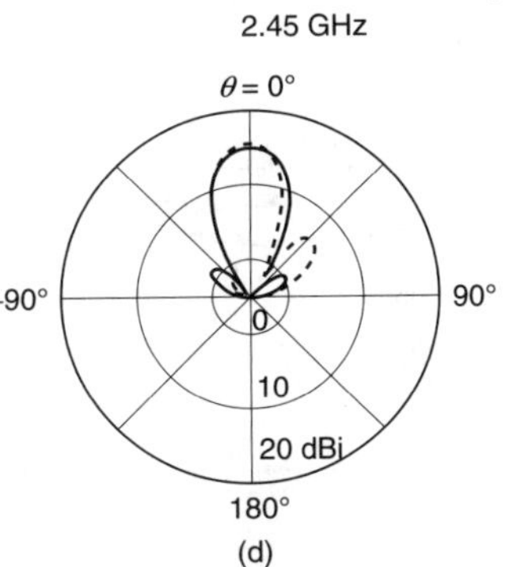

(c) (d)

Figure 7.16 Dual-polarized grid antenna array: (*a*) photo of antenna, (*b*) antenna specifications, (*c*) return loss and isolation, and (*d*) radiation patterns at 2.45 GHz (Photo courtesy of Compex Systems Pte Ltd.)

a dual-polarized P2P antenna that features a gain of 14 dBi over 2.4–2.5 GHz. The antenna design uses low-cost FR-4 material for both radiators and ground plane. It is fed directly by two coaxial cables, thus avoiding the need for any expensive RF connectors.

Figure 7.17 shows a dual-polarization antenna array that operates from 5.45–5.85 GHz and has a gain of 16 dBi. The 3-dB beamwidth in the E- and H-planes is 18°. The undesirable beam-squinting and cross-polarization increase as the array becomes larger in size. This antenna is fed by two folded stubs soldered to the SMA connectors.

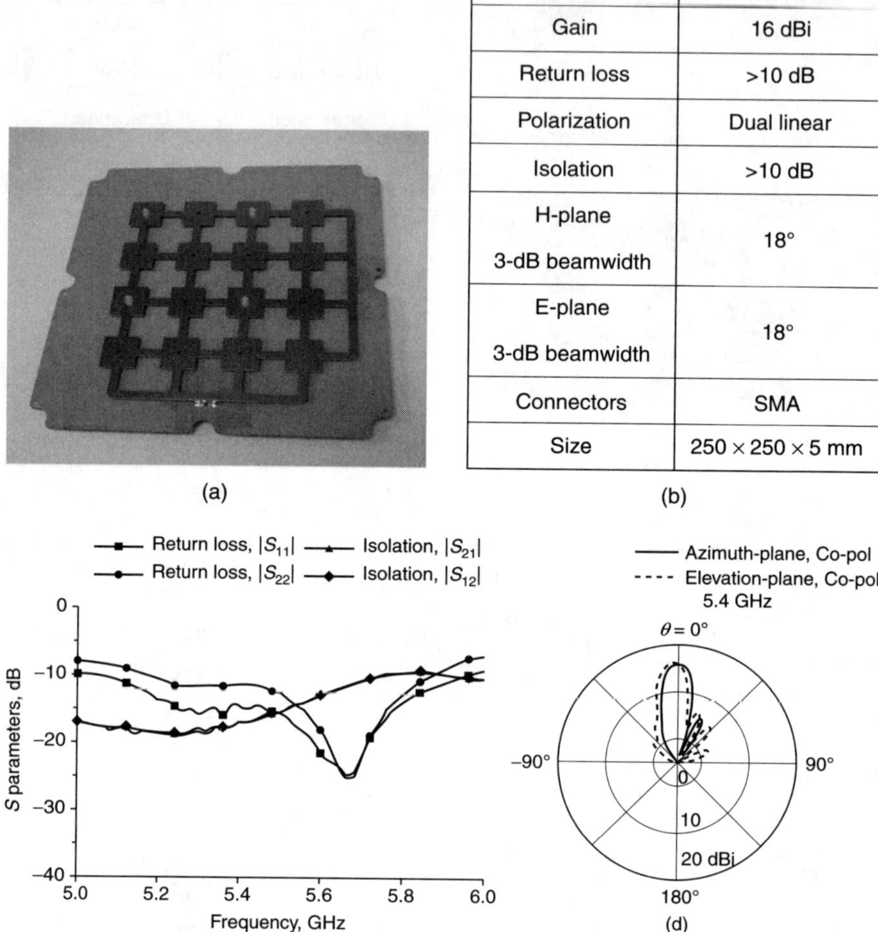

Frequency	5.45–5.85 GHz
Gain	16 dBi
Return loss	>10 dB
Polarization	Dual linear
Isolation	>10 dB
H-plane 3-dB beamwidth	18°
E-plane 3-dB beamwidth	18°
Connectors	SMA
Size	250 × 250 × 5 mm

(a)   (b)   (c)   (d)

Figure 7.17 A 16-dBi dual-polarity antenna array: (*a*) photo of antenna, (*b*) antenna specifications, (*c*) return loss and isolation, (*d*) radiation patterns at 5.4 GHz, (*e*) radiation patterns at 5.6 GHz, (*f*) radiation patterns at 5.8 GHz, and (*g*) broadband folded stub impedance matching structure (Photo courtesy of Compex Systems Pte Ltd.)

The length of the feeding lines has been optimized for impedance matching. Good impedance matching is easily achieved by using an N-type connector; the larger inner conductor of the N-type connector (diameter: 3 mm) results in a lower current density and, hence, a lower Q-value. On the other hand, SMA connectors and IPEX connectors have a smaller inner conductor, with a diameter of 1.3 mm and 0.5 mm, respectively. To achieve broadband performance, the folded stub structure shown in the figure is configured to simulate the larger N-type inner connector.

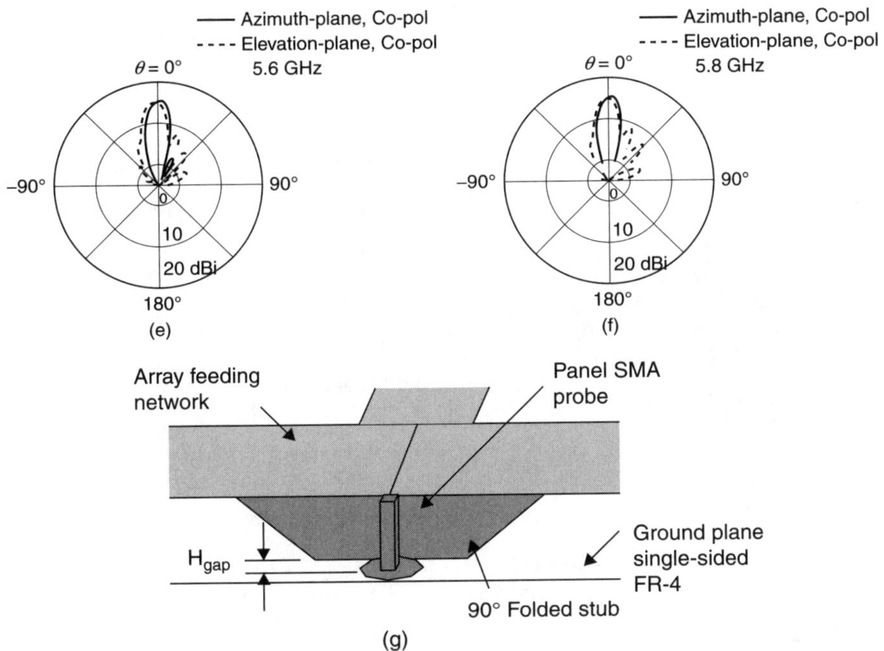

Figure 7.17 A 16-dBi dual-polarity antenna array: (*a*) photo of antenna, (*b*) antenna specifications, (*c*) return loss and isolation, (*d*) radiation patterns at 5.4 GHz, (*e*) radiation patterns at 5.6 GHz, (*f*) radiation patterns at 5.8 GHz, and (*g*) broadband folded stub impedance matching structure (Photo courtesy of Compex Systems Pte Ltd.) (*Continued*)

The H_{gap} of 0.1–0.5 mm can be easily controlled by soldering the folded stub on top of a piece of cardboard or adhesive tape.

7.4.5 Outdoor/Indoor P2MP HotSpot/HotZone Antenna

Figure 7.18 shows an embedded IEEE 802.11n P2MP antenna operating from 2.4–2.5 GHz. It is fed using the folded stub impedance matching structure shown in Figure 7.17*g*, which is soldered through a plate via connected to an IPEX connector using a co-planar waveguide with a ground plane on a four-layered PCB board. The construction of the four-layered PCB board is as follows: layer 1, surface mount components with RF mini PCI radio card on a 7 mils FR-4 substrate; layer 2, baseband signal lines on a 49 mils FR-4 substrate; layer 3, power lines on a 7 mils FR-4 substrate; and layer 4, antenna ground plane. The IPEX connector is soldered onto a co-planar waveguide with a finite ground plane structure, and the folded stub is inserted and soldered to the plated via. The same PCB can also be used for another dual-polarized antenna array operating from 5.45–5.85 GHz, as shown in Figure 7.19. In doing so, manufacturing and inventory cost can be significantly reduced.

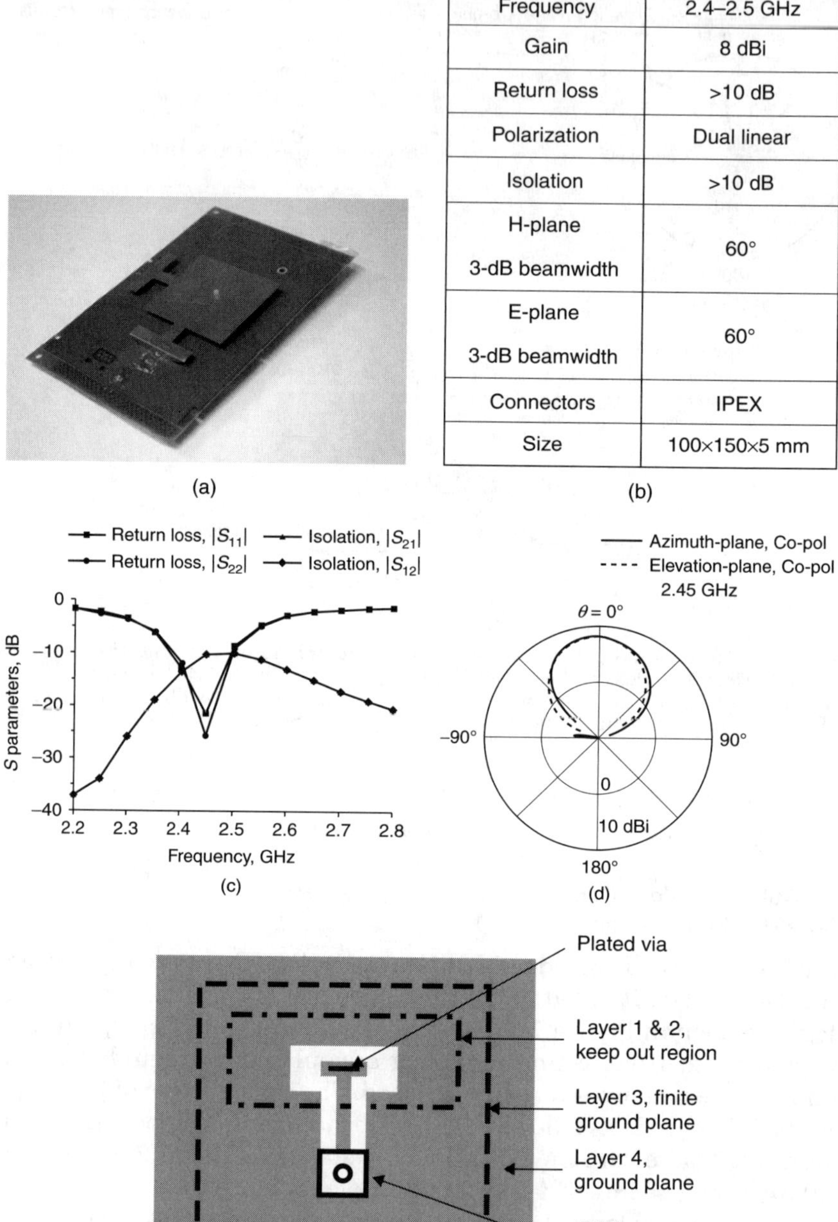

Frequency	2.4–2.5 GHz
Gain	8 dBi
Return loss	>10 dB
Polarization	Dual linear
Isolation	>10 dB
H-plane 3-dB beamwidth	60°
E-plane 3-dB beamwidth	60°
Connectors	IPEX
Size	100×150×5 mm

Figure 7.18 An 8-dBi dual-polarized antenna: (*a*) photo of antenna, (*b*) antenna specifications, (*c*) return loss and isolation, (*d*) radiation patterns at 2.45 GHz, and (*e*) IPEX connector and co-planar waveguide with ground plane feed (Photo courtesy of Compex Systems Pte Ltd.)

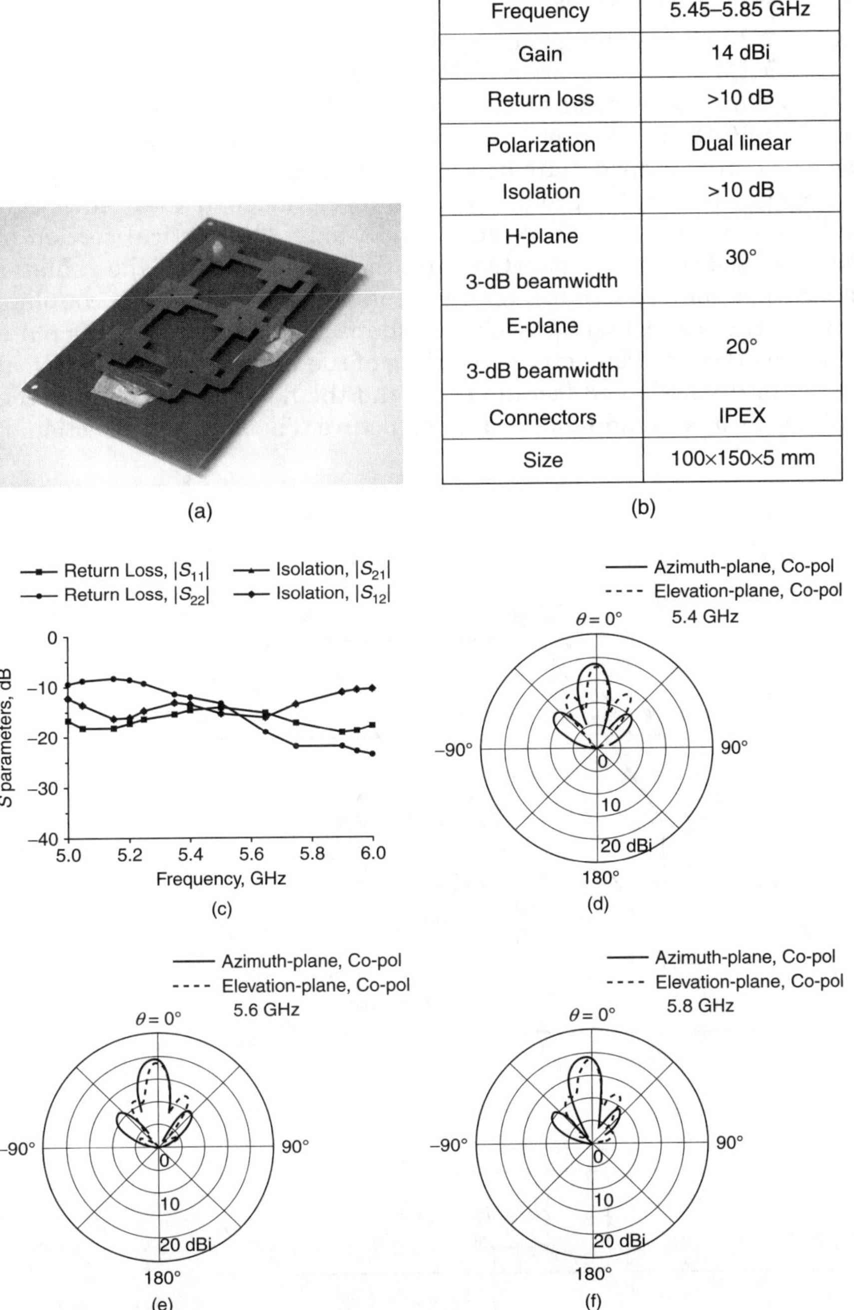

The antenna specifications table (b):

Frequency	5.45–5.85 GHz
Gain	14 dBi
Return loss	>10 dB
Polarization	Dual linear
Isolation	>10 dB
H-plane 3-dB beamwidth	30°
E-plane 3-dB beamwidth	20°
Connectors	IPEX
Size	100×150×5 mm

Figure 7.19 A 14 dBi dual-polarized antenna array: (*a*) photo of antenna, (*b*) antenna specifications, (c) return loss and isolation, (*d*) radiation patterns at 5.4 GHz, (*e*) radiation patterns at 5.6 GHz, and (*f*) radiation patterns at 5.8 GHz (Photo courtesy of Compex Systems Pte Ltd.)

7.4.6 MIMO Antenna Array

Figure 7.20 shows the geometry of the broadband antenna element for the MIMO antenna array.[40] The antenna consists of an L-shaped radiator, and its vertical section is shorted to the ground plane. It is fed via a 50-Ω coaxial probe with a diameter of 1.3 mm, which is connected to an L-shaped plate at the bottom of the radiator. The L-plate is suspended above the ground plane and positioned in a way such that the geometry has symmetry about the x-axis. The vertical section of the L-shaped plate is separated from the vertical wall of the radiator and the horizontal section of the radiator. The electromagnetic coupling between the vertical sections of the L-shaped radiator and feeding plate as well as the electromagnetic coupling of the upper end of the vertical section of the L-shaped feeding plate and the horizontal section of the radiator form a broadband feeding structure. The horizontal section of

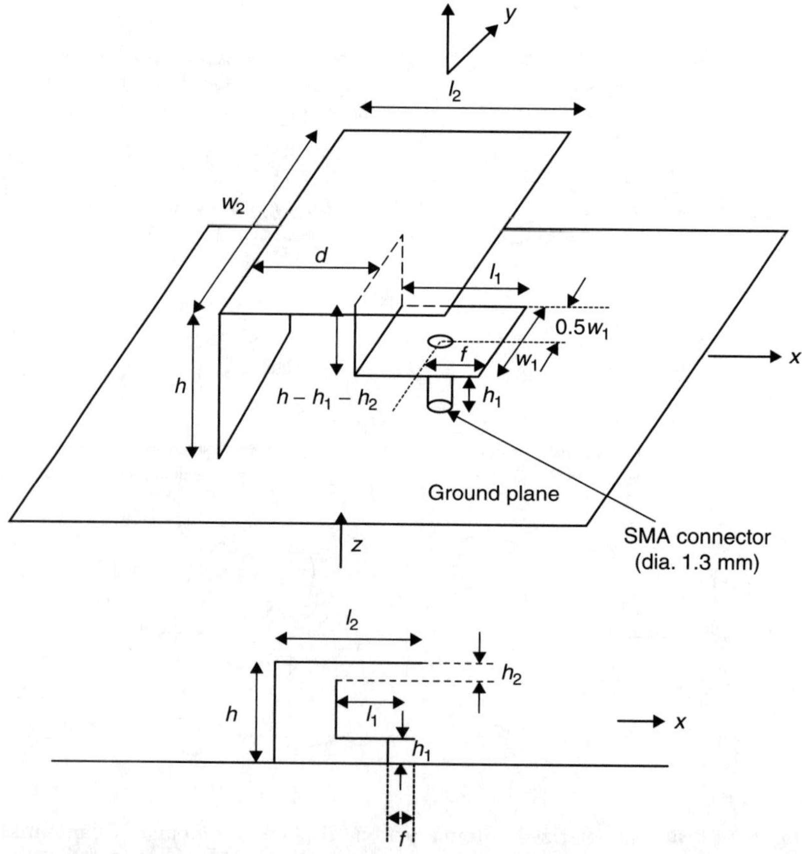

Figure 7.20 Antenna element geometry

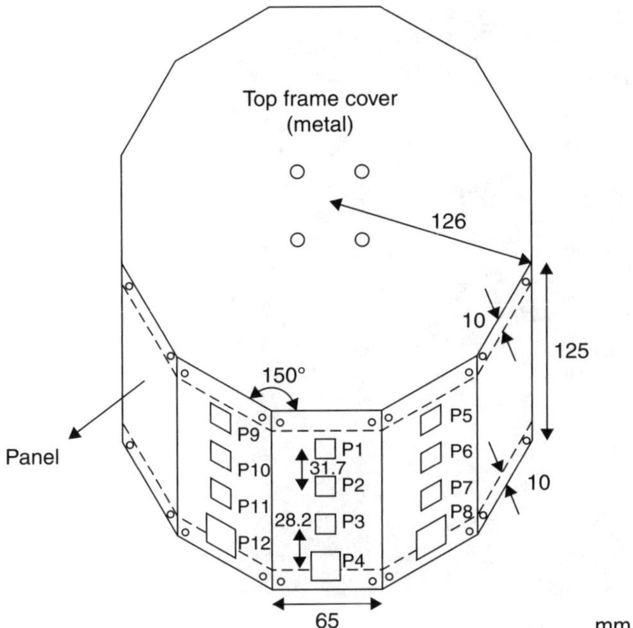

Figure 7.21 MIMO antenna array configuration

the L-shaped feeding plate increases the capacitance at the feed point to compensate for the increase in the inductance due to the probe across a broad bandwidth.

Figure 7.21 shows the antenna array arranged in a twelve-sectored configuration. Each sector is composed of three antennas operating at 5 GHz and one antenna operating at 2.4 GHz. The size of each ground plane panel is 125 mm × 65 mm and the radius of the antenna array is 126 mm. The feed-to-feed distance between the 5 GHz elements is 31.7 mm and the distance of the 5-GHz element to the 2.4-GHz element is 28.2 mm. The antenna prototype is shown in Figure 7.22. The dimensions of the 2.4-GHz and 5-GHz antenna elements are given in Table 7.3.

Figure 7.23 shows the measured impedance matching of the antenna elements. As can be seen, Antennas P1, P2, and P3 are well-matched within the 5–6-GHz band whereas Antenna P4 is matched in 2.4–2.5-GHz band. The higher return loss for Antenna P3 is due to the presence of the 2.4-GHz antenna P4.

TABLE 7.3 Antenna Elements Dimensions (in mm)

	d	l_1	l_2	w_1	w_2	h	h_1	h_2	f
2.4 GHz	4.5	13	21.5	5	20	9	3	2	3
5 GHz	7	6	10	6	20	9	3.5	1.5	3

Figure 7.22   MIMO antenna prototype

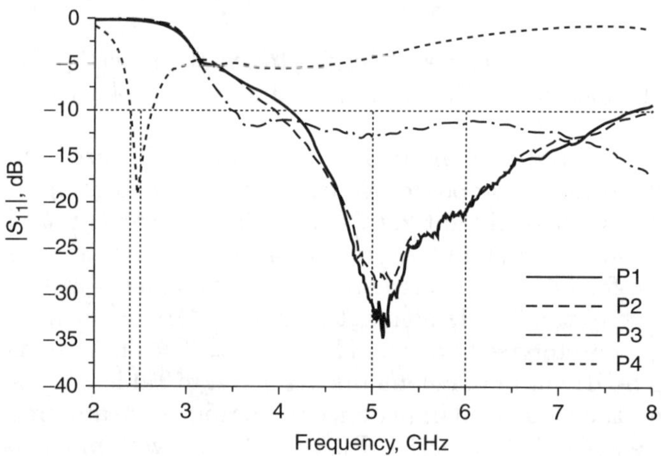

Figure 7.23   Measured return loss

Figure 7.24 shows the mutual coupling between the antenna elements. For brevity, only the mutual coupling between Antennas P1/P2, P1/P3, P2/P3, and P2/P6 are shown. The mutual coupling between the other elements, which are not given here, are relatively weak as they are spaced further apart. Also, the mutual coupling performance between the Antennas P4/P8 and P8/P12 are also provided. It can be

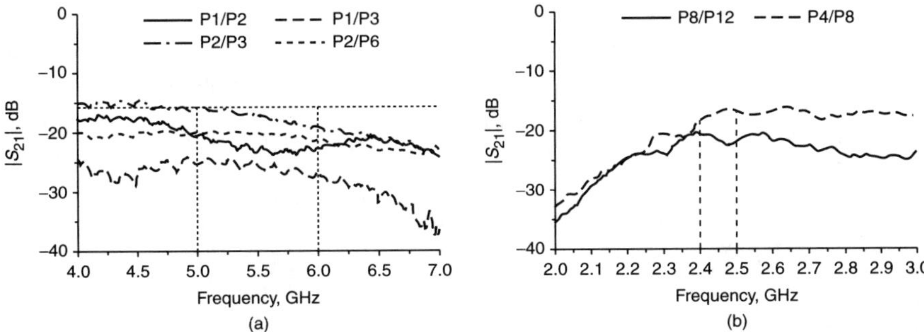

Figure 7.24 Measured mutual coupling: (a) elements at 5 GHz and (b) elements at 2.4 GHz

seen that the coupling between the 5-GHz elements are less than −15.9 dB, and −16.5 dB between the 2.4-GHz elements. Figure 7.25 plots the corresponding correlation performance of the antenna array calculated using Eq. 7.12. The correlation is less than −25 dB and the higher correlation experienced by Antennas P2/P3 compared to Antennas P1/P2 (although they are separated by the same distance) is due to the higher mutual coupling. Also, the higher return loss by Antenna P3 results in a higher correlation for Antennas P1/P3 and Antennas P2/P3. The 2.4-GHz antennas exhibit a higher correlation between P4/P8 compared to P8/P12 within the 2.4–2.5-GHz band, which is due to the higher mutual coupling between them.

The radiation patterns in the x-z and y-z planes of the Antenna P2 and Antenna P4 are shown in Figure 7.26 at 5.5 GHz and 2.45 GHz, respectively. Only the radiation patterns at the center frequencies are shown since they are stable across the entire bandwidth. The maximum gain in the x-z plane is around 3 dBi. The gain depends on the size of the ground plane. The beam is squinted 40–45° from the boresight

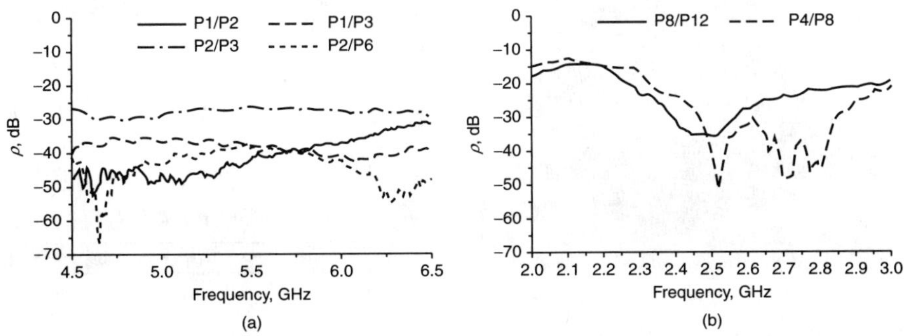

Figure 7.25 Measured correlation: (a) elements at 5 GHz and (b) elements at 2.4 GHz

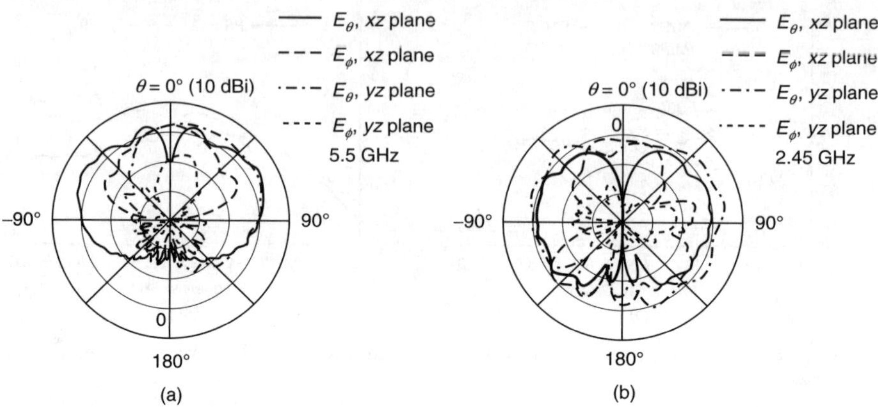

Figure 7.26 Measured radiation patterns in the x-z and y-z planes: (*a*) antenna P2 at 5.5 GHz and (*b*) antenna P4 at 2.45 GHz

due to the asymmetrical structure of the antenna. The squinted maximum beam is conducive to the indoor applications when the antenna is installed on the ceiling.

7.4.7 Three-Element Dual-Band MIMO Antenna

Figure 7.27 shows a dual-band (2.4 GHz/5.1–5.9 GHz) MIMO PIFA array arranged in a triangular fashion as well as its impedance, correlation, and radiation performance. Both the feeding pin at the edge and the shorting pin in the middle are inserted through the 20-mils single-sided FR-4. The ground planes are soldered together for rigidity.

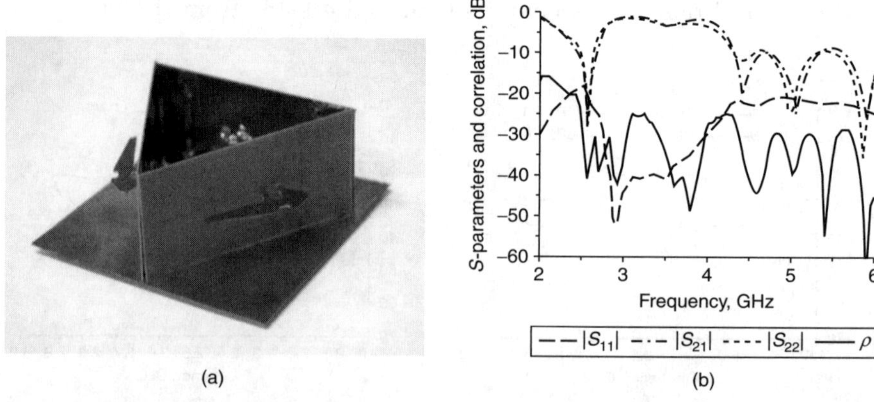

Figure 7.27 Dual-band MIMO antenna: (*a*) photo of antenna and (*b*) measured return loss, isolation, and correlation

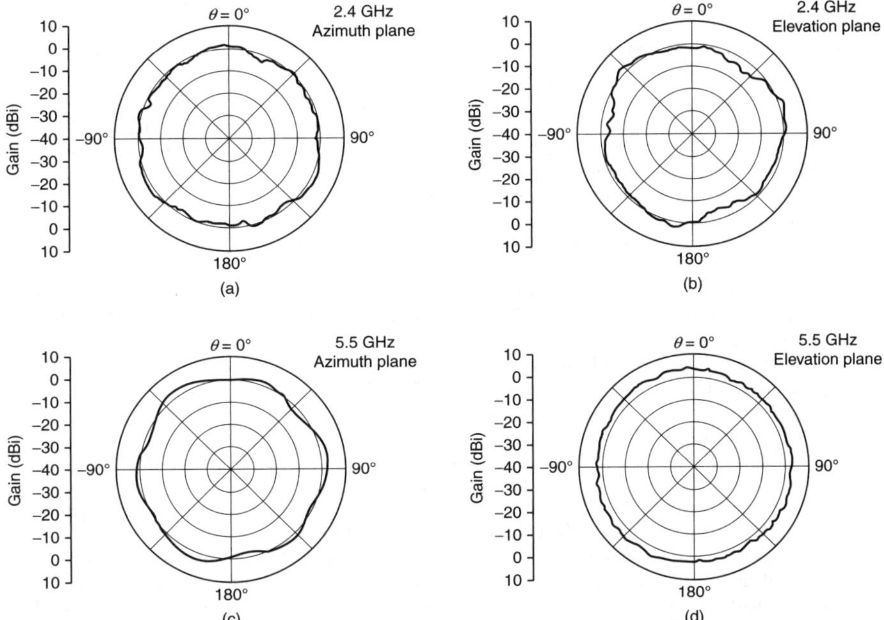

Figure 7.28 Total coverage: (*a*) azimuth plane at 2.4 GHz, (*b*) elevation plane at 2.4 GHz, (*c*) azimuth plane at 5.5 GHz, and (*d*) elevation plane at 5.5 GHz

The square ground plane below is optional. The antenna is well-matched at 2.4–2.5 GHz and 4.4–6 GHz. The isolation is kept greater than 17 dB across the well-matched bandwidths and the correlation is also found to be less than –20 dB. From the combined radiation patterns of the three elements, an omnidirectional coverage in the azimuth and elevation planes can be achieved, as shown in Figure 7.28.

7.5 Conclusion

This chapter has demonstrated a variety of WLAN antenna designs for applications in base stations, subscriber stations, and access points. Their cost effectiveness and performance were emphasized. Several state-of-the-art designs in the market were also presented. In the case studies, the critical issues regarding such designs and methods to overcome them were discussed.

As the WLAN/WiFi market continues to grow, this has spurred the demand for higher performance, more compact, and lower cost antenna designs. MIMO antennas (IEEE 802.11n) present many challenges, such as embedding multiple antennas into a single access point without increasing the overall size while increasing the overall data throughput.

References

1. IEEE 802.11, LAN/MAN Wireless LANS, IEEE Standards Organization, http://standards.ieee.org/getieee802/802.11.html.
2. IEEE 802.16, Broadband Wireless Metropolitan Area Network (Wireless MAN), IEEE Standards Organization, http://standards.ieee.org/getieee802/802.16.html.
3. IEEE 802.15, Wireless Personal Area Networks, IEEE Standards Organization, http://standards.ieee.org/getieee802/802.15.html.
4. Federal Communications Commission, http://www.fcc.gov/.
5. ETSI, http://www.etsi.org/.
6. Part 15, Radio Frequency Devices, Federal Communications Commission, http://www.fcc.gov/oet/info/rules/part15/part15-9-20-07.pdf.
7. L. Zheng and D. N. C. Tse, "Diversity and multiplexing: a fundamental trade-off in multiple-antenna channels," *IEEE Trans. Inform. Theory*, vol. 49, no. 5 (May 2003): 1073–1096.
8. I. E. Telatar, "Capacity of multi-antenna Gaussian channels," *European Trans. Tel.*, vol. 10, no. 6 (November/December 1999): 585–595.
9. T. M. Cover and J. A. Thomas, *Elements of Information Theory*, New York: John Wiley and Sons, 1991.
10. G. G. Rayleigh and J. M. Cioffi, "Spatio-temporal coding for wireless communication," *IEEE Trans. Commun.*, vol. 46, (March 1998): 357–366.
11. T. K. Moon and W. C. Stirling, *Mathematical Methods and Algorithms for Signal Processing,* Englewood Cliffs, NJ: Prentice Hall, 2000.
12. M. A. Khalighi, J. Brossier, G. Jourdain, and K. Raoof, "Water-filling capacity of Rayleigh MIMO channels," *Proc. IEEE 12th Int. Symp. on Personal, Indoor and Mobile Radio Comm.*, vol. 1, San Diego, CA (September 30–October 3, 2001): 155–158.
13. C. Waldschmidt, T. Fugen, and W. Wiesbeck, "Spiral and dipole antennas for indoor MIMO systems," *IEEE Antennas Wireless Propagat. Lett.*, vol. 1, no. 1 (2002): 176–178.
14. S. Blanch, J. Romeu, and I. Corbella, "Exact representation of antenna system diversity performance from input parameter description," *Electronics Lett.*, vol. 39, no. 9 (May 2003): 705–707.
15. T. Svantesson, "Correlation and channel capacity of MIMO systems employing multimode antennas," *IEEE Trans. Veh. Technol.*, vol. 51, no. 6 (November 2002): 1304–1312.
16. C. C. Martin, J. H. Winters, and N. R. Sollenberger, "MIMO radio channel measurements: Performance comparison of antenna configurations," *Proc. IEEE 54th Veh. Technol. Conf.*, vol. 2, Atlantic City, NJ (October 7–11, 2001): 1225–1229.
17. S. Sandhu, R. U. Nabar, D. A. Gore, and A. Paulraj, "Near-optimal selection of transmit antennas for a MIMO channel based on Shannon capacity," *Proc. 34th Asilomar Conf. Signals, Systems and Computers*, vol. 1, Pacific Grove, CA (October 29–November 1, 2000): 567–571.
18. D. Gore and A. Paulraj, "Space-time block coding with optimal antenna selection," *Proc. IEEE Int. Conf. Acoustics, Speech, and Signal Processing*, vol. 4, Salt Lake City, UT (May 7–11, 2001): 2441–2444.
19. D. A. Gore and A. Paulraj, "MIMO antenna subset selection with space-time coding," *IEEE Trans. Signal Processing*, vol. 50 (October 2002): 2580–2588.
20. A. F. Molisch, J. F. Winters, and M. Z. Win, "Capacity of MIMO systems with antenna selection," *Proc. IEEE Int. Conf. Commun.*, vol. 2, Helsinki, Finland (June 11–14, 2001): 570–574.
21. R. A. Andrews, P. P. Mitra, and R. deCarvalho, "Tripling the capacity of wireless communications using electromagnetic polarization," *Nature*, vol. 409 (January 2001): 316–318.
22. T. Svantesson, "On capacity and correlation of multi-antenna systems employing multiple polarizations," *Proc. IEEE Antennas and Propagation Society Int. Symp.*, vol. 3, San Antonio, TX (June 16–21, 2002): 202–205.
23. T. Svantesson, M. A. Jensen, and J. W. Wallace, "Analysis of electromagnetic field polarizations in multi-antenna systems," *IEEE Trans. Wireless Commun.*, vol. 3 (March 2004): 641–646.

24. R. G. Vaughan, "Polarization diversity in mobile communications," *IEEE Trans. Veh. Technol.*, vol. 39 (August 1990): 177–186.
25. J. W. Wallace, M. A. Jensen, A. L. Swindlehurst, and B. D. Jeffs, "Experimental characterization of the MIMO wireless channel: Data acquisition and analysis," *IEEE Trans. Wireless Commun.*, vol. 2 (March 2003): 335–343.
26. J. P. Kermoal, L. Schumacher, F. Frederiksen, and P. E. Mogensen, "Polarization diversity in MIMO radio channels: Experimental validation of a stochastic model and performance assessment," *Proc. IEEE 54th Veh. Technol. Conf.*, vol. 1, Atlantic City, NJ (October 7–11, 2001): 22–26.
27. P. Kyritsi and D. C. Cox, "Propagation characteristics of horizontally and vertically polarized electric fields in an indoor environment: Simple model and results," *Proc. IEEE 54th Veh. Technol. Conf.*, vol. 3, Atlantic City, NJ (October 7–11, 2001): 1422–1426.
28. P. Kyritsi, D. C. Cox, R. A. Valenzuela, and P. W. Wolniansky, "Effect of antenna polarization on the capacity of a multiple element system in an indoor environment," *IEEE J. Select Areas Commun.*, vol. 20 (August 2002): 1227–1239.
29. J. L. Allen and B. L. Diamond, "Mutual coupling in array antennas," Lincoln Laboratory, M.I.T., Tech. Rep. 424 (ESD-TR-66–443), 1966.
30. C. A. Balanis, *Antenna Theory: Analysis and Design*, New York: John Wiley and Sons, 1997.
31. J. W. Wallace and M. A. Jensen, "The capacity of MIMO wireless systems with mutual coupling," *Proc. IEEE 56th Veh. Technol. Conf.*, vol. 2, Vancouver, British, Columbia, Canada (September 24–28, 2002): 696–700.
32. J. W. Wallace and M. A. Jensen, "Termination diversity performance of coupled antennas: Network theory analysis," *IEEE Trans. Antennas Propagat.*, vol. 52 (January 2004): 98–105.
33. R. Janaswamy, "Effect of element mutual coupling on the capacity of fixed length linear arrays," *IEEE Antennas Wireless Propagat. Lett.*, vol. 1, no. 1 (2002): 157–160.
34. International Electrotechnical Commission, http://www.iec.ch/.
35. Z. N. Chen and M. Y. W. Chia, "A novel center-fed suspended plate antenna," *IEEE Trans. Antennas Propagat.*, vol. 51, no. 6 (June 2003): 1407–1410.
36. Z. N. Chen and M. Y. W. Chia, "Center-fed microstrip patch antenna," *IEEE Trans. Antennas Propagat.*, vol. 51, no. 3 (2003): 483–487.
37. Z. N. Chen, "Suspended plate antennas with shorting strips and slots," *IEEE Trans. Antennas Propagat.*, vol. 52, no. 10 (October 2004): 2525–2531.
38. W. K. Toh and Z. N. Chen, "A tunable dual-band planar antenna," *Elect. Lett.*, vol. 4, no. 1 (January 2008): 8–9.
39. W. K. Toh and Z. N. Chen, "On a broadband elevated suspended-plate antenna with consistent gain," *IEEE Antennas and Propagation Magazine*, vol. 50, no. 2 (April 2008): 95–105.
40. T. S. P. See and Z. N. Chen, "An electromagnetically coupled UWB plate antenna," *IEEE Trans. Antennas Propagat.*, vol. 56, no. 5 (May 2008): 1476–1479.

Antennas for Wireless Personal Area Network (WPAN) Applications: RFID/UWB Positioning

Xianming Qing and Zhi Ning Chen

Institute for Infocomm Research

8.1 Introduction

A wireless personal area network (WPAN) is a short range network for wirelessly interconnecting devices centered around an individual person's workspace.[1,2] Typically, a WPAN uses technology that allows communication within about 10 meters. This chapter addresses the antenna designs for two WPAN technologies: radio frequency identification (RFID) for assets tracking and ultra-wideband (UWB) for target positioning.

RFID technology has been developing rapidly in recent years, and its applications are found in many service industries, distribution logistics, manufacturing companies, and goods flow systems.[3,4] The reader antenna is one of the key components in an RFID system. The detection range/accuracy of an RFID system is directly dependent on the performance of the reader antenna. In addition, an optimized reader antenna design will benefit the RFID system with enhanced range, high accuracy, reduced fabrication cost, as well as simplified system configuration and implementation.

UWB radios employ very short pulses that spread their energy over a broad range of the frequency spectrum.[5,6] Due to the inherently fine temporal resolution of UWB, arrived multipath components can be sharply timed with a receiver to provide accurate time of arrival (TOA) estimation. This characteristic makes UWB ideal for high-precision radio positioning applications. UWB has been proposed as an underlying physical layer communication technology in the standard for low-speed WPAN (IEEE 802.15.4a). The position of a desired node can be determined in a variety of ways, such as angle of arrival (AOA), time of arrival (TOA), time difference of arrival (TDOA), or received signal strength (RSS), wherein antenna performance is vital to positioning accuracy.

This chapter consists of three parts. The first part will briefly introduce WPAN technologies, standards, and applications. The specific application scenarios: RFID and UWB positioning are detailed with system configurations, classification, spectra, regulations, and standardization. The second part will focus on discussing RFID reader antennas. Design considerations of antennas from the low frequency band (LF, $\leq$ 400 KHz) to the microwave band (MW, up to 24.125 GHz) are discussed. The case study in this part demonstrates RFID reader antenna designs at high frequency (HF, 13.56 MHz) and ultrahigh frequency (UHF, 840–960 MHz). Detailed antenna configurations, measurement setups, and measured results are explored. Furthermore, the proximity effects of metallic objects on the performance of the HF reader antenna in terms of resonant frequency, impedance matching, and relative magnetic field intensity are presented. The last part of the chapter introduces an antenna design for an indoor mono-station UWB positioning system. The requirements and design considerations of an antenna for such a UWB positioning system are discussed. As an example, a six-element sectored antenna array is presented. The antenna array configuration, the antenna element design, and the methods to control the beamwidth and restrain interference among antenna elements are discussed.

8.1.1 Wireless Personal Area Network (WPAN)

WPAN has been growing rapidly since 1998 when a consortium led by Ericsson Mobile Communications, Nokia Mobile phones, IBM, Intel, and Toshiba Corporation started organizing a large group of companies with the goal of developing the standards by which WPANs are developed and implemented. They formed a Bluetooth special interest group (SIG) that now includes more than 1,500 companies to develop and support Bluetooth technology standards as they are rolled into a wide array of products. The primary focus of the Bluetooth SIG has been developing standards and promoting WPANs for personal computing applications that combine cellular phones and personal computers (PC). By eliminating the need for cables, data capture products using WPAN technology provide great convenience, utility, safety, and cost reduction. Their low cost and desired data transfer rate make WPAN ideal for various applications such as linking peripheral devices (like printers, cellular phones, and home appliances) or a personal digital assistant (PDA) to a computer or to retail/industrial data capturing devices.

Typically, a WPAN features the following characteristics:

- Short coverage

- Low power consumption

- Low cost

- Unlicensed, Industrial, Scientific and Medical (ISM) /UWB bands

- Interference immunity

- Standardized interfaces

Various technologies such as Infrared, Bluetooth, HomeRF, ZigBee, UWB, RFID, near-field communication (NFC), and so on, can be used for WPAN applications.[7] Some of the technologies are matured with the commercial products that have been in development for years. The others are under development and the commercial products are expected in the near future.

Infrared   Infrared technology is probably the best known and most matured WPAN technology.[8] Communication takes place between two devices that are in line of sight of each other. This means the network linkage will fail if an object obstructs the infrared path between the two devices. Infrared technology is thus suitable for WPANs where both communicating devices can be placed close together and kept relatively static. The technology can be found in a range of devices, including PDAs, mobile phones, laptops, and remote controls. The advantages of using infrared technology include

- Relatively high data rate when compared to other WPAN technologies (currently up to 4 Mbps)

- Free from the interference caused by radio technologies

- Low device cost

- Lack of complicated addressing issues between devices

Bluetooth   Bluetooth[9] is a short-range wireless technology endorsed by major technology companies and has now been accepted as the IEEE 802.15.1 standard. Bluetooth uses frequency hopping spread spectrum radio technology. It chops up the data being sent and transmits chunks of it on up to 75 different frequencies. In its basic mode, the modulation is

Gaussian frequency shift keying (GFSK). It can achieve a high data rate of 1 Mbps. Bluetooth provides a way to connect and exchange information between devices such as mobile phones, telephones, laptops, personal computers, printers, global positioning system (GPS) receivers, digital cameras, video game consoles, and so on, over the unlicensed ISM 2.4-GHz band. Unlike infrared technology, having a clear line of sight between Bluetooth devices is unnecessary. The typical range for Bluetooth is short, only up to 10 meters.

Radio Frequency Identification Radio frequency identification (RFID), which was developed during World War II, is a technology that provides wireless identification and tracking capability and is more robust than the bar code. An RFID system can read and process the data from a mobile device (i.e., a tag) by using an RFID reader according to a particular application's need. The data transmitted by the tag may provide the information about identification or location, or specific information about the product tagged, such as price, color, date of purchase, and so on. The use of RFID technology in tracking and access applications first appeared during the 1980s. Today, RFID technology is found in a great number of applications such as E-passport, transportation payments, product tracking, animal identification, library inventory systems, human implants for health monitoring, and so on.

Typically, a passive RFID system can detect tags within a distance of less than 1.5 meters for near-field configurations and up to 10 meters for far-field configurations.

Near-Field Communication (NFC) Near-field communication (NFC) is a short-range HF wireless communication technology that enables the exchange of data between devices at a distance of up to 20 cm. The technology is a simple extension of the ISO 14443 proximity-card standard (contactless card, RFID) that combines the interface of a smartcard and a reader into a single device.[10] An NFC device can communicate with both existing ISO 14443 smartcards and readers, as well as with other NFC devices, and is thereby compatible with existing contactless infrastructure for public transportation and payment.

NFC devices can operate in passive or active mode. Passive mode enables a powerless NFC device such as a smart card to communicate with a powered device such as a PDA. This mode will also allow accessing of data when one of the NFC-enabled devices is turned off. NFC data rates are relatively slow compared with other WPAN technologies; the maximum data transfer rate of NFC (424 Kbps) is slower than that of Bluetooth. NFC also has a short-range of less than 20 cm, which provides a degree of security and makes it suitable for use in crowded areas where correlating a signal with its transmitting physical device might otherwise prove impossible.

ZigBee ZigBee[9] is the name of a specification for a suite of high-level communication protocols using small, low-power digital radios based on the IEEE 802.15.4 standard for WPANs, such as wireless headphones connected to cellular phones via a short-range radio. The technology is intended to be simpler and cheaper than other WPANs. ZigBee is targeted for WPAN applications that require a low data rate, long battery life, and secure networking. The data transfer rates of ZigBee depend on the frequency used, for example, 250 Kbps at 2.4 GHz, 40 Kbps at 915 MHz, and 20 Kbps at 868 MHz.

Ultra-Wideband (UWB) Ultra-wideband (UWB) is a radio technology that can be employed at very low energy levels for short-range communications by using a large portion of the radio spectrum. This method uses pulse-coded information with sharp carrier pulses at a bunch of center frequencies in a logical connex. UWB technology has traditional applications in radar. Most recent applications target sensors for data collection and precision positioning applications.

Due to the extremely low emission levels currently allowed by regulatory agencies, UWB systems are more suitable for short-range and indoor applications. Owing to the short duration of UWB pulses, realizing extremely high data rates in UWB-based systems is easier. Furthermore, the data rate can be readily traded for the range by simply aggregating pulse energy per data bit using either simple integration or coding techniques. High data rate UWB radios enable efficient data transfer from digital camcorders, wireless printing of digital pictures from a camera without an intervening personal computer, and the transfer of files among cellular phones and other handheld devices like personal digital audio and video players. UWB can also be applied to real-time positioning systems. Its high precision capabilities combined with very low power make it ideal for certain radio frequency sensitive environments such as hospitals and healthcare institutes.

Millimeter-Wave (mmW) The 60-GHz millimeter-wave band has been allocated worldwide for short-range wireless communications.[11] The millimeter-wave WPAN will operate in the 57–64 GHz unlicensed frequency band in the U.S. and the 59–66 GHz band in Japan and Europe. The millimeter-wave WPAN allows high coexistence with all other microwave systems in the 802.15 family of WPANs. In addition, the millimeter-wave WPAN allows very high data-rate applications (over 2 Gbps) such as high-speed Internet access, streaming content download (video on demand, HDTV, home theater, and so on), real-time streaming, and wireless data bus to replace the use of cables.

The physical (PHY) layer for WPANs operating in the 60-GHz bands has been developed by the IEEE 802.15 WPAN Millimeter-Wave Alternative PHY Task Group 3c.[12]

8.1.2 Radio Frequency Identification (RFID)

8.1.2.1 System Configuration A typical RFID system, as shown in Figure 8.1, is composed of at least a reader (interrogator), a reader antenna, a tag (transponder), a host computer, and middleware that includes software and a database. Generally, the reader emits a signal at a certain frequency. When an RFID tag passes within the reading zone of the reader antenna, the reader detects and interrogates the tag for its content information. This process is conducted by electric/magnetic field coupling or electromagnetic wave capturing via the antenna. The reader is capable of storing and transferring the object information, which may include stock number, current location, status, and a wide variety of other potential information, to a host computer system. The tags are attached to a wide variety of objects for the purpose of identification and tracking. Each tag has a manufacturer-installed unique identification code as well as additional available memory.

An RFID reader is a device that can read/write data from/to compatible RFID tags. The act of writing the tag data by a reader is called *creating a tag*. The process of creating a tag and uniquely associating it with an object is called *commissioning the tag*. Similarly, *decommissioning* a tag means to disassociate the tag from a tagged object and optionally destroy it. The time during which a reader emits RF energy to read tags is called a *duty cycle,* which is internationally regulated. Figure 8.2 shows several examples of commercial readers available in the market.

An RFID tag, or a transponder, is a device that can store and transmit data to a reader in a contactless manner through radio waves. Most of

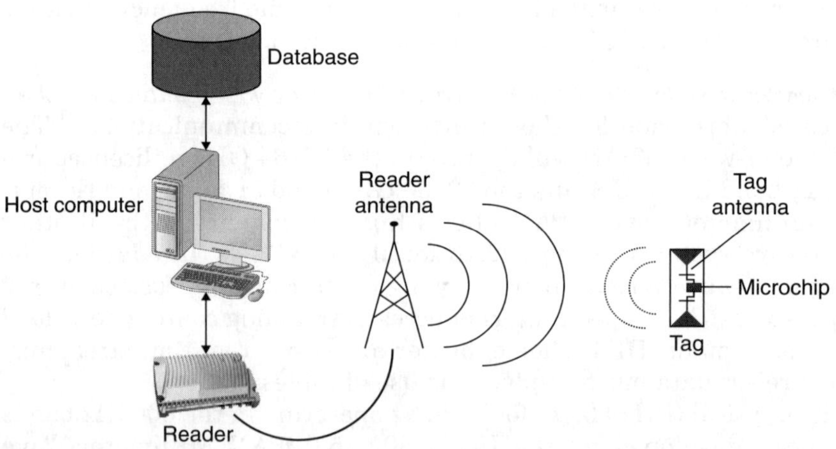

Figure 8.1 Configuration of a typical RFID system

Figure 8.2 RFID readers

the tags are with a "chip," which comprises two main components: a small application-specific integrated circuit (ASIC) and an antenna. Some commercial RFID tags are shown in Figure 8.3.

An RFID reader communicates with a tag through the reader antenna, which broadcasts the RF signals from a reader transmitter to its surroundings and receives the response from the tags. Because the frequency available for RFID applications varies within a broad frequency range from a low frequency (LF) of 125–134.2 kHz, a high frequency (HF) of 13.56 MHz, an ultra high frequency (UHF) of 840–960 MHz to microwave frequencies (MWF) of 2.4 GHz, 5.8 GHz, and 24 GHz, the

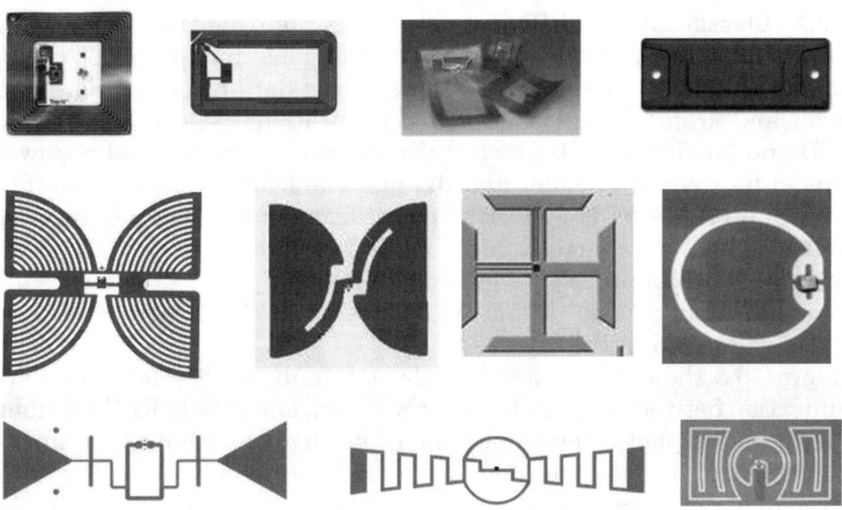

Figure 8.3 RFID tags

LF/HF reader antennas

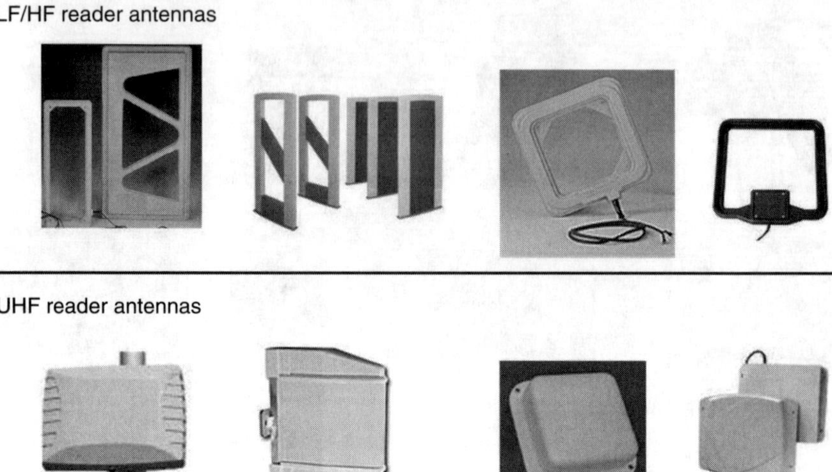

UHF reader antennas

Figure 8.4   RFID reader antennas

reader antennas used for RFID applications features distinct variations. As shown in Figure 8.4, for instance, loop antennas have been used at VLF/HF frequencies; the dimensions of the antennas can be up to a few meters. The antennas for UHF/MWF RFID applications are normally smaller patch antennas that are often required to be broadband and circularly polarized.

8.1.2.2 Classification   RFID systems are distinguished from each other by system usage, operating frequency, reading distance, protocol, power transfer to the tag, the procedure for sending data from the tag to the reader, and so on.[13] The most general classification is so-called near-field RFID and far-field RFID, which is determined by the method of power transfer between the reader and the tag. The RFID systems adopt different approaches for transferring power from an RFID reader to a tag, namely inductive/capacitive coupling and electromagnetic (EM) wave capturing. Both approaches can transfer enough power to a remote tag to sustain its operation—typically between 9 µW to 1 mW, depending on the ASIC used in the tag.

Figure 8.5 shows the power transfer mechanism as well as the communication between the reader and the tag in a near-field RFID system. When a tag is placed near the reader, the tag draws energy from the magnetic field that is generated by the reader antenna. The tag's resulting feedback on the reader antenna brings about the change in the magnetic field. This change is subsequently detected by the reader.

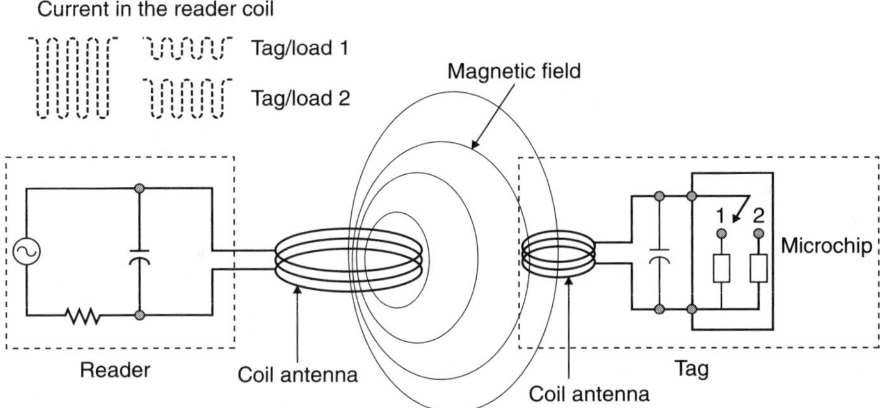

Figure 8.5 Power transfer and communication between reader and tag in a near-field RFID system

The data can be transferred from the tag to the reader if the timing within which the load resistor or capacitor is switched on and off is controlled. The tag varies its load impedance by switching either the load resistor or capacitor on and off. Hence, there is resistive-load modulation and capacitive-load modulation.[4]

The main limitation of near-field RFID systems is their limited reading distance. The reading distance of an HF near-field RFID system is typically less than 1.5 meters. A further limitation of the near-field RFID system is that detection is sensitive to the orientation of the tag since the magnetic field generated by the reader antenna is orientation dependent. The tag can be detected when it is positioned to allow enough magnetic flux going through it. If a tag is positioned parallel to the magnetic field, it cannot be detected by the reader because no magnetic flux is going through it.

In far-field RFID systems, the power transfer from the reader to the tag as well as the communication between the reader and the tag are carried out by transmitting and receiving electromagnetic waves. The tags used in far-field RFID systems capture the electromagnetic waves radiated from the reader antenna. The tag antenna receives the energy and develops an alternating potential difference that appears across the ports of the ASIC. A diode can rectify this potential and link it to a capacitor, which will result in an accumulation of energy in order to power up its electronics. The tags are located in the far-field zone of the reader antenna and the information is transmitted back to the reader by using backscattering modulation.

As shown in Figure 8.6, the reader antenna emits the energy that will be received by the tag, and part of the energy is then reflected from the

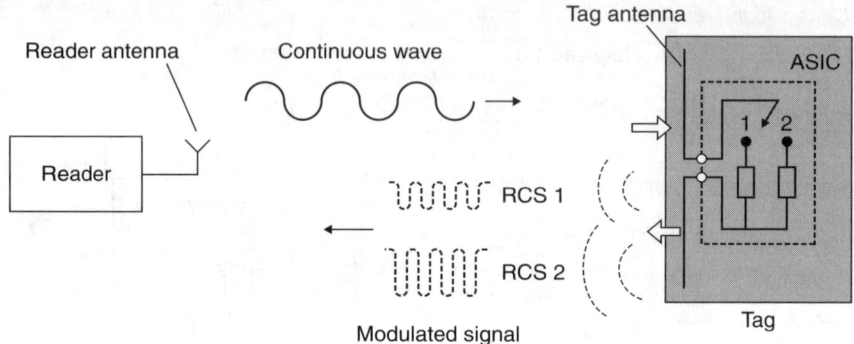

Figure 8.6 Power transfer and communication between reader and tag in a far-field RFID system

tag and detected by the reader. The variation of the tag's load (ASIC) impedance causes the intended impedance mismatch between the tag antenna and the load, which results in the variation of the backscattering signals. In other words, the amplitude of the backscattering from the tag can be influenced by altering the status of the ASIC, and thus the tag can reflect more or less of the incoming signal in a pattern that encodes the tag's ID.

8.1.2.3 Frequencies, Regulations, and Standardization The range and the scalability of the RFID system depend heavily on the operating frequency. The operating frequency significantly affects reading distance, data exchange speed, interoperability, and surface penetration. The need to ensure that the RFID systems co-exist with other radio systems such as mobile phones, maritime systems, and aeronautical radio systems significantly restricts the available spectrum for RFID applications to the ISM frequency bands. Figure 8.7 shows the major frequency spectra available for RFID applications.

No global public body governs the frequencies used for RFID systems. In principle, every country can set its own frequency ranges for

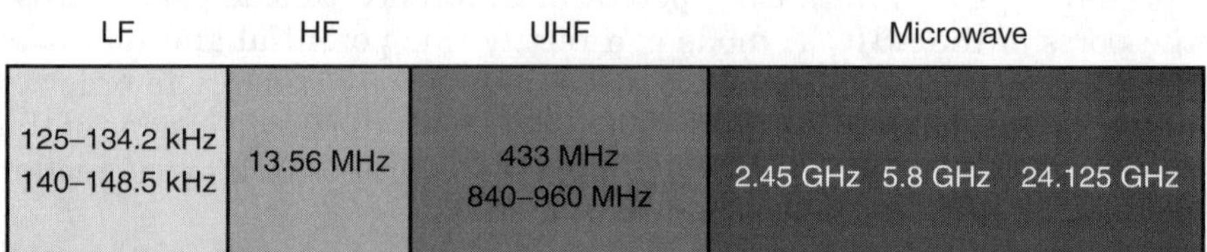

Figure 8.7 Major frequency ranges available for RFID applications

RFID applications. The main bodies governing the frequency allocation for RFID include

- *USA:* Federal Communications Commission (FCC)

- *Canada:* Canadian Radio-television and Telecommunications Commission (CRTC)

- *Europe:* European Representative Office (ERO), European Conference of Postal and Telecommunications Administrations (CEPT), European Telecommunications Standards Institute (ETSI), and national administrations

- *Japan:* Ministry of Internal Affairs and Communications (MIC)

- *China:* Ministry of Industry and Information Technology, Ministry of Science and Technology

- *South Africa:* Independent Communications Authority of South Africa (ICASA)

- *South Korea*: Ministry of Commerce, Industry and Energy

- *Australia:* Australian Communications and Media Authority

- *New Zealand:* Ministry of Economic Development

- *Singapore:* Infocomm Development Authority of Singapore

- *Brazil:* Agência Nacional de Telecomunicações (Anatel)

Low frequency bands of 125–134.2 kHz and 140–148.5 kHz and high frequency bands of 13.56 MHz can be used for RFID applications globally without any license. However, the UHF frequencies accepted for RFID applications vary from region to region. These frequency bands include the 840.5–844.5 MHz and 920.5–924.5 MHz bands in China, the 864–868 MHz band in Europe and Asia Pacific, the 902–928 MHz band in most parts of the world, and the 952–955 MHz band in Japan. In addition, each country/region regulates transmission power to minimize the possible interference to other radio users. The frequencies and power limitations of UHF RFID systems in different countries/regions are tabulated in Table 8.1.[14]

Standards are critical for RFID applications. A great deal of work has been going on over the past decade to develop the standards for different RFID frequencies and applications.[15] Several approved and proposed RFID standards deal with the *air interface protocol* (the way tags and readers communicate with each other), *data content* (the way in which the data is organized or formatted), *conformance* (the way to test whether products meet the standard), and *applications* (how standards are used on shipping labels, for example).

The International Organization for Standardization (ISO) has created the standards for animal tracking; for instance, ISO 11784 defines how

TABLE 8.1 Frequencies and Power Limitations of UHF RFID Systems in Different Countries/Regions

Country/Region	Status[1,2]	Frequency[3] (MHz)	Power[4]	Technique[5]	Comments
Argentina	OK	902–928	4W EIRP	FHSS	
Australia	OK	920–926	4W EIRP		4W EIRP is available through a license managed by GS1 Australia.
Austria	OK	865.6–867.6	4W EIRP	LBT	
Belgium	OK	865.6–867.6	2W ERP		
Bosnia Herzegovina	OK	865.6–867.6	2W ERP	LBT	
Brazil	OK	902–907.5	4W EIRP	FHSS	
		915–928	4W EIRP	FHSS	
Bulgaria	IP	865.6–867.6	2W ERP	LBT	New regulations should be in place shortly.
Canada	OK	902–928	4W EIRP	FHSS	
Chile	OK	902–928	4W EIRP	FHSS	
China	OK	840.5–844.5	2W ERP	FHSS	
		920.5–924.5	2W ERP	FHSS	
Costa Rica	OK	902–928	4W EIRP	FHSS	
Croatia	IP	865.6–867.6	2W ERP	LBT	New regulations should be in place shortly.
Cyprus	IP	865.6–867.6	2W ERP	LBT	New regulations should be in place shortly.
Czech Republic	OK	865.6–867.6	2W ERP	LBT	
Denmark	OK	865.6–867.6	2W ERP	LBT	
Dominican Republic	OK	902–928	4W EIRP	FHSS	
Estonia	OK	865.6–867.6	2W ERP	LBT	
Finland	OK	865.6–867.6	2W ERP	LBT	
France	OK	865.6–867.6	2W ERP	LBT	
Germany	OK	865.6–867.6	2W ERP	LBT	
Greece	IP	865.6–867.6	2W ERP	LBT	New regulations should be in place shortly.
Hong Kong, China	OK	865–868	2W ERP		
		920–925	4W EIRP		
Hungary	OK	865.6–867.6	2W ERP	LBT	
Iceland	OK	865.6–867.6	2W ERP	LBT	
India	OK	865–867	4W ERP		
Indonesia	IP				The 923-925 MHz band is being considered.
Iran, Islamic Rep.	OK	865–868 MHz	2W ERP		

TABLE 8.1   Frequencies and Power Limitations of UHF RFID Systems in Different Countries/Regions (*Continued*)

Country/Region	Status[1,2]	Frequency[3] (MHz)	Power[4]	Technique[5]	Comments
Ireland	IP	865.6–867.6	2W ERP	LBT	New regulations should be in place shortly.
Israel	OK	915–917 MHz	2W ERP		
Italy	OK	865.6-867.6	2W ERP	LBT	
Japan	OK	952–954	4W EIRP	LBT	License required for using 952–954 MHz at 4W EIRP.
		952–955	4W EIRP	LBT	952–955 MHz available for unlicensed use at 20 mW EIRP.
Korea, Rep.	OK	908.5–910	4W EIRP	LBT	
		910–914	4W EIRP	FHSS	
Latvia	IP	865.6–867.6	2W ERP	LBT	New regulations should be in place shortly.
Lithuania	IP	865.6–867.6	2W ERP	LBT	New regulations should be in place shortly.
Luxembourg	OK	865.6–867.6	2W ERP	LBT	
Malaysia	OK	866–869			Allocation under consideration. 868 MHz is available at 50 mW power.
		919–923	2W ERP		Unlicensed use allowed up to 2 W ERP. Use up to 4W ERP allowed under license.
Malta	OK	865.6–867.6	2W ERP	LBT	
Mexico	OK	902–928	4W EIRP	FHSS	
Moldova	OK	865.6–867.6	2W ERP	LBT	
Netherlands	OK	865.6–867.6	2W ERP	LBT	
New Zealand	OK	864–868	4W EIRP		
Norway	IP	865.6–867.6	2W ERP	LBT	
Peru	OK	902–928	4W EIRP	FHSS	
Philippines	IP	918–920	0.5W ERP		In progress.
Poland	OK	865.6–867.6	2W ERP	LBT	
Portugal	OK	865.6–867.6	2W ERP	LBT	
Puerto Rico	OK	902–928	4W EIRP	FHSS	
Romania	OK	865.6–867.6	2W ERP	LBT	
Russian Federation	IP	865.6–867.6	2W ERP	LBT	Licensed use only.
Serbia and Montenegro	OK	865.6–867.6	2W ERP	LBT	

(Continued)

TABLE 8.1   Frequencies and Power Limitations of UHF RFID Systems in Different Countries/Regions (*Continued*)

Country/Region	Status[1,2]	Frequency[3] (MHz)	Power[4]	Technique[5]	Comments
Singapore	OK	866–869	0.5W ERP		
		920–925	2W ERP		
Slovak Republic	OK	865.6–867.6	2W ERP	LBT	
Slovenia	OK	865.6–867.6	2W ERP	LBT	
South Africa	OK	865.6–867.6	2W ERP	LBT	
		915.4–919	4W ERP	FHSS	
		919.2–921	4W ERP	Non-modulating	
Sweden	OK	865.6–867.6	2W ERP	LBT	
Switzerland	OK	865.6–867.6	2W ERP	LBT	
Taiwan	OK	922–928	1W ERP	FHSS	Indoor.
		922–928	0.5W ERP	FHSS	Outdoor.
Thailand	OK	920–925	4W ERP	FHSS	
Tunisia	IP	865.6–867.6	2W ERP	LBT	Plans adopting European regulations.
Turkey	OK	865.6–867.6	2W ERP	LBT	
United Arab Emirates	OK	865.6–867.6	2W ERP	LBT	Aligned with European regulations.
United Kingdom	OK	865.6–867.6	2W ERP	LBT	
United States	OK	902–928	4W EIRP	FHSS	
Uruguay	OK	902–928	4W EIRP	FHSS	
Vietnam	OK	866–869	0.5W ERP		
		920–925	2W ERP		License required for power greater than 0.5W ERP.

Note:

[1] OK indicates regulations are in place or will be in place shortly.

[2] IP indicates appropriate regulations expected within 6 to 12 months.

[3] Indicates the frequency band(s) authorized in the country for RFID applications. The objective is to get a band available in the 840 to 960 MHz spectrum.

[4] Indicates the maximum power available to RFID applications. The power is expressed either as Effective Isotropic Radiated Power (EIRP) or Effective Radiated Power (ERP). Please note that 2 Watts ERP is equivalent to 3.2 Watts EIRP.

[5] Indicates the reader-to-tag communication technique. FHSS stands for frequency hopping spread spectrum and LBT stands for listen before talk.

data is structured on the tag and ISO 11785 defines the air interface protocol. ISO has created a standard for the air interface protocol for RFID tags used in payment systems and contactless smart cards (ISO 14443) as well as in vicinity cards (ISO 15693). It has also established the standards for testing the conformance of RFID tags and readers (ISO 18047) and for testing the performance of RFID tags and readers (ISO 18046).

For automatic identification and item management, ISO has developed the standards, known as the ISO 18000 series, for covering the air interface protocol for systems likely to be used to track goods in the supply chain. They cover the major frequencies used in RFID systems around the world. The seven parts are

- *18000–1:* Generic parameters for air interfaces for globally accepted frequencies

- *18000–2:* Air interface for 135 KHz

- *18000–3:* Air interface for 13.56 MHz

- *18000–4:* Air interface for 2.45 GHz

- *18000–5:* Air interface for 5.8 GHz

- *18000–6:* Air interface for 860 MHz to 930 MHz

- *18000–7:* Air interface at 433.92 MHz

Besides ISO, EPCglobal is another well-known standards body involved in the development of RFID standards for automatic identification and item management. EPCglobal is a joint venture between GS1 (formerly known as EAN International) and GS1 US (formerly the Uniform Code Council, Inc.). The current focus of the group is to create a worldwide standard for RFID and use the Internet to share data via the EPCglobal network.

Class 1 Generation 2 UHF Air Interface Protocol Standard "Gen 2," commonly known as the *Gen 2 standard,* defines the physical and logical requirements for a passive-backscatter, interrogator-talks-first (ITF) RFID system operating in the frequency range of 860–960 MHz.

8.1.3   Ultra-Wideband (UWB) Positioning

8.1.3.1   UWB Spectrum and Regulation   UWB technology has been used in military applications for many years. The current development of UWB technology is focused on consumer electronics and communications. UWB-based radio systems are targeted to feature low power, low cost, and extremely low interference. A significant difference between conventional radio systems and the UWB-based systems is that the conventional systems transmit information by varying the power level, frequency, and/or phase of a sinusoidal wave whereas the UWB-based systems transmit information by generating radio energy at specific time instants and occupying a large bandwidth, thus enabling pulse-positioning or time modulation. The information can also be modulated by UWB signals (*pulses*) by encoding the polarity as well as the amplitude of the pulse and/or by using orthogonal pulses. UWB pulses can be sent sporadically at relatively low pulse rates to support time/position modulation. They can also be sent at high pulse rates up to the inverse of the UWB pulse bandwidth.

Regarding the regulation of UWB radio, many organizations and government entities set rules and recommendations for UWB usage. The Internal Telecommunication Union (ITU) is an international body where governments and the private sector work together on issues pertinent to telecommunication networks. The group that undertakes the work on UWB was created in 2002 to study the compatibility between UWB and other communication services. Most of the telecommunications services that occupy the allocated spectrum would prefer to keep UWB out of their frequency ranges. There are some regional organizations involved in these activities as well; for example, the European Conference of Postal & Telecommunications Administrations (ECPT) has created a task group under the Electronic Communications Committee (ECC) to draft a proposal regarding the usage of UWB in Europe.

At the national level, each country has its own regulatory body to set the rules. The United States is the first country to legalize the UWB spectrum for commercial applications. The rules are meant to protect existing radio services, particularly those national security–relevant services, such as aviation systems and global positioning systems (GPS). In 2004, the FCC authorized the unlicensed use of UWB in 3.1–10.6 GHz band.[16] The FCC defined UWB in terms of a transmission from an antenna for which the emitted signal bandwidth exceeds a bandwidth of 500 MHz or 20% of the center frequency. The power spectral density emission limit by UWB emitters operating over the band of 3.1–10.6 GHz is limited to –41.3 dBm/MHz, which is much lower than that of other radio systems (as shown in Figure 8.8).

FCC regulation of UWB is widely referenced by other countries/ organizations. The defined frequency range of 3.1–10.6 GHz has been adopted in most countries while the emission limit has been modified

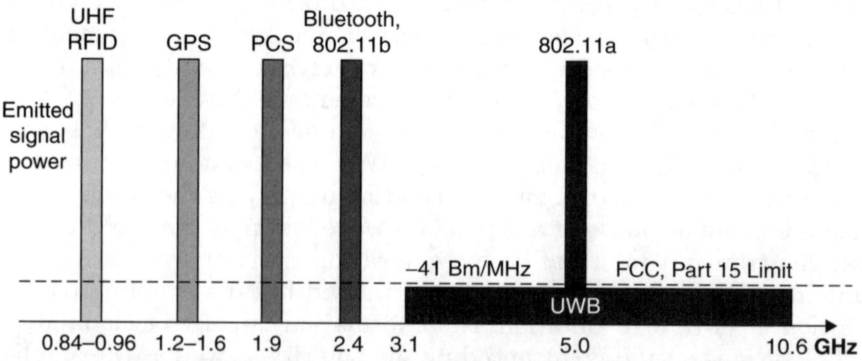

Figure 8.8 Comparison of the spectra and emission power levels for various radio systems

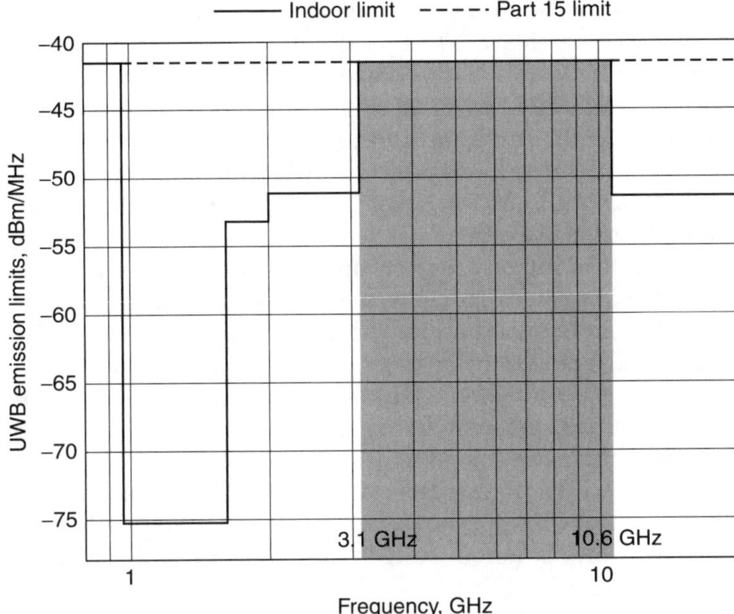

Figure 8.9 FCC UWB emission limits (indoor)

for different countries/regions. Figure 8.9 illustrates the UWB emission power-level mask by FCC for indoor applications.[16]

8.1.3.2 IEEE 802.15.4a The IEEE 802.15.4 standard provides a framework for low data-rate communication systems, typically in sensor networks.[17,18] The original 802.15.4 standard released in 2003 adopted a wideband physical layer using a Direct Sequence Spread Spectrum technique (DSSS). The standard provided the specifications for operation in three different frequency bands: the 868-MHz band in Europe, the 915-MHz band in the United States, and the 2.4-GHz ISM band worldwide. In order to allow the coexistence of several networks in the same location, a frequency division multiplexing (FDM) approach was adopted by dividing each of the bands into channels; the access strategy within a single network was based on time division. The channel scheme is as follows:

- One channel (Channel 0) was defined in the 868-MHz band

- Ten channels (Channels 1–10) were defined in the 915-MHz band, with a channel spacing of 2 MHz

- Sixteen channels (Channels 11–26) were defined in the 2.4-GHz band, with a channel spacing of 5 MHz

The channels defined in the first two bands were intended for very low bit-rate operations. The rates are 20 Kbps and 40 Kbps per channel for the 868-MHz and the 915-MHz bands, respectively. The 2.4-GHz ISM band, on the other hand, allowed bit rates up to 250 Kbps per channel due to the larger bandwidth allocated to each channel. The revised 820.15.4-2006 standard introduced new modulation schemes for the channels in the 868-MHz and 915-MHz bands, allowing bit rates of 250 Kbps per channel for these two bands.

The 802.15.4a standard introduces new options for the physical layer in order to support higher data rates and accurate ranging capability, enabling new applications based on distance and the positions of the devices in the network. The 802.15.4a physical layer is based on two different technologies: UWB and chirp signals. The direct sequence UWB (DS-UWB) PHY was adopted because it is spectrally efficient and very robust even at low transmitted power. In addition, DS-UWB can support precision ranging. The Chirp Spread Spectrum (CSS) PHY was added to the standard because the CSS supports communications to the devices moving at high speeds and at longer ranges than any of the other PHYs in the IEEE 802.15.4 standard.

The UWB signal adopted in 802.15.4a uses two frequency bands within the range of frequencies available by the regulation of the FCC: a lower band across 3.1–4.8 GHz and an upper band across 6.2–9.7 GHz, which is illustrated in Figure 8.10. The chirp solution has the advantage of working in the 2.4-GHz ISM band, and thus allows the deployment of 802.15.4a networks almost worldwide, including the countries where UWB emissions are not yet allowed. The standard for the chirp signals defines that 14 channels spaced at 5 MHz are in the frequency range between 2410 and 2486 MHz. As a consequence of the lower bandwidth

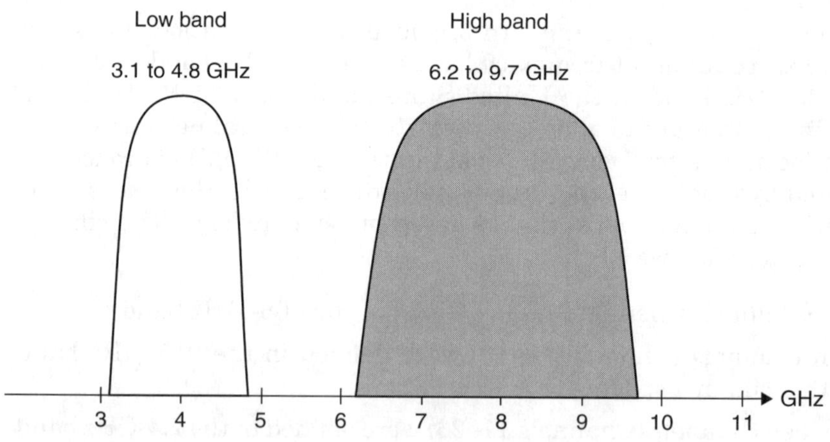

Figure 8.10 DS-UWB bands

compared to UWB, the bit rate achievable with the chirp signals is at most 1 Mbps. Furthermore, the chirp approach does not support ranging although accurate ranging solutions based on chirp signals were proposed.

8.1.3.3　DS-UWB　For WPAN applications using UWB technology, the direct sequence UWB (DS-UWB) has been proposed and is currently being developed for UWB positioning applications. The DS-UWB scheme is based on the original proposal by Freescale Semiconductor, Inc. (formerly known as XtremeSpectrum) and supported by the UWB Forum. The proposed system uses DS-CDMA with variable code length to provide data rates of 28 to 1320 Mbps over the frequency band of 3.1–4.8 GHz or 6.2–9.7 GHz.

The DS-UWB system operates using pulses that spread out their spectrum over a wide bandwidth, typically many hundreds of MHz or even several GHz. Each of the DS-UWB pulses has an extremely short duration, which is typically between 10 and 1000 picoseconds. In view of the wide bandwidth over which the DS-UWB transmissions are spread, the actual energy density is exceedingly low. Typically, a DS-UWB transmitter might transmit less than -71 dBm/MHz when integrated over the total bandwidth of the transmission. The extremely spectral density suggests that DS-UWB transmission does not cause interference with other radio transmissions operating in the existing bands and using conventional carrier-based techniques.

There are a number of ways in which DS-UWB transmissions can be modulated to enable data to be carried. Two of the most popular forms of the modulation used for DS-UWB are pulse position modulation (PPM) and binary phase shift keying (BPSK). PPM encodes the information by modifying the time interval and hence the position of the pulses whereas BPSK reverses the phase of the pulses to signify the data to be transmitted.

DS-UWB systems may feature the following merits:

- Low fading due to a wide bandwidth of greater than 1.7 GHz

- Excellent combination of high performance and low system complexity for WPAN applications

- Low power level compared to any other technologies for WPAN applications.

8.1.3.4　UWB Positioning Methods　Today, a multitude of applications and services can benefit from indoor (in-building) positioning and navigation, such as logistics, routing, sales, asset tracking, personal safety, emergency response, and so on. Conventional GPS receivers do not work properly inside buildings due to the absence of line of sight to satellites. Cellular positioning systems generally fail to provide a satisfactory degree of accuracy. The delivered position cannot be used for determining whether a target person stays inside or outside the building. Any commercial and social interactions that are being conducted indoors are thus not able to take advantage of outdoor positioning systems like GPS.

Generally, indoor positioning is a big challenge since the multipath signals tend to ruin position accuracy. Conventional narrow-band systems are subjected to multipath Rayleigh fading where the signals from many delayed reflections combine at the receiving antenna according to their random relative phases. This results in possible summation or cancellation, depending on the specific propagation path to a given location. In mobile systems, the multipath effect leads to dramatic signal strength fluctuations as a function of distance, which poses a problem to positioning accuracy.

UWB radios, however, can be substantially resistant to such effects by using very short pulses. In addition, the short pulse enables accurate TOA measurements with good multipath tolerance because the UWB radios can distinguish the directly transmitted signal and the multipath signal from ceilings, walls, and floors easily. A comparison of the multipath effects between a conventional RF system and a UWB-based system is shown in Figure 8.11.

Conventional UWB positioning systems[19,20,21] adopt multiple references (more than three) to locate the position of a mobile UWB tag. The references located at different locations usually need to be synchronized and connected to a common point. By using either the TOA or the TDOA

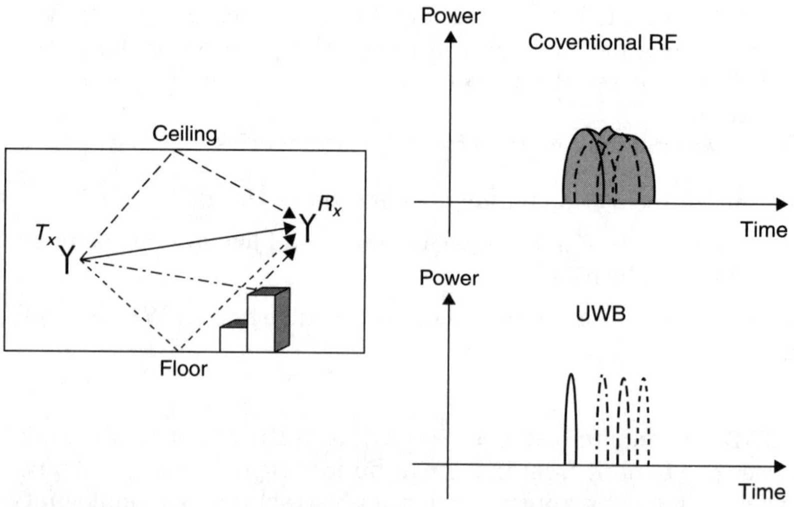

Figure 8.11 Multipath effect within an indoor environment

method, the tag's position is calculated by its geometric relationship to these reference points. Figure 8.12 illustrates a system that is used to track the position of mobile tags. Several UWB radio units are positioned at preselected locations to configure a coverage area. These reference UWB radio units are first synchronized together. When a mobile tag enters the covered area, it is synchronized with the reference UWB radio units, and the time-tagging range measurement is conducted by any reference UWB radio unit. The mobile tag's position can then be calculated in a three-dimensional coordinate system.

The other one method of UWB positioning is to emulate a radar system in which both the range and direction of a UWB tag are measured by a single base station.[22] As shown in Figure 8.13, in such a

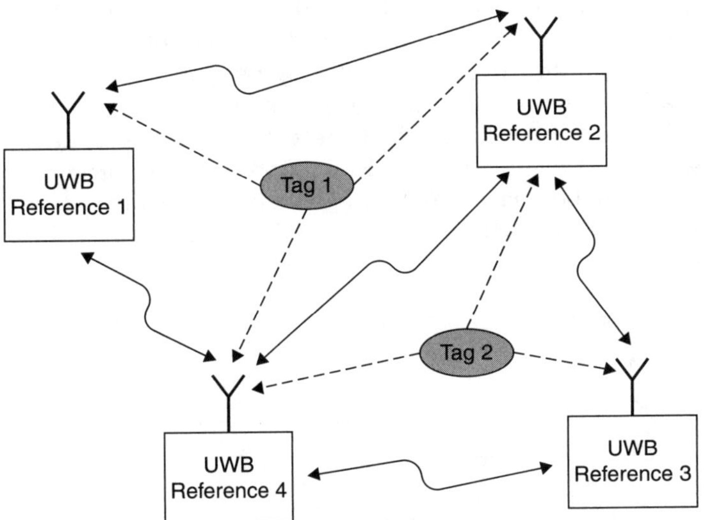

Figure 8.12 Schematic diagram of a UWB positioning system using multiple synchronized references

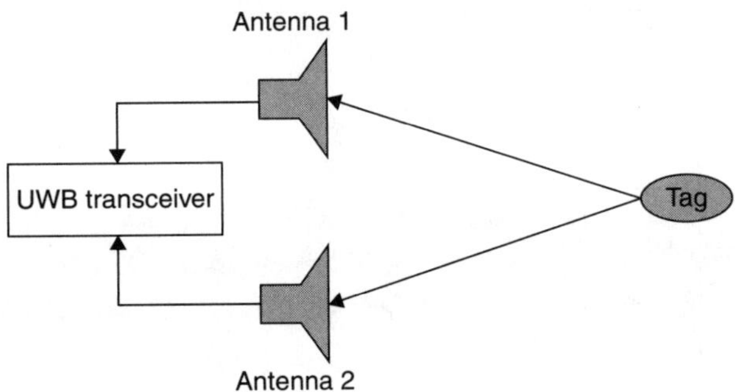

Figure 8.13 Configuration of a UWB monopulse radar system

system, range measurement is easily carried out by measuring the TOA of the reflected pulses from the tag. The direction measurement is generally conducted by measuring the time difference of the pulses received by two separated antennas. Since the time difference is very small, the achieved accuracy of the direction of arrival (DOA) using such a method is limited if the separation between the two antennas is small.

An alternative way to find the DOA of a UWB pulse is to measure the amplitude difference of the UWB pulses received by a directional antenna array.[23] Figure 8.14 illustrates the concept behind this method. A sectored antenna array is used for transmitting/receiving UWB signals; the main beam of the directional antenna elements is aligned at different directions; and a 360° circumference can be covered by using multiple antenna elements. A UWB signal from the tag along certain direction (such as θ shown in Figure 8.14) is able to be received by all the antenna elements with different magnitudes. No doubt, antenna Element #5 and Element #6 will receive a stronger signal $R_5(\theta) + R_6(\theta)$, which will be greater than any other pair of antenna elements, such as $R_4(\theta) + R_5(\theta)$, or $R_1(\theta) + R_6(\theta)$. The DOA of the signal can be estimated to be between antenna Element #5 and Element #6. A more accurate direction calculation can be carried out by iteratively comparing the preknown gain response of the antenna elements, $G_5(\theta)$ and $G_6(\theta)$, within

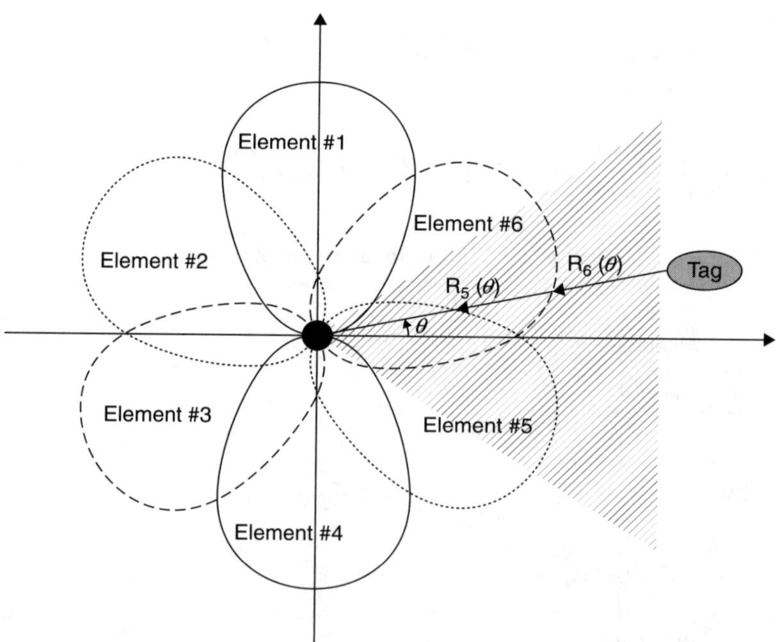

Figure 8.14   Monopulse UWB DOA estimation using sectored antenna array

the sector of interest to the received signals until the true direction is determined. A monostation UWB positioning system based on the mentioned technique is able to achieve a high accuracy on the order of a centimeter in range and a few degrees in direction, which are desirable for indoor positioning applications.[23]

8.2　Antenna Design for RFID Readers

8.2.1　Design Considerations

Reader antennas play an important role in RFID applications. RFID systems using optimized reader antennas feature enhanced detection range, higher detection accuracy, simpler system configuration, easier implementation, and lower costs. In general, the loop antennas used in LF/HF RFID systems are required to generate desired magnetic field strength within a specific coupling zone. While at UHF and MWF bands, patch antennas are normally adopted as reader antennas, which prefer to be circularly polarized, broadband, and high gain. In addition, common factors such as reliability, size, weight, and cost must also be considered in all RFID antenna designs.

8.2.1.1　LF/HF Reader Antenna Design Considerations　Loop antennas have been widely used in LF/HF RFID systems as reader antennas for many years since most LF/HF RFID systems use inductive coupling for reader/tag power transferring and communications. System performance in an application with specified EIRP and preselected tags is mainly determined by the characteristics of the reader antenna, which is relevant to various factors such as coupling zone, magnetic field strength, quality factor, impedance matching, environment, size, and so on.

Coupling Zones　Figure 8.15 demonstrates the coupling between the tags and the reader antenna in a near-field RFID system. The maximum coupling is achieved when the tags are positioned orthogonal to the magnetic field generated by the reader antenna and thus the tag can be detected. When a tag is positioned parallel to the reader antenna, namely in x-y plane, it can be detected at boresight ($\pm$ z-direction) by the reader antenna with the largest reading range since maximum magnetic flux is going through the tag antenna. At the sides of the reader antenna ($\pm$ x and $\pm$ y-directions), the tag can be detected as well with a certain reading range. If the tag is placed with an offset from the boresight, the magnetic flux going through the tag antenna may be too weak to power up the ASIC such that the tag may not be detected at certain positions.

On the other hand, if the tag is positioned perpendicular to the reader antenna, namely in the y-z plane, a different coupling zone is observed. The tag cannot be detected at the boresight ($\pm$ z-direction) and sides

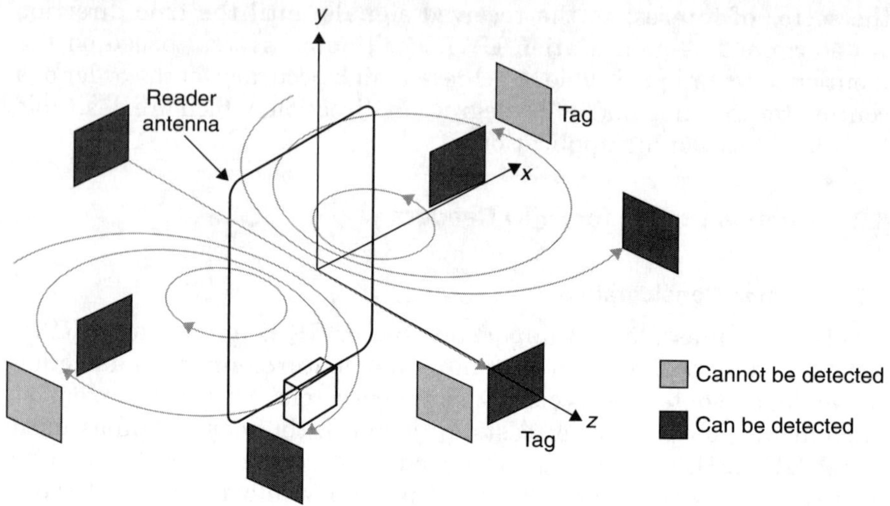

Figure 8.15 Coupling between the tags and the reader antenna

($\pm x$ and $\pm y$-directions) of the reader antenna because no magnetic flux is going through the tag antenna. When the tag is placed at an offset position from the boresight or sides of the reader antenna, however, the magnetic flux going through the tag antenna may be large enough to power up the ASIC and thus the tag can be detected with a certain reading range. These reading zones are shown in Figure 8.16.

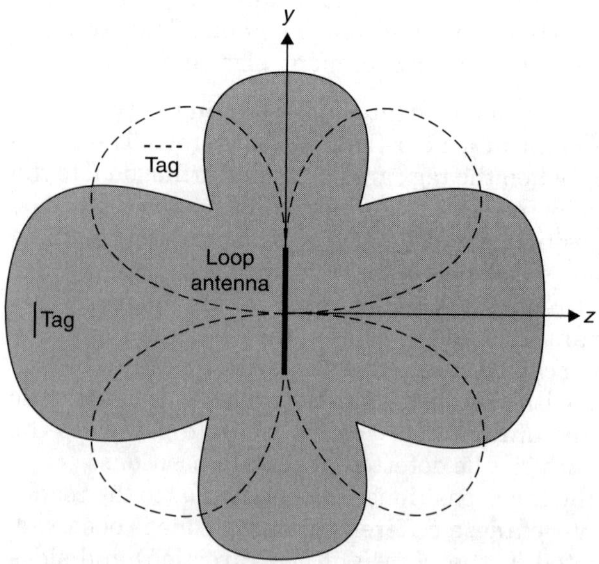

Figure 8.16 Tag coupling zones (top view)

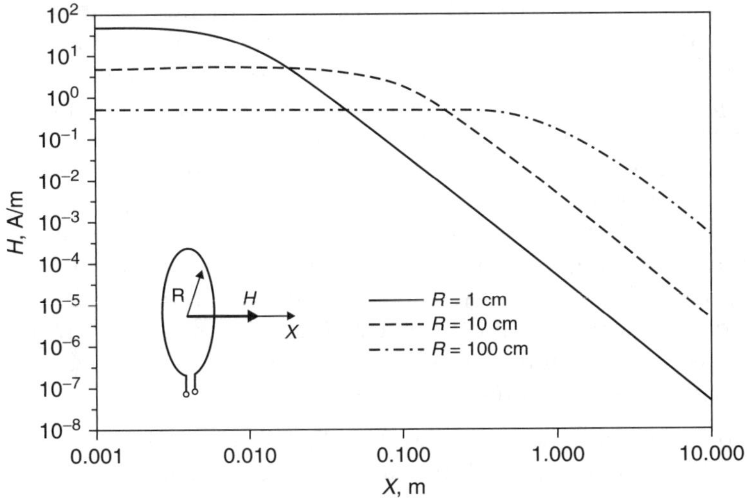

Figure 8.17 Magnetic field strength of a single circular loop antenna with different radius

Antenna Size and Magnetic Field Strength　Figure 8.17 shows the calculated magnetic field strength, H, of three single circular loop antennas at a distance of 0–10 m. It is observed that the field strength flattens out at short distances $(X < R)$ from the antenna and then falls rapidly when $X > R$. In free space, field strength decay is approximately 60 dB per decade in the near-field zone.

Generally, a reader antenna with a larger size offers a longer reading range. However, an accurate assessment of a system's maximum reading range requires knowing the interrogation field strength, H_{min}, of the tag[4] and the output power of the RFID reader. When the selected antenna size is too large, the field strength H may be too weak to supply the tag with sufficient operating energy, even at a distance $X = 0$. If the selected antenna size is too small, the magnetic field falls too rapidly to maintain enough strength to power up the tag when the distance increases. The "optimal" antenna size is subject to specific application scenarios.

Resonance and Impedance Matching　In order to transfer maximum energy, the electrical small loop antenna in inductively coupled RFID systems should be tuned to resonate at the carrier frequency, for example, 134.2 kHz or 13.56 MHz. At resonance, the inductive reactance (X_L) of the antenna is compensated by the capacitive reactance (X_C), which can be attained from the antenna itself or an additional circuit. The resonant frequency can be calculated as follows:

$$f = \frac{1}{2\pi\sqrt{L \cdot C}}$$

(8.1)

Obviously, the inductance (L) and capacitance (C) are interrelated. At a specific resonant frequency, if an antenna is too large, the inductance is thus very large and the required capacitance is too small to be realized. In practical applications, there are some techniques to reduce inductance such as connecting two antennas in parallel. In addition, antenna inductance can be calculated accurately using commercial simulation tools or measured using impedance analyzers or vector network analyzers.

To achieve optimal system performance, the antenna should have good impedance matching to the RFID reader with certain input impedance, for instance, 50 Ω. Generally, the return loss of a reader antenna is required to be less than –20 dB (or a voltage standing wave ratio, VSWR < 1.2). Numerous matching techniques can be used to achieve the desired impedance matching for LF/HF loop antennas.

Quality Factor (Q factor) The dimensionless figure of merit called the Quality Factor (Q) is another important parameter in the design of LF/HF RFID antennas. The Q factor represents the amount of electrical resistance of a system at resonance. In general, a high Q antenna not only transfers higher energy at resonance but also has a narrow bandwidth that limits the possible out-of-band interference. However, a very high Q may conflict with the bandpass characteristics of the RFID reader. If the antenna has a very narrow bandwidth due to a very high Q, the signals in the sidebands may be cut off. On the other hand, if the antenna features very low Q, i.e., a very broad bandwidth, the sidebands will pass through but random ambient noise occurs in that frequency range as well. Too much noise will degrade the quality of the received signals. As a tradeoff, the Q factor of a single loop antenna connected to a 50-Ω load (i.e., the reader) should be about 20 or less.[24] The Q factor can be determined by the 3-dB bandwidth of the magnetic field response of the antenna and calculated by

$$Q = \frac{F_0}{\Delta f} \tag{8.2}$$

where, F_0 is the resonant frequency (for example, 13.56 MHz) and Δf is the 3-dB bandwidth of the field response.

Effect of Environment In practical RFID applications, reader antennas are always positioned in the proximity of other materials such as metals and liquids. These situations occur when the antenna is mounted on a conveyor belt, a bookshelf, embedded into a table, or an aquatic body. RFID system performance may be degraded when these materials are in the antenna's coupling zone. Among these materials, metallic objects have a significant effect on antenna performance. The proximity of the metallic objects lowers antenna inductance and thus shifts up the resonant frequency. Here, a variable capacitor is recommended for use in the matching circuit for tuning purposes.

In summary, an LF/HF reader antenna design is dependent on the specific application's requirements. The size and shape of the antenna can be optimized with prior knowledge of the reader output power and selected tag. The antenna can also be co-designed with the environment by considering the proximity effect of objects. In addition, a simple and cost-effective antenna configuration is preferred.

8.2.1.2 UHF/MWF Reader Antenna Design Considerations

Compared to the loop antennas in LF/HF RFID applications, UHF/MWF reader antennas have a number of variations wherein the most commonly used is the circularly polarized patch antenna. Linear antennas such as dipole antennas have also been adopted in some specific applications. Generally, the major considerations in UHF/MW RFID reader antenna design include frequency range/bandwidth, polarization, gain, axial ratio, impedance matching, size, cost, and mechanical robustness.

Frequency Range/Bandwidth As mentioned in Section 8.1.2.3, the frequencies for UHF RFID applications vary from country to country. Therefore, a broadband reader antenna that features desirable performance across the entire UHF RFID band ranging from 840 to 960 MHz (13.3%) would be preferable for system configuration, implementation, and cost reduction. The MWF RFID bands are allocated at 2.4 GHz (2.400–2.4385, 1.59%), 5.8 GHz (5.725–5.875 GHz, 2.59%), and 24 GHz (24.00–24.25 GHz, 1.04%).

Polarization Generally, reader antennas at UHF/MWF bands are required to be circularly polarized for detecting arbitrarily oriented tags. The axial ratio (AR) is typically required to be less than 3 dB within the required beamwidth. Usually, a properly designed circularly polarized antenna can achieve less than 1 dB AR at boresight.

Besides the circularly polarized antenna, linearly polarized antennas such as patch and dipole antennas can also be applied to specific RFID applications where the orientation of the tags is fixed.

Gain For increasing reading range, a higher gain antenna is preferable for RFID applications. However, an RFID system with a high gain antenna may conflict with EIRP regulations. For an EIRP limitation of 4 W (36 dBm), the reader antenna gain cannot be more than 6 dBic when a 1 W (30 dBm) reader is used. Instead, a consistent gain response over the entire band is more important, which provides the most convenience in terms of system configuration, implementation, and cost reduction.

Impedance Matching The output power of the reader at the UHF/MWF bands can be up to 1 W (30 dBm); hence, the connected antenna must have very good impedance matching with the reader. Poor matching causes a large reflected power back to the receiver module of the reader, thus degrading system performance. Typically, the antenna is required to have a return loss of less than −20 dB, or VSWR ≤ 1.2. For some systems, a return loss of less than −15 dB is acceptable.

Size and Cost   The size of the reader antenna is not a critical concern for most UHF/MWF RFID applications. Of course, a compact and low-profile antenna is preferable. The conventional single-fed circularly polarized patch antenna has a simple and compact structure, is easy to manufacture, and is low cost. However, it has an inherently narrow axial ratio and impedance bandwidth of 1–2%. On the other hand, the multiple feeding techniques result in complicated structures, manufacturing difficulty, and increased antenna size, but a greater axial ratio bandwidth.

Reliability   The antenna must be a reliable component in a system and able to sustain variations due to temperature, humidity, stress, and so on, because the antenna may be mounted outdoors in some applications.

8.2.2   Case Study

8.2.2.1   HF Antenna

Antenna Configuration and Simulation   Figure 8.18 illustrates the geometry of a single loop antenna. The loop antenna is configured on a printed circuit board (PCB) with the dimensions 280 mm × 280 mm. The FR4 PCB has dielectric parameters of $\varepsilon_r = 4.4$, $\tan\delta = 0.02$, and thickness of 0.508 mm. The dimension of the square loop antenna is 250 mm × 250 mm, and the width of the trace is 10 mm. The antenna is connected to a sub-miniature version A (SMA) coaxial connector; the inner/outer conductors of the SMA connector are soldered to the two output stubs of the antenna. A capacitance impedance matching circuit is designed to

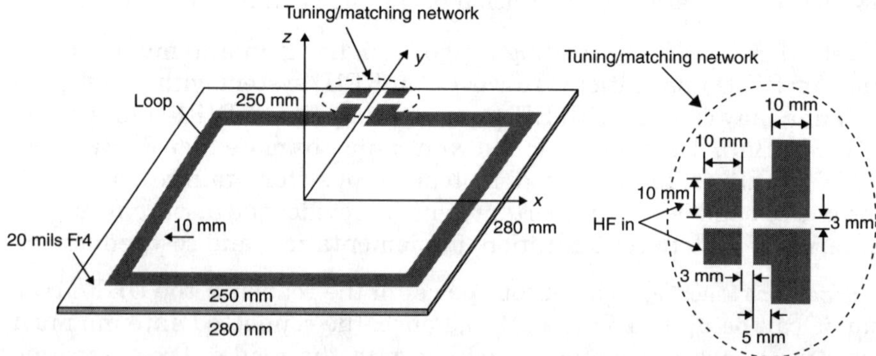

Figure 8.18   HF single loop antenna

tune the antenna to the resonant frequency of 13.56 MHz and match the antenna to a 50-Ω feed line.

The inductance of the loop antenna can be derived from the input impedance of the antenna, which can be simulated by using commercial simulation tools. Figure 8.19 shows the simulated input impedance of the loop antenna. The real part of the input impedance is around 1.5 Ω with slight variation. The imaginary part of the input impedance increases from 54 to 72 Ω over a range of 11.5–15.5 MHz. The inductance of the loop can be extracted from imaginary part of the impedance, as shown in Figure 8.20. An inductance of 728 nH is obtained at 13.56 MHz with a variation of $\pm$ 1 nH over the 11.5–15.5 MHz frequency band.

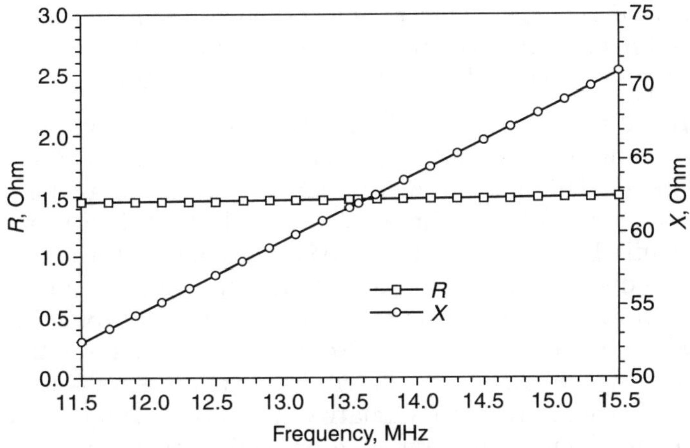

Figure 8.19 Simulated input impedance of the loop antenna by IE3D[25]

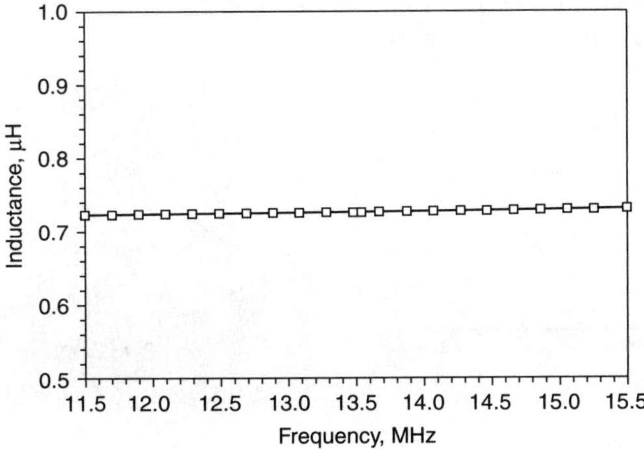

Figure 8.20 Simulated inductance of the loop antenna

The loop antenna is designed to resonate at 13.56 MHz. The required capacitance for forming the resonant circuit can be calculated by using Eq. 8.1. Considering $L = 728$ nH and $f = 13.56$ MHz, the required capacitance is $C = 189.2$ pF.

Tuning/Impedance Matching Circuit   Here, the capacitance matching method is demonstrated for achieving the desired resonant frequency and good impedance matching. As shown in Figure 8.21, the requested capacitance can be realized by using several series/parallel capacitors with $C_{p1} = 390$ pF, $C_{p2} = 120$ pF, $C_{p3} = 41.95$ pF, $C_{s1} = 47$ pF, and $C_{s2} = 82$ pF. The adoption of the variable tuning capacitor C_{p3} (7–50 pF) enables the resonant frequency to be tuned from 13.28 MHz to 15.02 MHz, which offers great convenience for antenna tuning in practical applications where the antenna is often de-tuned by an adjacent object such as a metallic plate. Because additional insulation resistance of the capacitor contributes to achieving an adequate Q factor, the extra resistor is not needed.

Antenna Prototype and Results   Figure 8.22 shows the single loop antenna prototype. The measured return loss and magnetic field response are shown in Figure 8.23. The measurement was carried out by using a vector network analyzer (VNA, Agilent 8753E) and an H-field probe (Langer EMV-Technik LF-R 400).[26] The probe is positioned at the center of the antenna at a distance of 50 mm. In addition, the probe is placed parallel to the loop antenna to allow maximum magnetic flux going through it. The antenna and the probe are connected to Port 1 and 2 of the VNA, respectively. The impedance matching of the antenna is indicated by $|S_{11}|$. The resonant frequency, relative magnetic field intensity, field distribution, and Q factor are extracted from the measured $|S_{21}|$. In the measurement, the VNA is calibrated using a standard calibration procedure at the bandwidth of interest. Using an Omron HF reader (VS 702S) with 4 W output power and Phillips SLI tag, the antenna can achieve a reading range up to 0.45 meter.

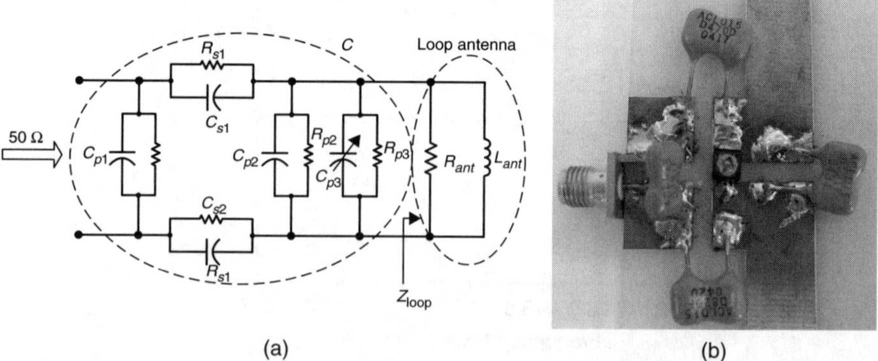

Figure 8.21   Tuning/impedance matching circuit: (*a*) equivalent circuit and (*b*) practical circuit

Figure 8.22 HF loop antenna prototype

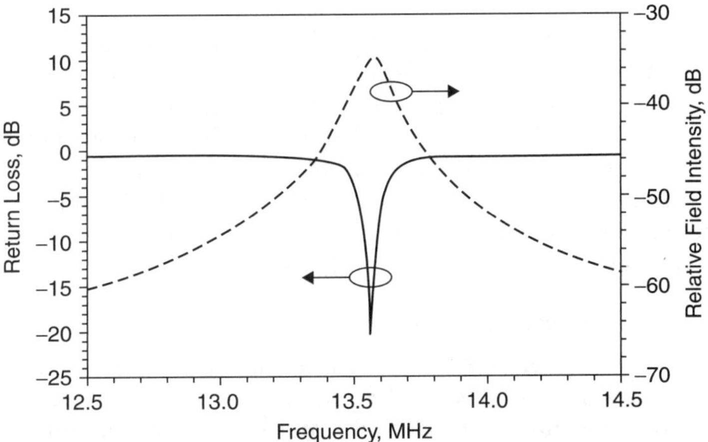

Figure 8.23 Measured return loss and magnetic field response of the loop antenna

Proximity Effects of Metal Environment RFID applications for large-scale item-level management in retail and libraries have greatly spurred the development of so-called *RFID Smart Shelf*. To detect the tagged items, a large number of reader antennas must be positioned inside the shelves, which are usually made of metal. Metallic environments severely affect the performance of the HF loop antenna by detuning the resonant frequency and reducing the magnetic field intensity.[27]

Figure 8.24 depicts the configuration of the loop antenna with typical metal environments in practical applications. Figure 8.24*a* shows the antenna with a back-placement metal plate, wherein a metal plate is placed parallel to the antenna. The dimensions of the metal plate are

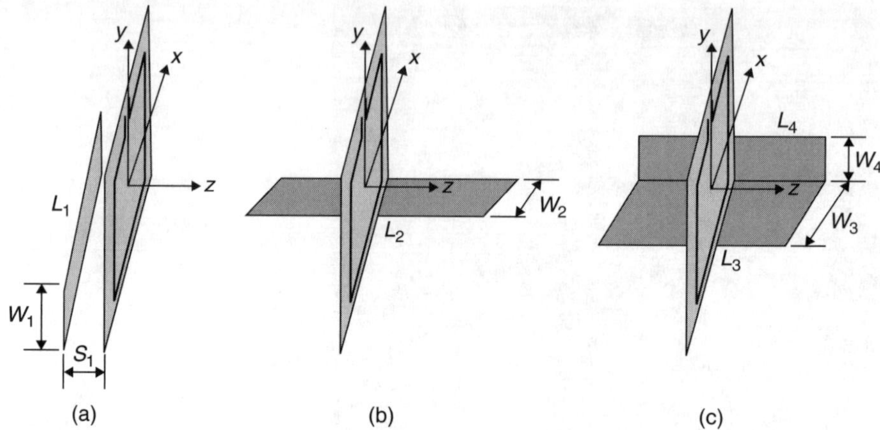

Figure 8.24 Typical antenna/metal plate configurations: (a) antenna with back-placement metal plate, (b) antenna with bottom-placement metal plate, and (c) antenna with bottom-placement and side-placement metal plates.

indicated by L_1 and W_1 and the separation between the antenna and the metal plate is S_1. Figure 8.24b demonstrates the configuration of the antenna with a bottom-placement metal plate, where a metal plate with dimensions of $L_2 \times W_2$ is placed at the bottom of the antenna. Figure 8.24c illustrates the antenna with bottom-placement and side-placement metal plates. The metal structure is composed of two metal plates that are perpendicular to each other. The dimensions of the metal plates are indicated as (L_3, W_3) and (L_4, W_4), respectively.

■ **Antenna with back-placement metal plate**

The length of the metal plate shown in Figure 8.24a is selected to be the same as the antenna, namely $L_1 = 280$ mm. The width of the metal plate, W_1, is changed from 80 mm to 280 mm, whereas the separation (S_1) between the antenna and the metal plate is fixed at 10 mm. Figures 8.25 and 8.26 show the measured return loss and relative magnetic field intensity of the antenna, respectively. The presence of the metal plate shifts up the resonant frequency of the antenna. The bigger the metal plate, the greater the frequency shift. The resonant frequency shifts to 14.00 MHz when $W_1 = 80$ mm and 17.53 MHz when $W_1 = 280$ mm. Good impedance matching is maintained at the corresponding resonant frequency. However, a RFID system using such an antenna configuration cannot receive the signal at the designated frequency of 13.56 MHz since the antenna is no longer well-matched at 13.56 MHz; hence, the reading range will be greatly reduced. The relative magnetic field intensity is measured at the side of the antenna without the metal plate at a distance of 50 mm from the center of the antenna. The magnetic field decreases when the size of the metal plate increases. As the metal plate

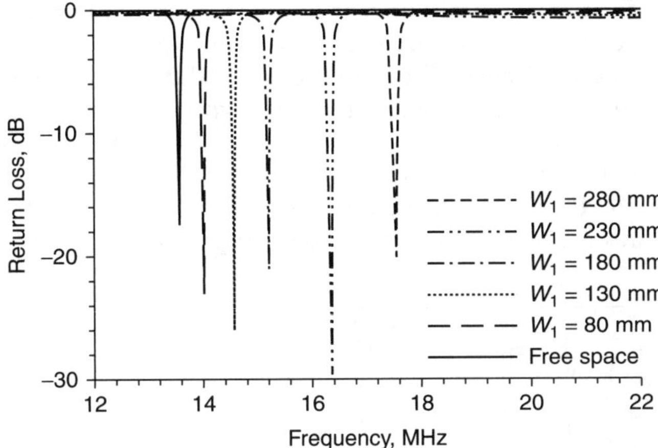

Figure 8.25 Measured return losses of the antenna with back-placement metal plates. Dimensions of the metal plates: $L_1 = 280$ mm, $W_1 = 80 - 280$ mm; separation: $S_1 = 10$ mm

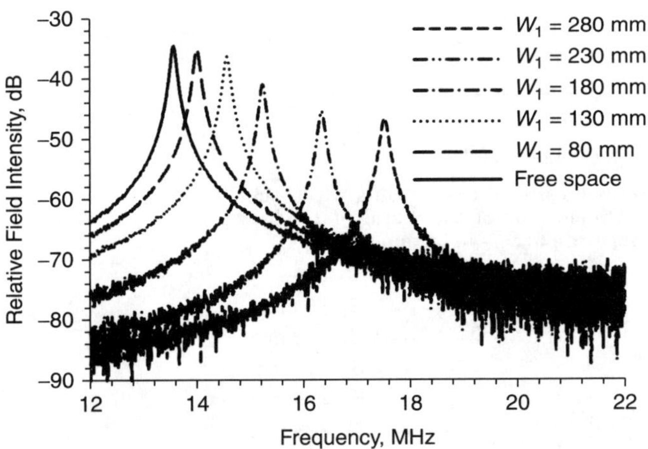

Figure 8.26 Measured magnetic field responses of the antenna with back-placement metal plates. Dimensions of the metal plates: $L_1 = 280$ mm, $W_1 = 80 - 280$ mm; separation: $S_1 = 10$ mm

becomes smaller ($L_1 = 280$ mm, $W_1 = 80$ mm), the relative magnetic field intensity (-35.0 dB) of the antenna is almost the same as that in free space (-34.8 dB). For the bigger metal plate ($L_1 = 280$ mm, $W_1 = 280$ mm), the relative magnetic field intensity drops to -46.4 dB.

When the size of the metal plate is further increased, the frequency shift is found to be very slight and the magnetic field intensity remains almost unchanged. Further study suggests that the frequency shift and the weakening of the magnetic field intensity are dependent on disturbance of the magnetic flux by the metal plate. The metal plate with a smaller size partially blocks the magnetic flux whereas a larger metal plate (such as 280 mm × 280 mm) blocks almost all of the magnetic flux on one side. Further increasing the size of the metal plate only slightly increases the blocking of the magnetic flux such that the effects on the resonant frequency and field intensity are very limited.[27]

Figures 8.27 and 8.28 show the measured return loss and relative magnetic field intensity of the antenna with various metal plate

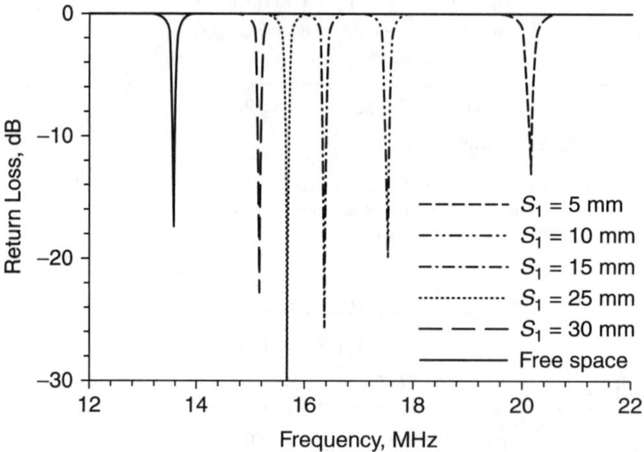

Figure 8.27 Measured return losses of the antenna with backplacement metal plate. Dimensions of the metal plate: $L_1 = 280$ mm, $W_1 = 280$ mm; separation: $S_1 = 5$–35 mm

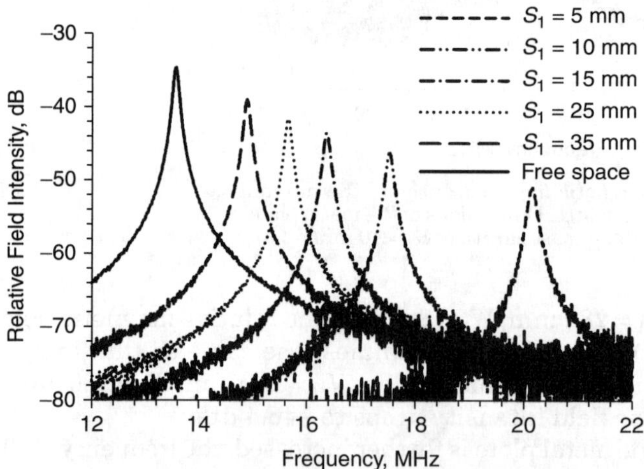

Figure 8.28 Measured magnetic field responses of the antenna with back-placement metal plate. Dimensions of the metal plate: $L_1 = 280$ mm, $W_1 = 280$ mm; separation: $S_1 = 5$–35 mm

(280 mm × 280 mm) separations. Placing the metal plate closer to the antenna results in a higher resonant frequency and weaker magnetic field intensity. The resonant frequency shifts to 20.13 MHz and the relative magnetic field intensity reduces to −51.2 dB when the metal plate is placed 5 mm away from the antenna. The frequency shift is up to 6.57 MHz and the relative magnetic field intensity degrades by 16.4 dB compared to the antenna without any metal plate.

■ **Antenna with bottom-placement metal plate**

Referring to Figure 8.24*b*, a metal plate with a length of L_2 and a width of W_2 is placed at the bottom of the antenna. The length of the metal plate, L_2, is selected as 280 mm whereas the width, W_2, changes from 80 mm to 280 mm. The impedance matching characteristic is similar to the back-placement case just described except for the frequency shift. The measured magnetic field responses are shown in Figure 8.29. The maximum frequency shift is 0.235 MHz when the metal plate is 280 mm × 280 mm, but the magnetic field intensity hardly changes because the presence of the metal plate results in little disturbance to the magnetic flux generated by the antenna.

■ **Antenna with bottom-placement and side-placement metal plates**

Referring to Figure 8.24*c*, the configuration of a loop antenna with a bottom-placement metal plate and a perpendicular side-placement metal plate can be seen in many practical scenarios such as in libraries, department stores, and warehouses. Here, the antenna is located

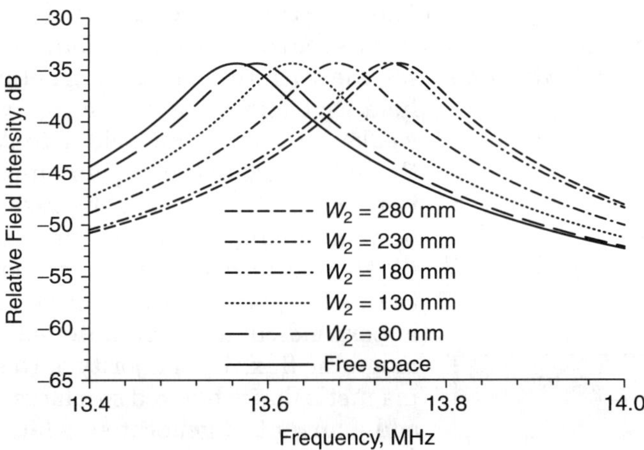

Figure 8.29 Measured magnetic field responses of the antenna with bottom-placement metal plates. Dimensions of the metal plates: L_2 = 280 mm, W_2 = 80–280 mm

TABLE 8.2 Measured Resonant Frequency and Relative Magnetic Field Intensity of the Antenna with Bottom-Placement and Side-Placement Metal Plates

Dimensions of the metal plates $L_3 \times L_4 \times W_3 \times W_4$ (mm)	280 × 280 × 280 × 0	280 × 280 × 280 × 80	280 × 280 × 280 × 130	280 × 280 × 280 × 180	280 × 280 × 280 × 230	280 × 280 × 280 × 280	280 × 500 × 280 × 80	280 × 900 × 280 × 80
Resonant frequency (MHz)	13.80	13.89	13.95	13.95	14.00	14.07	13.89	13.90
Relative magnetic field intensity (dB)	−34.53	−35.12	−35.13	−35.32	−35.40	−35.41	−35.20	−35.22

at the center of the metal plates. The width of the bottom-placement metal plates, W_3, is kept at 280 mm, whereas the length, L_3, varies from 280 mm to 900 mm. The width of the side-placement metal plate, W_4, varies from 80 mm to 280 mm and the length, L_4, from 280 mm to 900 mm. The measured results of the resonant frequency and relative magnetic field intensity are tabulated in Table 8.2. Compared to the case with a bottom-placement metal plate, the frequency shift is slightly higher whereas the variation of the relative magnetic field intensity is less than 1 dB.

It has been found that the resonant frequency of a loop antenna in the proximity of a metal plate is always shifted up with a reduction in the magnetic field intensity. The back-placement plate has a more severe effect on antenna performance than the bottom-placement and/or side-placement metal plate. Referring to Figure 8.30, a metal plate is positioned close to a loop antenna. To satisfy the boundary conditions, the normal components of a magnetic field must be zero on the surface of the metal plate. An additional current, known as the *eddy current,* is induced within the metal plate,[4] which opposes the variation of the magnetic field that induces it and severely weakens the magnetic field so that the inductance of the loop antenna is reduced. As a result, the resonant frequency is shifted up. As illustrated in Figure 8.31*a* when the metal plate is back-placed, it is perpendicular to the antenna's magnetic flux. The majority of the magnetic flux is blocked so a larger eddy current is generated, which results in the dramatic reduction in inductance of the loop antenna. Therefore, a significant frequency

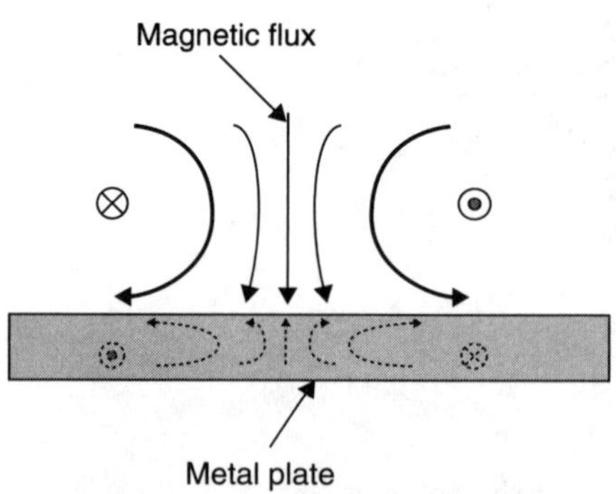

Figure 8.30 Magnetic fields on the surface of a metal plate

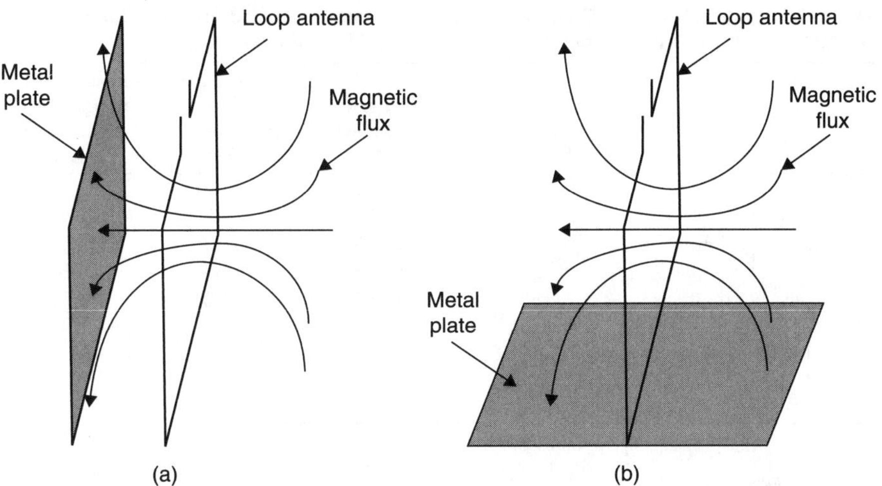

Figure 8.31 Metal plates and magnetic flux of loop antenna: (*a*) back-placement metal plate and (*b*) bottom-placement metal plate

shift is observed. For antennas with bottom-placement and/or side-placement metal plates, the eddy current effect is much less because the metal plate is oriented in parallel to the majority of the magnetic flux such that the magnetic flux is hardly disturbed (Figure 8.31*b*). Hence, the inductance and the resonant frequency of the loop antenna are hardly changed.

The frequency shift of the antennas with the different metal plate placements can be predicted by calculating the variation in antenna inductance. The simulated and measured results are summarized in Table 8.3.

8.2.2.2 UHF/MWF Antenna Circularly polarized (CP) radiation can be achieved when two orthogonal modes of equal amplitude are excited with a 90° phase difference.[28] In general, the feeding structures of CP antennas may be categorized into single and hybrid feeds. A single feed offers a CP antenna several advantages: a simple structure, easy to manufacture, and a small size. However, such antenna designs in their simple forms usually suffer from an inherently narrow axial ration (AR) and impedance bandwidths of 1–2%.[29] To improve bandwidth, a variety of CP antennas have been studied, wherein the bandwidth of AR and gain have been enhanced by, for example, modifying radiator shapes and feeding structures, as well as optimizing the antenna or array configurations.[30,31,32,33,34,35,36] Usually, a CP antenna with a hybrid feed features a larger axial ratio bandwidth but has a complicated structure, is expensive to manufacture, and is larger in size.

TABLE 8.3   Comparison of Calculated and Measured Resonant Frequencies of Loop Antennas with Different Placements and Sizes of Metal Plates

	S_1 (mm)	Inductance (calculated), nH	Resonant frequency (calculated), MHz	Resonant frequency (measured), MHz
	5	329.12	20.26	20.13
	10	435.14	17.62	17.53
	15	490.25	16.60	16.50
	25	556.55	15.58	15.63
	35	606.87	14.92	14.88
W_1 (mm)				
	80	690.37	13.99	14.00
	130	634.74	14.59	14.55
	180	574.39	15.34	15.23
	230	496.95	16.49	16.33
	280	435.14	17.62	17.53
W_2 (mm)				
	80	730.40	13.60	13.59
	130	721.88	13.68	13.64
	180	714.55	13.75	13.71
	230	711.44	13.78	13.76
	280	707.3	13.82	13.79

As an example, a sequentially fed stacked patch antenna is demonstrated to cover the UHF RFID band of 840–960 MHz with acceptable performance in terms of gain, axial ratio, and return loss, as well as having a simple configuration and low cost.

Antenna Configuration   Figure 8.32 shows the configuration of the antenna, which is made up of four layers of conducting plate/strip. The antenna is composed of two radiating patches, a suspended microstrip feed line, and a ground plane. The feed line is suspended above the ground plane at a height of h_1 and with an open-circuited end, which simplifies the antenna structure. The main radiating patch is placed at a height of h_2 above the feed line. It is sequentially fed by four probes that are connected to the microstrip line. The probes are of diameter $d = 2.2$ mm and uniformly positioned in an anti-clockwise direction to feed the patch sequentially. Left-hand circular polarization (LHCP) radiation will be generated by this configuration, and right-hand circular polarization (RHCP) radiation

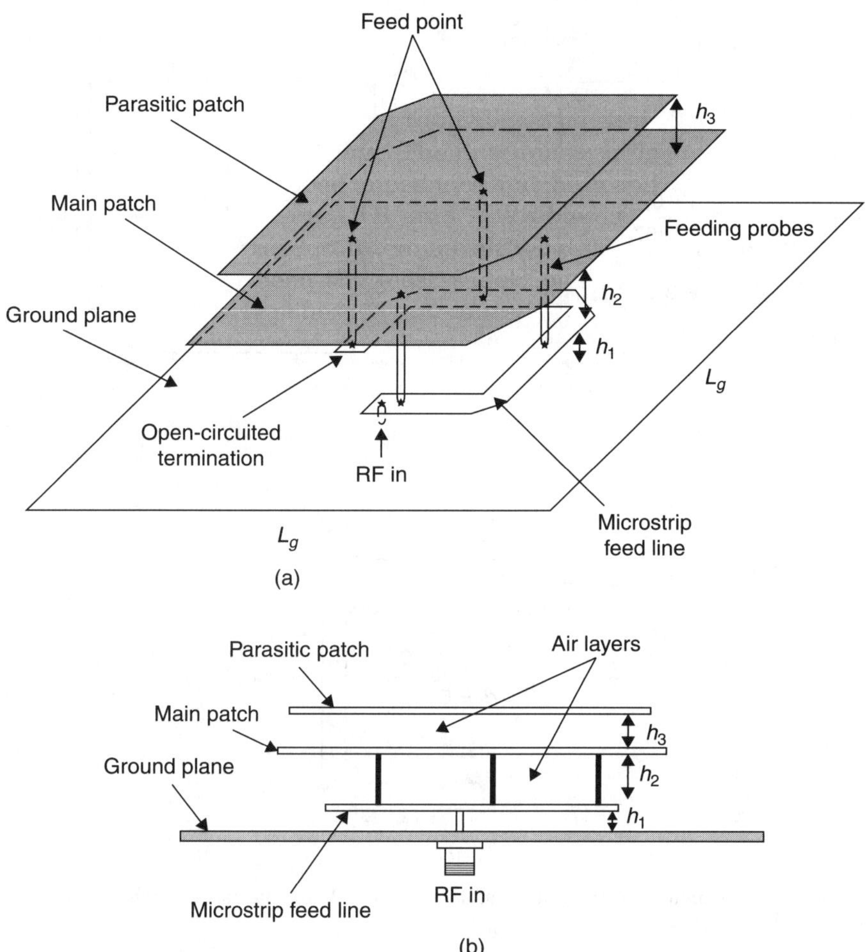

Figure 8.32 Configuration of the UHF circularly polarized antenna: (a) exploded view and (b) side view

can be achieved by locating the feed line and probes in a clockwise manner. A stacked patch, which is placed right above the main patch at a height of h_3, is used to enhance the impedance bandwidth. The two patches are truncated at two corners to create the additional degenerating modes for broadening the AR bandwidth. The suspended structure with the air substrate is conducive to achieving the high gain and broad impedance bandwidth as well as cost reduction.

The detailed dimensions of the patches, the suspended microstrip feed line, and the position of the feeding probes of the optimized antenna, are illustrated in Figure 8.33, where $h_1 = 5$ mm, $h_2 = 20$ mm, $h_3 = 10$ mm, and $L_g = 250$ mm.

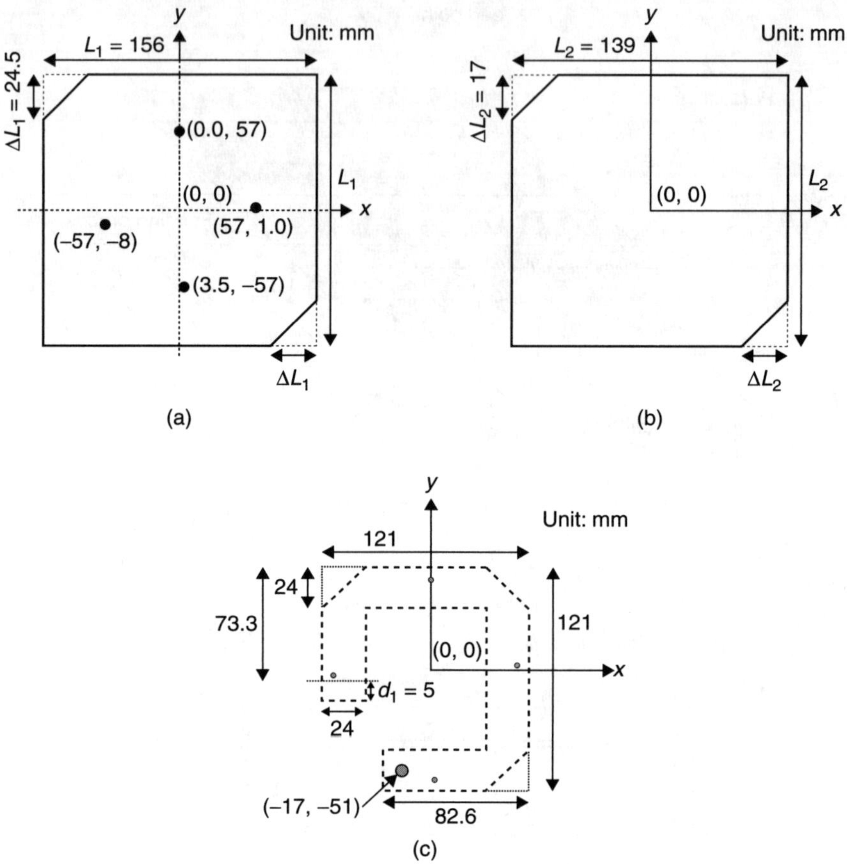

Figure 8.33 Dimensions of the antenna: (*a*) main patch, (*b*) parasitic patch, and (*c*) suspended microstrip line

Measured and Simulated Results Figure 8.34*a* shows the simulated and measured return loss of the antenna. The measured return loss is less than −15 dB over the frequency range of 760–967 MHz. Figure 8.34*b* exhibits the simulated and measured AR at boresight. A measured 3-dB AR bandwidth of 818–964 MHz is obtained. The measured boresight gain is more than 8.3 dBic over the bandwidth of 815–970 MHz with a peak gain of 9.3 dBic at 900 MHz, as illustrated in Figure 8.34*c*.

Figure 8.35 exhibits the measured CP radiation patterns at 840, 910, and 955 MHz in the *x-z* and *y-z* planes. The 3-dB AR beamwidth is more than 75° for all the patterns, which is summarized in Table 8.4. A large 3-dB AR beamwidth is desirable for wide angle coverage in RFID applications. The desired CP characteristic of the antenna is accredited to the sequential feed arrangement. This advantage stems from

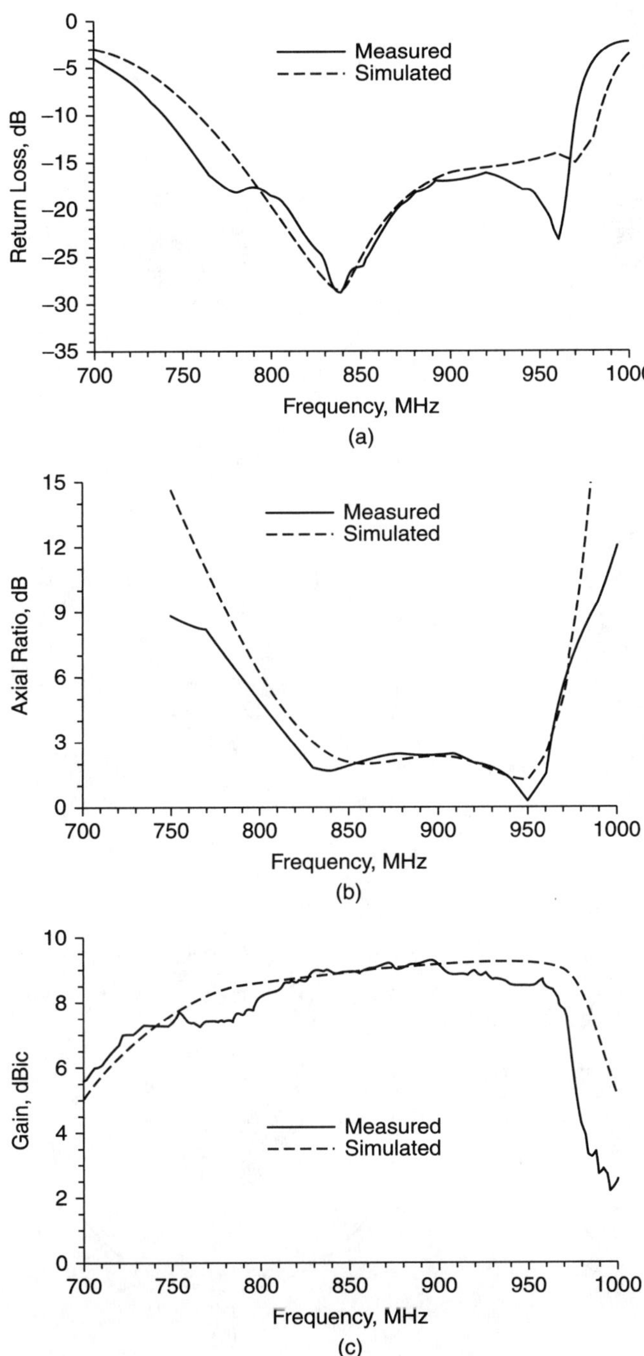

Figure 8.34 Simulated and measured results of the proposed antenna: (*a*) return loss, (*b*) axial ratio, and (*c*) gain

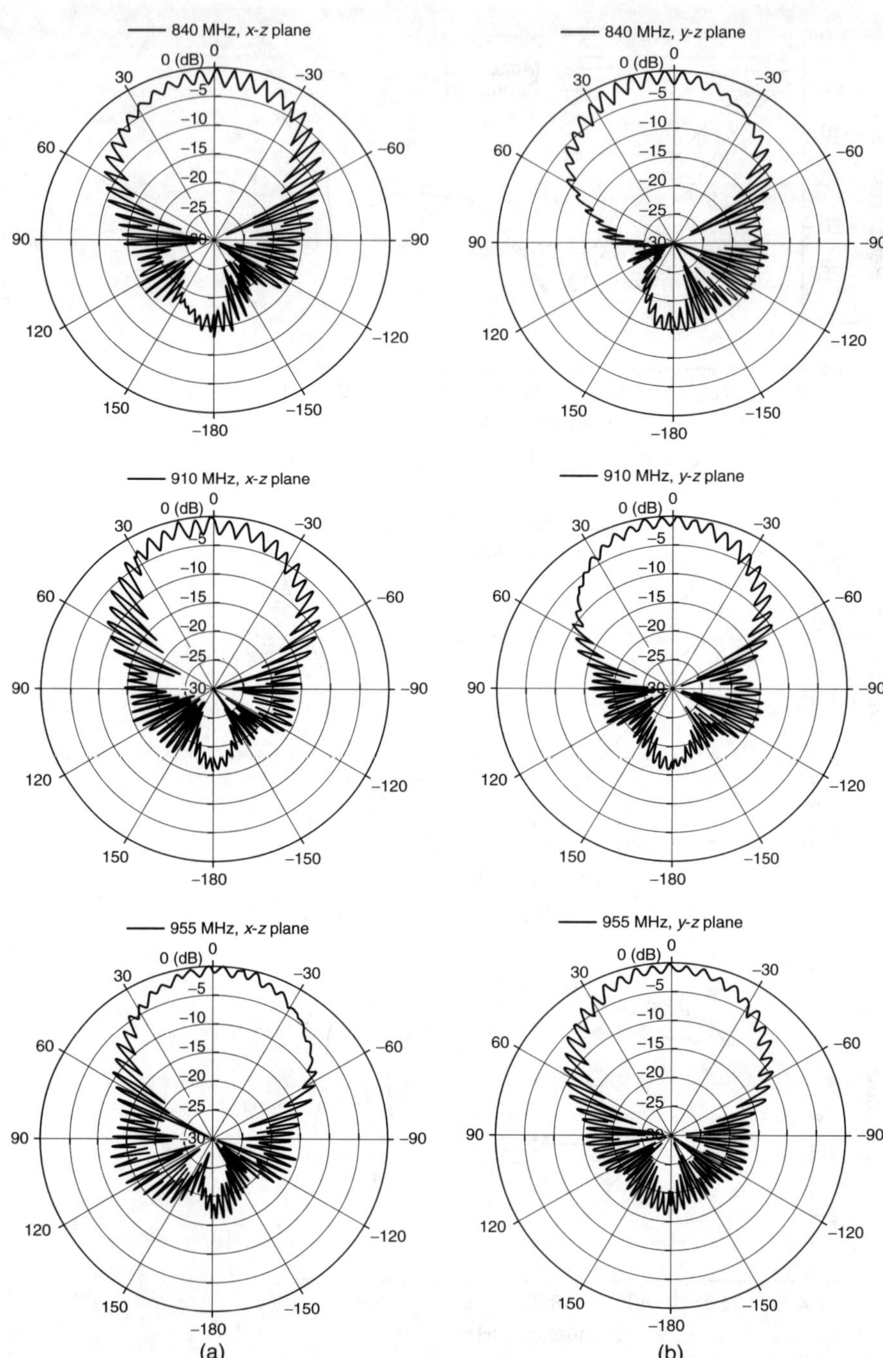

Figure 8.35 Measured radiation patterns at 840 MHz, 910 MHz, and 955 MHz: (a) x-z plane and (b) y-z plane

TABLE 8.4   Measured 3-dB Axial Ratio Beamwidth of the UHF RFID Reader Antenna

Frequency (MHz)	840	870	900	910	930	955
3-dB AR beamwidth (x-z plane)	85°	75°	75°	80°	85°	105°
3-dB AR beamwidth (y-z plane)	130°	115°	95°	100°	113°	89°

the symmetry of the feeding structure that cancels out the unwanted cross-polarization radiation. In addition, the front-to-back ratio of the antenna is greater than 15 dB in both the x-z and y-z planes at all the frequencies.

Parametric Studies   Without any dielectric substrate, antenna performance is determined only by the geometrical structure parameters. Moreover, antenna performance may be much more sensitive to some parameters than to others. Since the effects of the parameters, such as the size and height of the main patch, the size of the parasitic patch, and so on, are well known, they are excluded here. The parameters to be demonstrated include the truncation of the patches, the height of the parasitic patch, the size of feeding probes, the extension of the open-circuited microstrip line end, and the size of the ground plane. In order to understand the influence of the parameters on antenna performance, only one parameter is varied at a time unless otherwise indicated.

- **Truncation of the main patch, ΔL_1**

Figure 8.36 shows the effect of truncation of the main patch, ΔL_1, on the impedance matching, AR, and gain of the antenna. The ΔL_1 significantly affects the AR of the antenna. When the main patch is not truncated, i.e., $\Delta L_1 = 0$, the antenna exhibits the widest impedance and gain bandwidths but the narrowest bandwidth for the AR. Increasing ΔL_1 improves the AR bandwidth but narrows the gain bandwidth and degrades the impedance matching. However, over-truncating the patch (such as $\Delta L_1 = 55$ mm) will degrade all the bandwidths for impedance, AR, and gain.

- **Truncation of the parasitic patch, ΔL_2**

Unlike ΔL_1, ΔL_2 has a significant effect on the impedance and AR bandwidths but hardly any effect on the gain performance of the antenna unless the parasitic patch is over-truncated. When the parasitic patch is not truncated ($\Delta L_2 = 0$ mm), the antenna features good impedance matching but the AR bandwidth is reduced. The impedance and AR bandwidths of the antenna change for ΔL_2 of 10–20 mm but decrease when the patch is over-truncated. The results are illustrated in Figure 8.37.

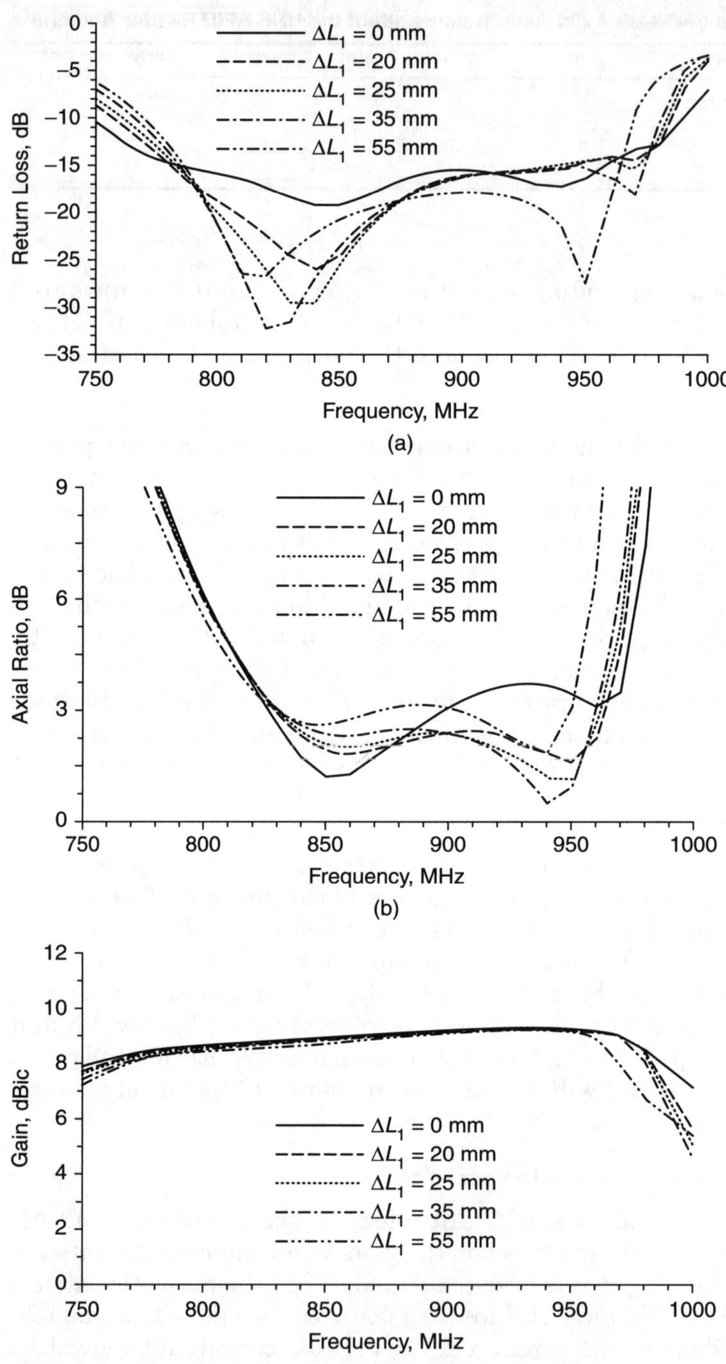

Figure 8.36 Effect of truncation of the main patch, ΔL_1, on antenna performance: (*a*) return loss, (*b*) axial ratio, and (*c*) gain

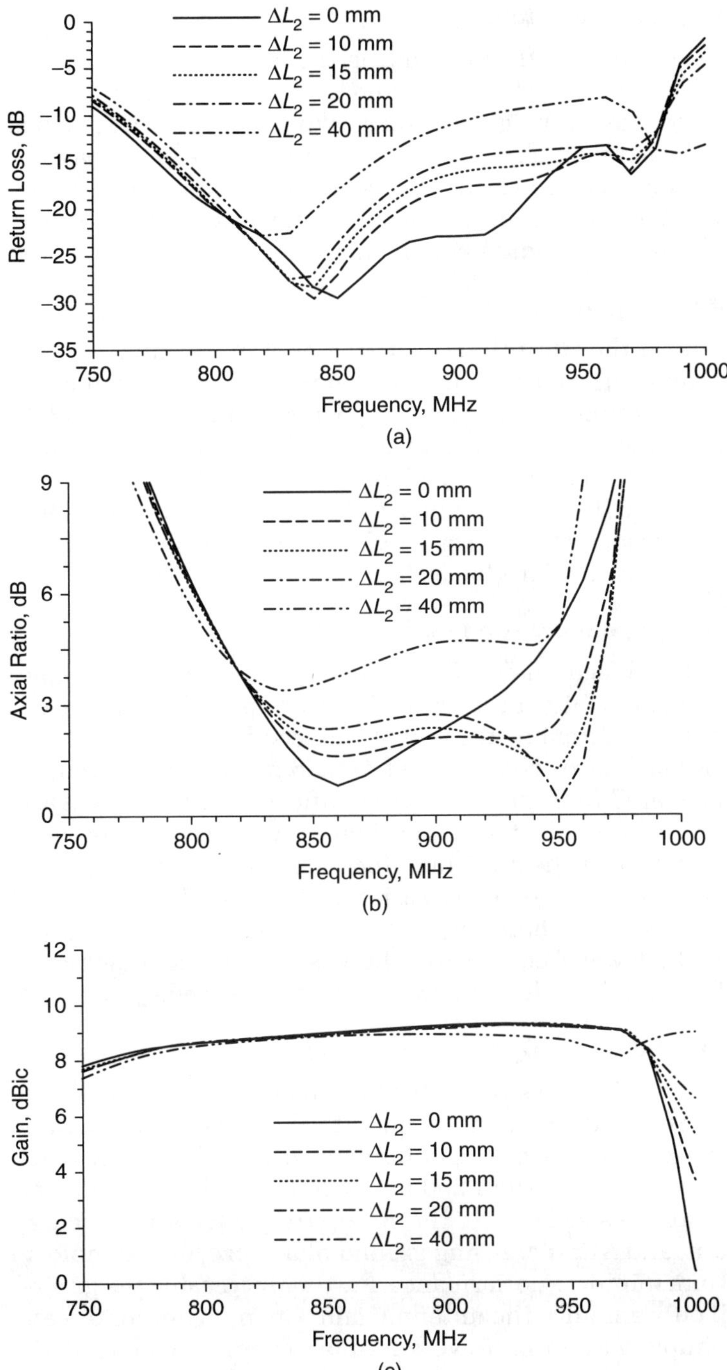

Figure 8.37 Effect of truncation of the parasitic patch, ΔL_2, on antenna performance: (a) return loss, (b) axial ratio, and (c) gain

■ **Height of the parasitic patch, h_3**

Figure 8.38 exhibits the effect of varying the height, h_3, of the parasitic patch on antenna performance. As shown, the operating frequency band is shifted down as the height increases. When the parasitic patch is placed close to the main patch, the strong coupling between the patches greatly reduces the AR bandwidth. The weakly coupled patches with a larger h_3 also suffer from reducing the AR bandwidth, although their impedance and gain bandwidths become larger.

■ **Diameter of feeding probes, d**

Figure 8.39 shows the effect that the diameter of the feeding probe, d, has on the antenna performance. Its effect on impedance matching and AR are severe. The thinner the probes, the poorer the impedance matching and AR performance. The long and thin feeding probes introduce a large inductance that degrades impedance matching. In addition, the large inductance also disturbs the phase characteristic at the feeding points and thus degrades AR performance. Feeding probes with a 2–3 mm diameter are recommended in the design.

■ **Extension of the open-circuited feed line, d_1**

The open-circuited feed line configuration simplifies antenna implementation and reduces fabrication cost. However, open-circuited termination may cause reflection at the end of the feed line and thus affect the antenna performance. The effect of extending the open-circuited feed line is illustrated in Figure 8.40. A significant effect on the AR is observed. Optimal AR performance is achieved when the last probe is positioned at the edge of the strip line. Increasing d_1 greatly degrades AR performance. When d_1 reaches 25 mm, AR is greater than 3 dB over the whole frequency band. d_1 affects impedance matching and gain mainly at the lower frequencies, whereas impedance matching is improved while the gain is degraded when d_1 is increased.

■ **Size of the ground plane, L_g**

The effect of ground plane size on the antenna performance is shown in Figure 8.41. As expected, the antenna with the larger ground plane has superior performance over the antennas with smaller ones. When the ground plane is smaller than 200 mm × 200 mm, the antenna performance degrades, especially at the lower frequencies in terms of impedance, gain, and AR. Increasing ground plane size, for example, to 250 mm × 250 mm improves performance. Further increasing the ground plane size will only enhance the antenna gain. Changing ground plane size offers a simple way to improve antenna performance but at the expense of increasing the overall antenna volume. Moreover, practical antenna designs are always subject to size constraints.

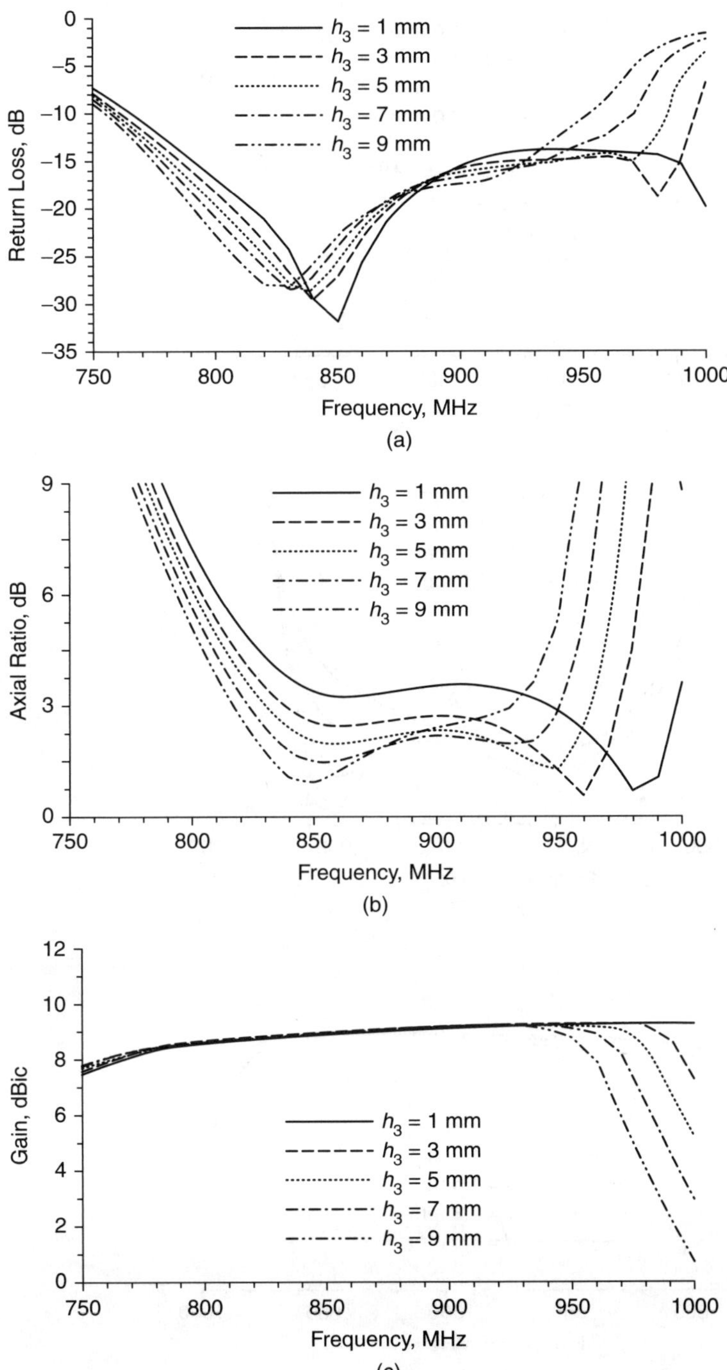

Figure 8.38 Effect of the height of the parasitic patch, h_3, on antenna performance: (*a*) return loss, (*b*) axial ratio, and (*c*) gain

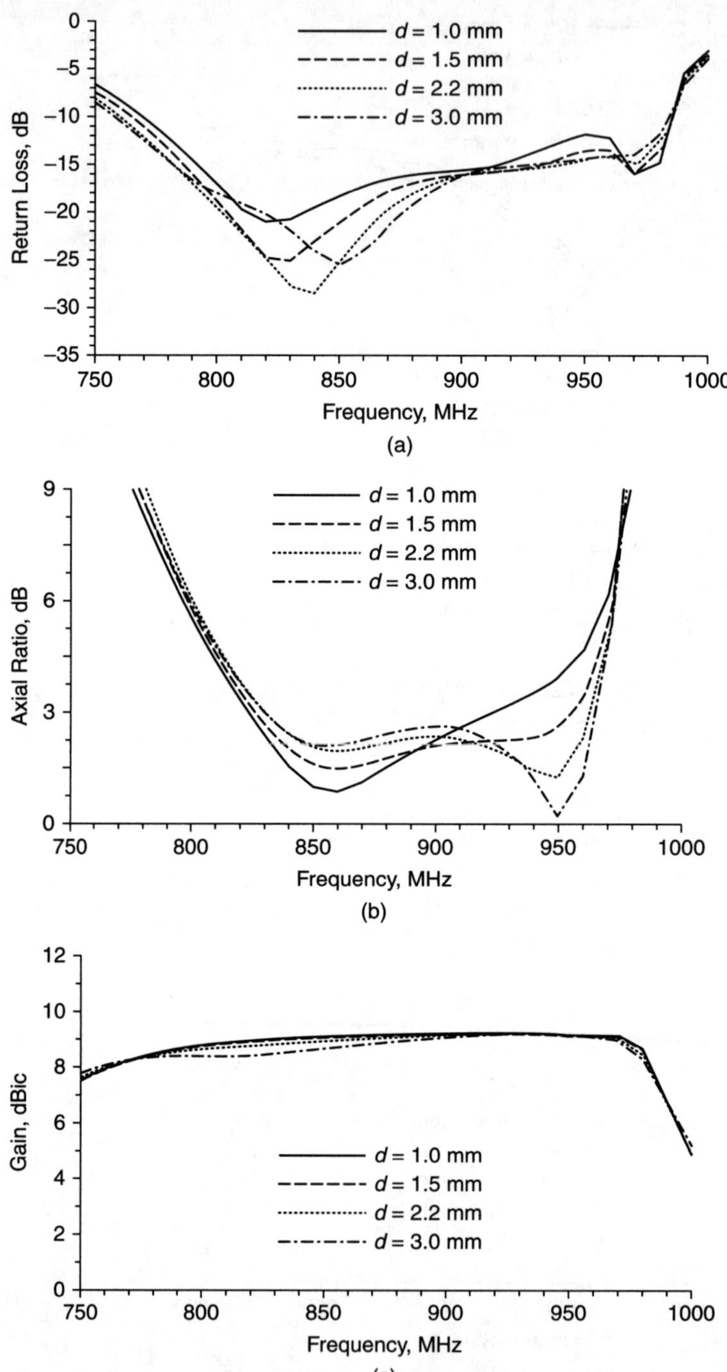

Figure 8.39 Effect of the probes' diameter, d, on antenna performance: (*a*) return loss, (*b*) axial ratio, and (*c*) gain

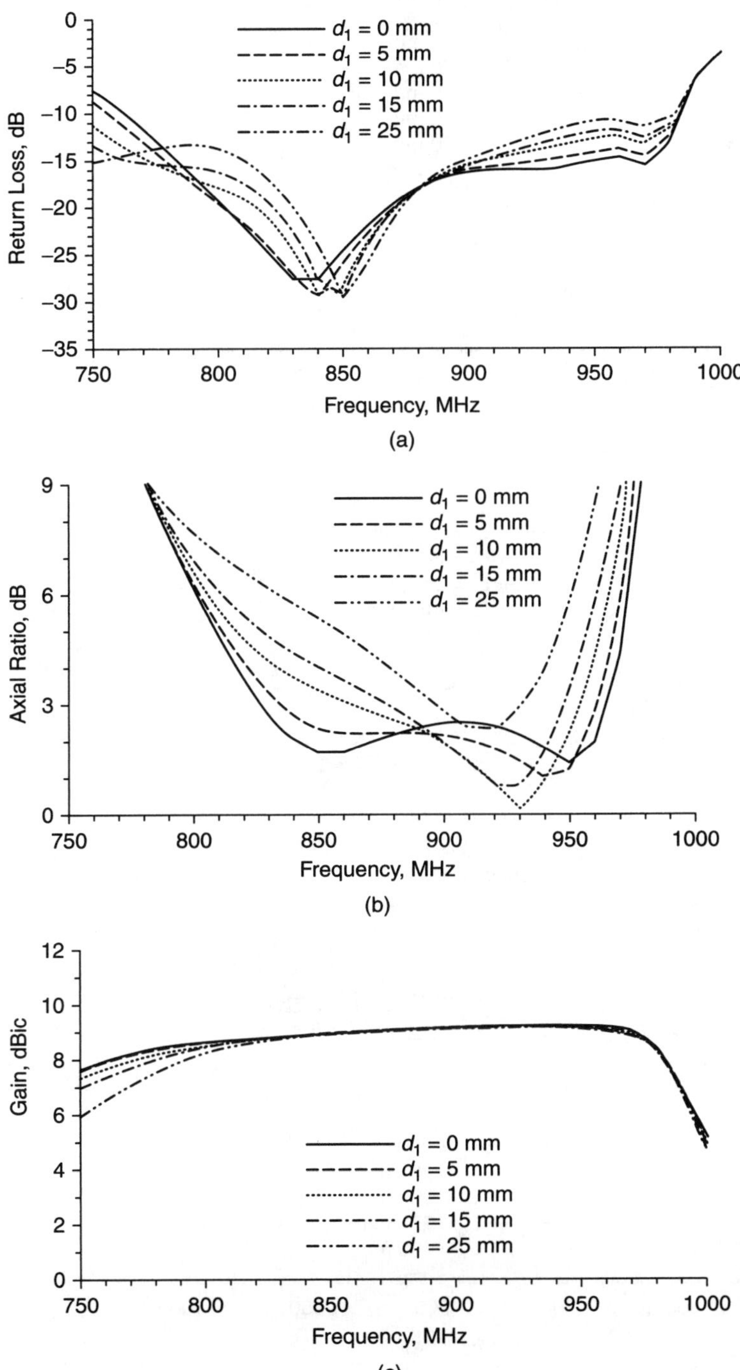

Figure 8.40 Effect of the extension of the open-circuited feed line, d_1, on antenna performance: (*a*) return loss, (*b*) axial ratio, and (*c*) gain

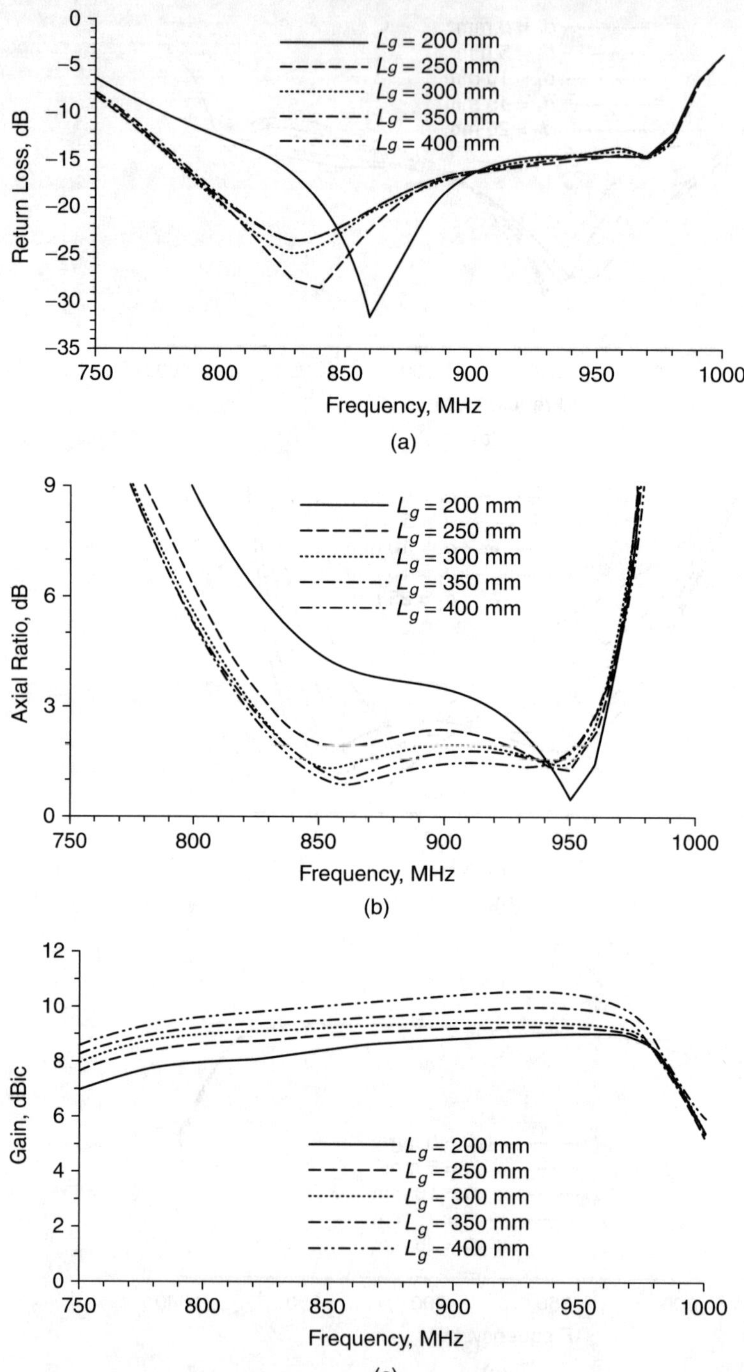

Figure 8.41 Effect of ground plane size, L_g, on antenna performance: (*a*) return loss, (*b*) axial ratio, and (*c*) gain

8.3 Antenna Design for Indoor Mono-Station UWB Positioning System

8.3.1 Design Considerations

The multiple antenna elements in the indoor mono-station UWB positioning system are generally placed symmetrically along a circumference. The patterns of the antenna elements are partially overlapped and thus the sectored antenna array can offer 360° coverage. Generally, more antenna elements offer higher accuracy at the price of a more complicated system and increased cost. There is a compromise between system complexity and accuracy. Figure 8.14 demonstrates a sectored antenna configuration consisting of six antenna elements. These antenna elements are configured in such a way that their maximum gain directions are separated 60° from each other. A 360° coverage can be achieved if the antenna elements feature pattern beamwidths of at least 60°. In some applications, if a smaller DOA span is of interest (e.g., 180° instead of 360°), fewer antenna elements, for instance, three, can be used.

8.3.2 Case Study: Six-Element Sectored Antenna Arrays

Antenna Configuration The configuration of the six-element sectored antenna array is shown in Figure 8.42, and the three-dimensional view of the antenna array is shown in Figure 8.42a. The antenna array consists of six identical antenna elements arranged in a hexagonal configuration. Two hexagonal covers (bottom/top cover) are used to support the antenna elements and create the enclosure for the antenna array. Figure 8.42b shows the top view of the antenna array without the top cover. A hexagonal reflector is centrally positioned behind the antenna elements. The length of each side of the hexagonal reflector is 45 mm and the distance from the two opposite pointed ends is 90 mm. The antenna array is arranged such that the pointed edge of the reflector is aligned behind and along the midpoint of each antenna element. Absorber slabs are employed to improve the isolation between the antenna elements. Each absorber slab is placed such that one side touches the pointed edge of the outer wall formed by two adjacent antenna elements and the other side touches the midpoint of the wall of the reflector. The antenna prototype is illustrated in Figure 8.42c.

The geometry of the wide slot antenna element is shown in Figure 8.43. The antenna element was etched onto a PCB (Rogers 4003) with $\varepsilon_r = 3.38$, $\tan\delta = 0.002$, and a thickness of 0.8128 mm. The rectangular wide slot measuring 32 mm × 21 mm is etched on the bottom surface of the PCB, while the fork-like feeding stubs are positioned on the opposite side of the PCB. The microstrip line, which has a width of 1.86 mm, is connected to

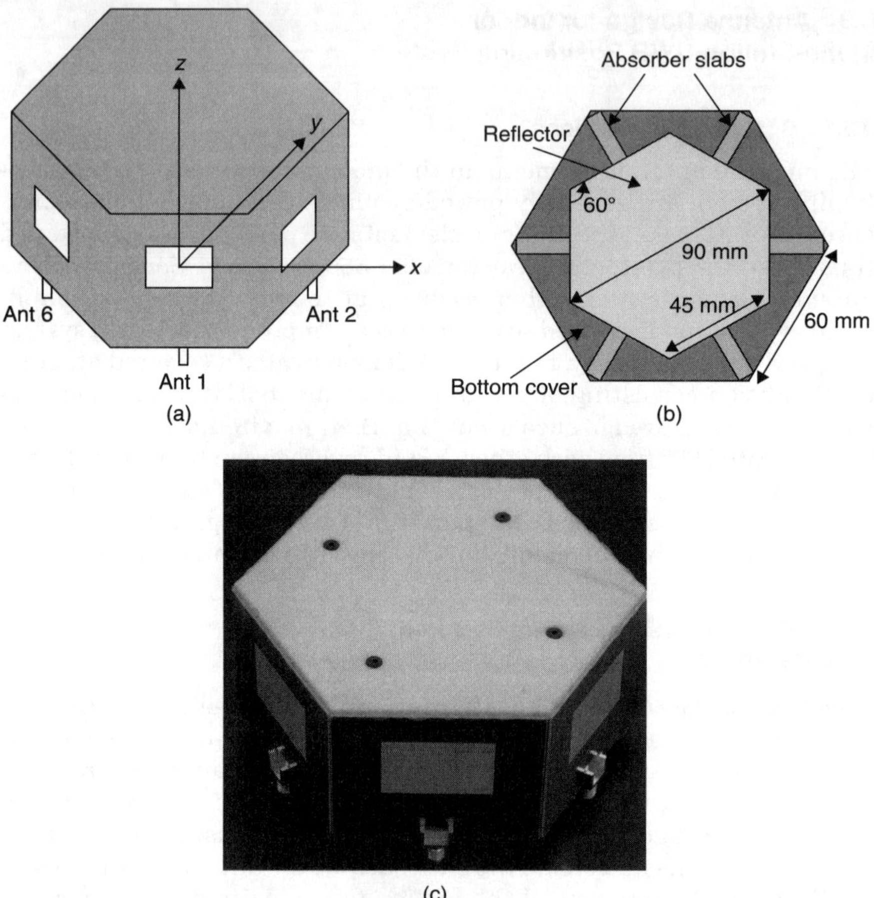

Figure 8.42 Six-element sectored antenna array: (a) three-dimensional view, (b) top view (with top cover open), and (c) antenna prototype

the horizontal arm measuring 13 mm × 1 mm. The identical strips measuring 1.86 mm × 9.5 mm extend from the ends of the horizontal arm. There is a gap of 0.77 mm between the horizontal arm and the bottom side of the slot. Impedance matching of the antenna element can be achieved by adjusting the dimensions of the fork-like feeding stubs as well as the dimensions of the slot.[37]

Measurement Results The measured impedance matching of one element of the antenna array is shown Figure 8.44. As shown, the antenna is well-matched over 2.9–6.0 GHz with a return loss of less than −10 dB. The impedance matching performance of the other elements is almost the same. The measured gain patterns (x-y and y-z planes) of the element at different frequencies are shown in Figures 8.45 and 8.46, respectively. Stable gain patterns are observed over the frequency band.

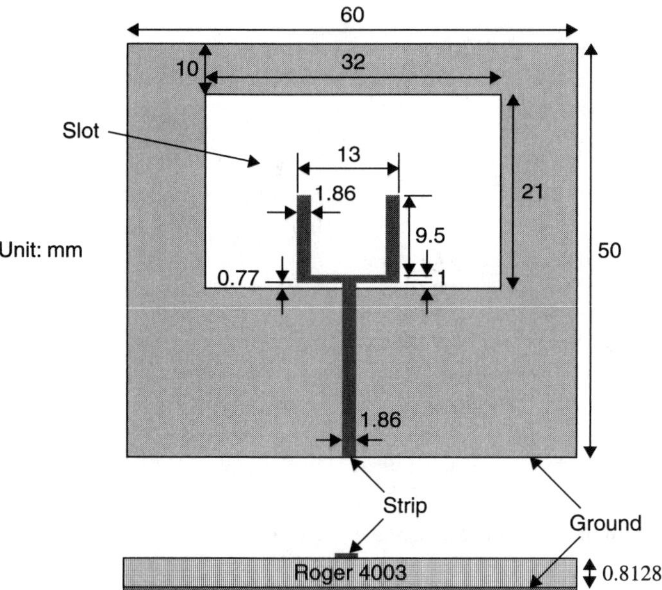

Figure 8.43   Geometry of antenna element

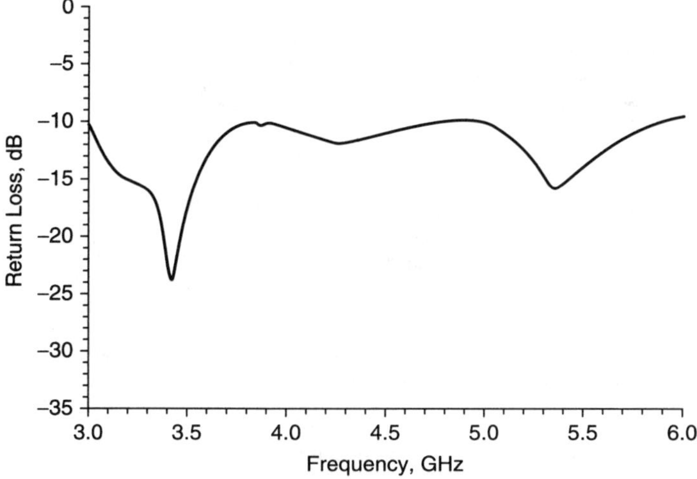

Figure 8.44   Measured return loss of the antenna element

For indoor mono-station UWB positioning applications, the peak amplitude pattern is more useful for DOA calculations since it represents the pulse response over the entire operating frequency band. Figure 8.47 shows a waveform of the UWB pulse that is transmitted by one of the antenna elements shown above and received by a horn antenna. The peak amplitude pattern is obtained by rotating the antenna element

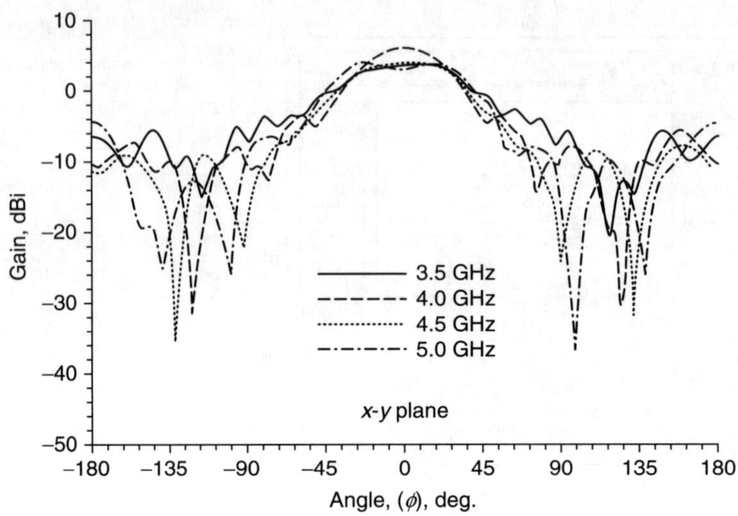

Figure 8.45 Measured gain pattern of the antenna element in *x-y* plane

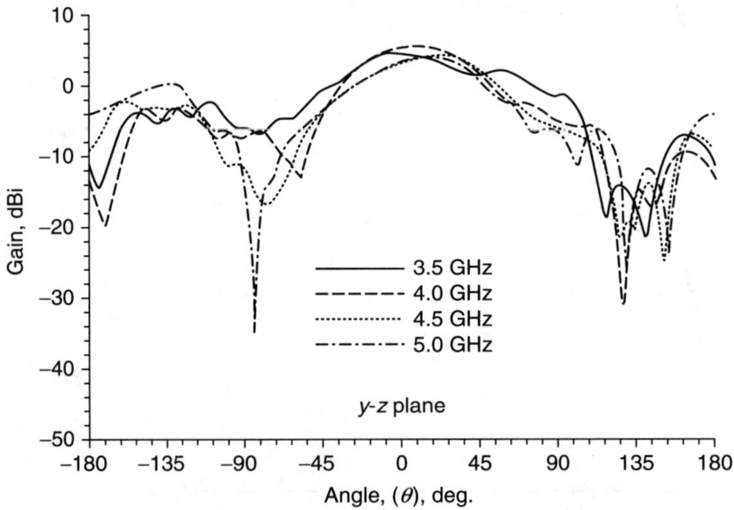

Figure 8.46 Measured gain pattern of the antenna element in *y-z* plane

and obtaining the peak value of the waveform at each direction. The normalized peak amplitude pattern of the antenna element is shown in Figure 8.48, where a half-power beamwidth of 69° is achieved.

The beamwidth of the peak amplitude pattern of the antenna elements can be controlled by changing the shape and size of the reflector. Table 8.5 exhibits the peak amplitude pattern beamwidth of the

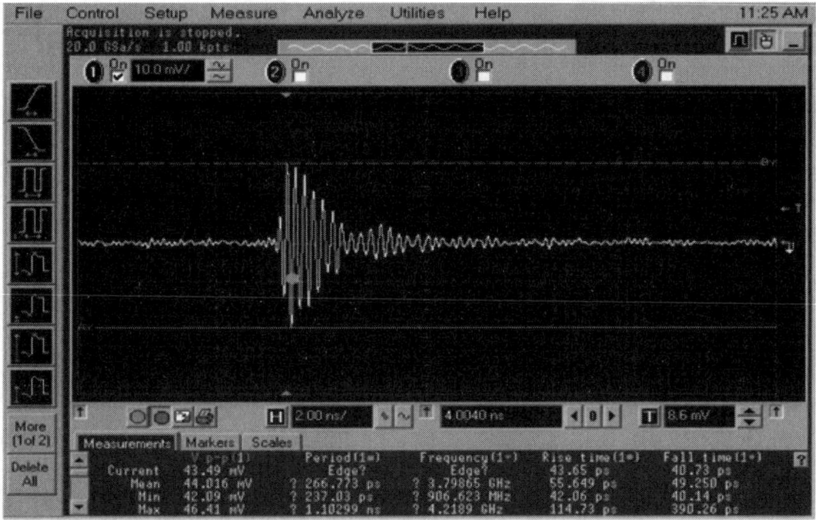

Figure 8.47   Received waveform at the boresight (Tx: sectored antenna element, Rx: horn antenna, $\phi = 0°$)

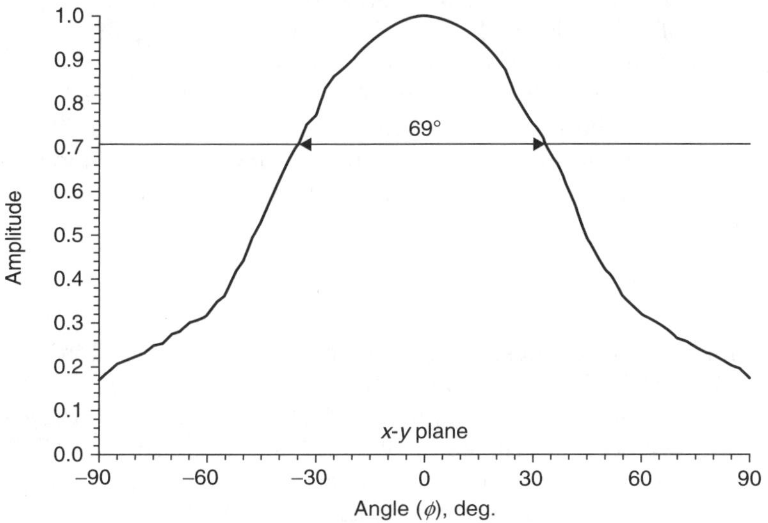

Figure 8.48   Measured peak amplitude pattern in the x-y plane

TABLE 8.5   Effect of Reflector Size on Peak Amplitude Pattern Beamwidth

Reflector	Hexagonal	Hexagonal	Hexagonal
d (mm)	90	80	70
Beamwidth	69°	73°	78°

TABLE 8.6   Effect of Reflector Shape on Peak amplitude Pattern Beamwidth

Reflector	Hexagonal	Star-shaped	Cylindrical
d (mm)	90	90	90
Beamwidth	69°	75°	80°

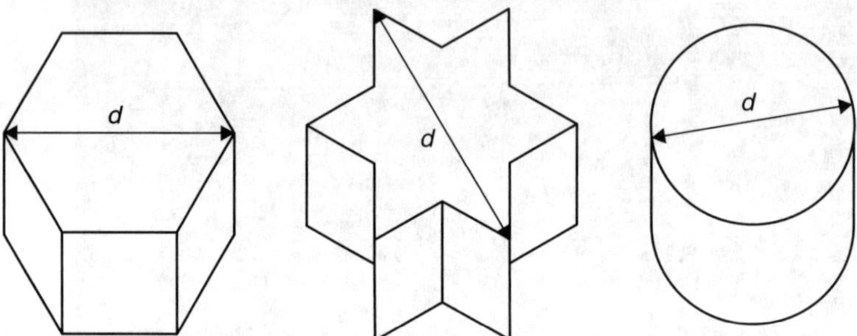

Figure 8.49   Different reflector configurations

antenna element against the size of the hexagonal reflector. When the diameter of the hexagonal reflector decreases from 90 mm to 70 mm, the beamwidth of the peak amplitude pattern increases from 69° to 78°. Table 8.6 shows the results of the peak amplitude pattern beamwidth with different reflectors illustrated in Figure 8.49. Compared to the hexagonal reflector, the star-shaped and cylindrical reflectors offer a larger beamwidth. As observed, it is possible to vary the shape and size of the reflector to obtain a desirable beamwidth to meet specific system specifications.

8.4   Conclusion

There are many technologies, such as Infrared, Bluetooth, HomeRF, ZigBee, UWB, RFID, NFC, and so on, that have been used for WPAN applications. Some of them, such as Infrared, Bluetooth, and RFID, are matured and have been developed for years. The others, such as UWB and NFC, are still being developed. This chapter has addressed the important issues related to antenna designs for two WPAN applications: radio frequency identification for assets tracking and ultra-wideband for target positioning.

The performance of the reader antenna is crucial for an RFID system. The optimal antenna design always offers the RFID system enhanced range, high accuracy, reduced fabrication cost, as well as simplified system configuration and implementation. Since the frequencies for RFID applications spans from a low frequency of 125 KHz to the microwave frequency of 24 GHz, the selection of the antenna as well as the design considerations for specific RFID application feature distinct differences. Loop antennas have been a popular choice of reader antennas for LF/HF RFID systems. The major design considerations for the antennas in such systems are to achieve the desired magnetic field strength over certain coverage. The antennas require resonance at the desired frequency with good impedance matching and adequate Q factor. For RFID applications at UHF and MWF bands, a number of antennas can be used as reader antennas, although the circularly polarized patch antenna has been the most widely used antenna. Generally, a broadband circularly polarized antenna with good axial ratio, gain, and impedance-matching performance is always preferable for RFID system configuration and implementation.

The technology of UWB positioning is still being developed. Monostation UWB positioning technology features a simple system configuration with high accuracy and it has great potential for indoor positioning applications. The six-element sectored antenna array exhibited in this chapter demonstrates the requirements and design considerations of the antenna array for such a system. The antenna configuration, antenna element design, and method of controlling the beamwidth of the antenna element are expected to be of beneficial reference when designing sectored antenna arrays for indoor UWB positioning applications.

References

1. M. Daniel, *Telecommunications Technology Handbook,* Boston, MA: Artech House, 2003.
2. M. Golio, *The RF and Microwave Handbook,* Boca Raton, FL: CRC Press, 2000.
3. R. Want, "An introduction to RFID technology," *IEEE Pervasive Computing,* vol. 5, no. 1 (January–March 2006): 25–33.
4. K. Finkenzeller, *RFID Handbook,* 2nd Ed. Chichester: John Wiley & Sons, 2004.
5. K. Siwiak and D. McKeown, *Ultra-wideband Radio Technology,* Chichester: John Wiley & Sons, 2004.
6. M. Ghavami, L. B. Michael, and R. Kohno, *Ultra-wideband Signals and Systems in Communication Engineering,* Chichester: John Wiley & Sons, 2004.
7. Personal Area Networks, http://foi.becta.org.uk/display.cfm?cfid=1476190&cftoken =29154&resID=15939.
8. J. L. Miller, *Principles of Infrared Technology: a Practical Guide to the State of the Art,* New York: Chapman & Hall, 1994.
9. H. Laboid, H. Afifi, and C. De Santis, *Wi-Fi, Bluetooth, Zigbee and WiMAX,* Dordrecht: Springer, 2007.
10. Ecma International, Standard ECMA-340, Near Field Communication Interface and Protocol (NFCIP-1), December 2004.
11. S. Q. Xiao, M. T. Zhou, and Y. Zhang, *Millimeter Wave Technology in Wireless PAN, LAN, and MAN,* Boca Raton, FL: CRC Press, 2008.
12. B. H. Walke, S. Mangold, and L. Berlemann, *IEEE 802 Wireless Systems: Protocols, Multi-hop Mash / relaying, Performance and Spectrum Coexistence,* Chichester: John Wiley & Sons, 2006.
13. T. Hassan and S. Chatterjee, "A taxonomy for RFID," *IEEE System Sciences, 39th Int. Conf,* vol. 8 (January 2006): 184b.
14. EPCglobal, Regulatory status for using RFID in the UHF spectrum, May 2007. http://www.epcglobalinc.org/tech/freq_reg/RFID_at_UHF_Regulations_20070504.pdf.

15. Anonymous, "A summary of RFID standards," *RFID Journal,* www.rfidjournal.com/article/articleview/1335/2/129/.
16. Federal Communications Commission, First report and order, revision of Part 15 of Commission's rule regarding ultra-wideband transmission system FCC 02-48, April 22, 2002.
17. IEEE 802.15 Working Group for WPAN, http://ieee802.org/15/index.html.
18. L. D. Nardis, M. G. D. Benedetto, "Overview of the IEEE 802.15.4/4a standards for low data rate wireless personal data networks," Positioning, Navigation and Communication, 2007. WPNC '07. 4th Workshop on (March 2007): 285–289.
19. R. J. Fontana, Ultra Wideband Precision Geolocation Systems, U.S. Patent No. 6,054,950, April 2000.
20. S. Krishnan, P. Sharma, G. Zhang, and H. W. Ong, "A UWB based location system for indoor robot navigation," *IEEE International Conference on Ultra-Wideband,* Singapore (September 2007): 77–82.
21. J. T. Matheney, L. M. Lee, and D. D. Mondul, Apparatus and method for locating objects in a Three-Dimensional Space, US patent No. 6,483,461, November 2002.
22. J. D. Taylor, *Introduction to Ultra-Wideband Radar Systems,* Boca Raton, FL: CRC Press, 1995.
23. X. Sun, Y. Ma, J. Xu, J. Zhang, and J. Wang, "A high accuracy mono-station UWB positioning system," *IEEE International Conference on Ultra-Wideband,* Hannover, Germany (September 2009): 201–204.
24. Texas Instruments, HF antenna design notes-Technical application report, Literature number: 11-08-26-003, September 2003.
25. IE3D version 12, Zeland Software, Inc., Fremont, CA, U.S.A.
26. http://www.langer-emv.com/prod_lf1_en.html.
27. X. Qing and Z. N. Chen, "Proximity effects of metallic environments on high frequency RFID reader antenna: study and applications," *IEEE Transactions on Antennas and Propagation,* vol. 55, no. 11 (November 2007): 3105–3111.
28. C. A. Balanis, *Antenna Theory: Analysis and Design,* 3rd Ed., New York: John Wiley & Sons, 2005.
29. Y. T. Lo and W. F. Richards, "Perturbation approach to design of circularly polarized microstrip antennas," *Electron. Lett.,* vol. 17, no. 11 (May 1981): 383–385.
30. T. Vlasits, E. Korolkiewicz, A. Sambell, and B. Robinson, "Performance of a cross aperture coupled single feed circularly polarised patch antenna," *Electron. Lett.,* vol. 32, no. 7 (March 1996): 612–613.
31. E. Aloni and R. Kastner, "Analysis of a dual circularly polarized microstrip antenna fed by crossed slots," *IEEE Transactions on Antennas and Propagation,* vol. 42, no. 8 (August 1994): 1053–1058.
32. H. Kim, B. M. Lee, and Y. J. Yoon, "A single-feeding circularly polarized microstrip antenna with the effect of hybrid feeding," *IEEE Antennas Wireless Propagat. Lett.,* vol. 2, no. 1 (April 2003): 74–77.
33. G. A. E. Vandenbosch and A. R. Van de Capelle, "Study of the capacitively fed microstrip antenna element," *IEEE Transactions on Antennas and Propagation,* vol. 42, no. 12 (December 1994): 1648–1652.
34. K. L. Wong and T. W. Chiou, "Broad-band single-patch circularly polarized microstrip antenna with dual capacitively coupled feeds," *IEEE Transactions on Antennas and Propagation,* vol. 49, no. 1 (January 2001): 41–44.
35. C. L. Mak, K. F. Lee, and K. M. Luk, "Broadband patch antenna with a T-shaped probe," *IEE Microwaves Antennas Propagation,* vol. 147, no. 2 (April 2000): 73–76.
36. H. W. Kwa, X. Qing, and Z. N. Chen, "Broadband Single-fed Single-Patch Circularly Polarized Antenna for UHF RFID Applications," *IEEE International Symposium on Antennas & Propagation,* San Diego, CA (July 5, 2008): 972–975.
37. X. Qing, Y. W. M. Chia, and X. H. Wu, "Wide slot Antenna for UWB Applications," *IEEE Antenna and Propagation Symposium,* Columbus, Ohio, USA (June 2003): 834–837.